INTRODUCTION TO
Probability and Statistics

INTRODUCTION TO
Probability and Statistics

SIXTH EDITION

Henry L. Alder   Edward B. Roessler
UNIVERSITY OF CALIFORNIA, DAVIS

W. H. FREEMAN AND COMPANY
San Francisco

Cover illustration: A computer-generated figure composed of bivariate normal distribution functions with positions, widths, and amplitudes chosen randomly. From *Computer Graphics: 118 Computer-Generated Designs* by Melvin L. Prueitt. Dover Publications, Inc., 1975.

Library of Congress Cataloging in Publication Data

Alder, Henry L
    Introduction to probability and statistics.
    Bibliography: p.
    Includes index.
    1. Probabilities. 2. Statistics.
I. Roessler, Edward Biffer, joint author.
II. Title.
QA273.A43 1976   519.2   76-13643
ISBN 0–7167–0467–6

Printed in the United States of America

AMS 1970 subject classifications: 60-01, 62-01

1 2 3 4 5 6 7 8 9

Contents

Preface to the Sixth Edition

This book has been written to serve as a textbook for a one-semester intro-ductory course in probability and statistics. It is also suitable for either a quarter course meeting four hours a week for ten weeks or a two-quarter course meeting three hours a week. The first edition of the text was a revision of the mimeographed and lithoprinted forms used for several years by in-structors on the Davis campus of the University of California and at some other institutions. The many hundreds of students in these classes came from almost all fields of specialization in the natural and social sciences, and a few from the humanities.

In the typical American college or university, introductory statistics courses are offered in many different departments, are directed toward a great number of different specializations, and require widely differing levels of mathematical preparation. This phenomenon might lead to the mistaken impression that basic statistics is essentially different in the various fields of application. In this respect, instruction in statistics seems to occupy a unique position; in almost every other science there exists one introductory course with a fairly well-defined amount of subject material and anyone requiring some background in that science takes that course. Thus, almost every college or university offers an introductory chemistry course in which all students who need some knowledge of chemistry enroll. There seem to be no reasons except purely administrative or traditional ones why statistics should not be treated similarly. Indeed, much benefit could be derived from centralizing the teaching of basic statistics in one course in one department, which, in most schools, would be either the mathematics or the statistics department. If only one such course is offered, much needless duplication can be avoided and a single standard of achievement can be maintained. It is for such a course that this book is designed.

Mathematical knowledge equivalent to two years of high school algebra is all that is required as background for this textbook. At one point in Chapter 12 trigonometry is used, but readers without training in that subject may omit this passage without loss of continuity. The limited level of mathematical preparation required for this book makes it necessary to state some theorems without proof. However, almost all theorems whose proofs depend only on mathematics as commonly taught in two years of high school algebra are proved. The reader may be surprised to find that the number of theorems not proved here, because of the limited mathematical preparation assumed, is relatively small, and he may find consolation in the fact that his understanding of the omitted proofs would not be increased appreciably even after a course or two in beginning calculus. To avoid the mathematical difficulties associated with limiting processes, with which the reader ordinarily will not become acquainted until he studies calculus, discussions have been restricted, wherever possible, to the finite case; in particular, all populations are assumed to be finite.

The students who enroll in the statistics course at Davis come from all the agricultural, biological, and medical sciences, business administration, economics, home economics, psychology, sociology, and geology. To allow for these differing fields of specialization, the examples and exercises have been chosen from all of these subject areas.

Exercises are an integral part of the text. It has been our experience that an introductory course in statistics is taught most effectively in the same way as are other introductory courses in mathematics, a number of homework problems being assigned after every lecture. Special care has been taken to insure that exercises involve a minimum of computation, and that none (except for a few in Chapter 15) requires machine calculation. This has often necessitated the inclusion of fictitious data, although those in many exercises represent the results of actual experiments.

Although most students taking the beginning course in statistics on the Davis campus enroll in it during their second year in college, some are freshmen To take this course as early as possible appears the best advice. There seems no valid reason why it could not be taken in high school, since it is certainly no more difficult than a course in trigonometry and, judging from the reaction of many students, may even be simpler. An early course in statistics enables the student to use his knowledge in the many courses in which some form of statistics is presently applied. In addition to other obvious benefits gained from an early exposure, a course in statistics is a convenient and effective means of reviewing high school algebra before taking other mathematics courses that demand a thorough knowledge of algebra. Because of this, we have included exercises which involve almost all topics ordinarily encountered in two years of high school algebra—students seem to find solution of practical, interesting problems a much less painful way of reviewing their algebra than traditional methods.

Chapters 1 through 13 constitute the basic core of information required in all fields where statistics is used. Chapters 14 and 15 are designed primarily for students in business administration and economics. Consequently it is recommended that instructors in general one-semester statistics courses, particularly those oriented towards the biological sciences, omit Chapters 14 and 15 and consider two possible alternatives: one, complete coverage of Chapters 1 through 13; the other, coverage of Chapters 1 through 10, 12, 13, 16, and 17. Instructors in one-semester courses oriented towards the social sciences—in particular, economics—may find it preferable to omit Chapters 16 and 17, but to include Chapters 11, 14, and 15. If the complete coverage of all chapters is desired, the text would be suitable for a one-semester course meeting four hours a week or a two-quarter course of ten weeks each meeting three hours a week.

With an increasing number of universities on the quarter system, more and more institutions now desire to cover the basic material of probability and statistics in a one-quarter course meeting four hours a week for ten weeks. Experience has shown that, for such a course, Chapters 1 through 13 of this text can be covered, with the following exclusions: Section 5.4; Section 7.2; Chapter 11; Sections 12.6, 12.7, 12.8, 12.9, 12,10, and 12.11; and Section 13.5. For such a course, it is recommended that Chapter 11 be given as a reading assignment after Chapter 10 has been covered.

The present edition differs from the Fifth Edition not only in a large number of minor improvements but also in four major changes.

1. Since the numbers used in the examples and exercises of this text usually represent actual data, wherever possible they have been brought up to date with latest available information.

2. Exercises have been added in a few chapters, and review exercises covering the material of Chapters 3 to 13 appear in a new Appendix consisting of a set of exercises, all of which are different from those in Chapters 3 to 13 and are designed to give the reader experience in deciding which statistical technique learned in Chapters 3 to 13 is the appropriate one to use in each of these exercises.

3. Small additions have been made in a few chapters to include material which, experience has indicated, is frequently needed or useful. In particular, discussion of the binomial distribution (Chapter 6) has been expanded by treating this distribution as a special case of a probability distribution of a discrete variable and similarly the normal distribution (Chapter 7) is treated as a special case of a distribution of a continuous variable. Also, in the chapter on regression and correlation (Chapter 12), we have added an elementary proof, which does not require calculus, for obtaining the line of regression, and we have added a section on exponential and power curves. In the chapter on chi-square distribution (Chapter 13), a discussion of the proper way of drawing conclusions from a contingency table has been added and, in the chapter on analysis of variance (Chapter 17), material has been included to clarify the conditions under which Duncan's new multiple range test should be applied.

4. A few changes have been made which delete, add, or alter terminology in order to conform to current practices and usage in statistics.

We are grateful to the Literary Executor of the late Sir Ronald A. Fisher, F.R.S., and to Dr. Frank Yates, F.R.S., and to Longman Group Ltd., London, for permission to reprint, in abridged form, Tables III, IV, and VII from their book *Statistical Tables for Biological, Agricultural, and Medical Research* (6th edition, 1974).

We gratefully acknowledge the many valuable suggestions and comments made by Professors Hubert A. Arnold, George A. Baker, and Curtis M. Fulton in the course of their use of an early draft of this book in their classes. We are indebted to Professor Jerry Foytik for his advice and helpful suggestions concerning the original draft and, in particular, for his comments on the material from economics covered in Chapter 15. We express special appreciation to Professor Gordon V. Shute of the City College of Chicago for his careful reading of the Fifth Edition and his many valuable suggestions. We are also grateful to Professors Dorothy L. Bernstein of Goucher College and Corwin L. Atwood of the University of California, Davis for their advice on certain parts of this new edition.

Henry L. Alder
Edward B. Roessler

August 1976

INTRODUCTION TO
Probability and Statistics

1
Introduction

Statistics is the science dealing with the *collection, organization, analysis,* and *interpretation* of numerical data.

"Collection of data" is the process of obtaining measurements or counts. Valid conclusions can result only from properly collected or from representative data. Although this is a very important part of statistical procedure, in the interest of conciseness we shall not discuss it but shall consider the treatment of data already available.

"Organization of data" is the task of presenting the collected measurements or counts in a form suitable for deriving logical conclusions. Representative methods of organizing and presenting data by means of tables and graphs are discussed in Chapter 2.

"Analysis of data" is the process of extracting from the given measurements or counts relevant information, from which a summarized and comprehensible numerical description can be formulated. The most important measures used for this purpose—the mean, the median, the range, the standard deviation, and others—are discussed in Chapter 4.

"Interpretation of data" is the task of drawing conclusions from the analysis of the data and usually involves the formulation of predictions concerning a large collection of objects from information available for a small collection of similar objects. The interpretation of data forms the main portion of the text.

Statistics, then, is a science that deals with problems capable of being answered to some degree by numerical information, that, is information obtained by counting or measuring. It matters little whether we are making insect counts for a biological study or surveying the number of workers or work-hours in an industrial plant. The duties of the statistician are first to select the kind of information needed, then to direct the proper and efficient collection and processing of that information, and, finally, to interpret the results. In interpreting results, especially where they are based on incomplete data, the statistician must apply principles and techniques that yield valid findings. He is often expected to make wise decisions in the face of uncertainty.

The word statistics has two greatly differing meanings. When used as indicated in the preceding paragraph, it is a scientific procedure used in the study and evaluation of numerical data. When used as the plural of statistic, it is synonymous with the term "numerical data." Thus if we say there are statistics in the *World Almanac* or in the *Statistical Abstract of the United States*, we mean there are numerical data in them. This is the older and more common meaning of the word. Originally, statistics were gathered for the purpose of providing governmental heads with data for managing the affairs of state. Such information expressed in numbers dates back to Aristotle and his treatises on the "matters of state." In fact, there is evidence that the words "statistics" and "state" are derived from the same root. From earliest times most civilized countries have compiled large scale "statistics" in order to ascertain, for military and fiscal reasons, the manpower and material strength of the nation. We read in the Bible of such censuses, and compilations for purposes of taxation were common practices in all parts of the Roman Empire.

The study of probability began during the Italian Renaissance when gamblers seeking to develop systems of winning at dice consulted such scholars as Girolamo Cardano (1501–1576) and the famous mathematician-astronomer Galileo Galilei (1564–1642); Galileo wrote a short essay in which he set forth the fundamental laws of probability that form the basis for the whole science of statistics.

In the sixteenth and seventeenth centuries, games of chance were especially popular with people of wealth and as more complicated games were introduced and larger sums of money were involved, the need for a rational method of calculating the chances in various games became increasingly important. A French intellectual who was also a passionate gambler, Chevalier de Méré, consulted the famous mathematician and philosopher Blaise Pascal (1623–1662), whose interest led Pascal to correspond with some of his mathematical friends, especially Pierre de Fermat (1601–1665). That correspondence forms the origin of modern probability theory and combinatorial analysis.

Other well-known mathematicians active in the study of the laws of chance were Gottfried Wilhelm Leibniz (1646–1716) and Jakob Bernoulli (1654–1705), who was the first of nine mathematicians in the famous Bernoulli family. All won measures of distinction, and Jakob, his brother Johann Bernoulli (1667–1748), and his nephews Nikolaus Bernoulli (1687–1759) and Daniel Bernoulli (1700–1782) achieved worldwide renown. The first extensive treatise on the theory of probability as a whole was written by Jakob Bernoulli, who expounded the principle of the Law of Large Numbers. Nikolaus Bernoulli applied the concept of probability to problems in law. Daniel Bernoulli applied the calculus of probability to epidemiology and the study of insurance.

During the same period, important advances were made in the collection of demographic data and the development of the body of knowledge now known as statistics. In England, John Graunt (1620–1674) made a semi-mathematical study of vital statistics and the statistics of insurance and economics. His work was extended by Sir William Petty (1623–1687), who studied the vital statistics of the population of the city of London, the first work of this kind ever undertaken, and by Edmund Halley (1656–1742), who developed mortality tables and is credited with originating the science of life statistics.

Abraham De Moivre (1667–1754) enunciated procedures for the probabilities of compound events, derived the theory of permutations and combinations from the principles of probability, and founded the science of life contingencies. In 1733 he discovered the equation of the normal curve upon which much of the theory of inductive statistics is based. The same bell-shaped curve is often referred to as the "Laplacian curve," the "Gaussian curve," or the "Gauss-Laplace curve," in honor of Marquis de Laplace (1749–1827) and Karl Friedrich Gauss (1777–1855), who independently rediscovered the equation. Gauss derived it from a study of errors in repeated measurements of the same quantity. He also originated the method of least squares and developed the theory of observational error. Laplace made great contributions to the application of statistics to astronomy and with Adrien Marie Legendre (1752–1833) introduced the use of partial differential equations into the study of probability. In 1815 the term "probable error" appeared for the first time in the writings of Friedrich Wilhelm Bessel (1784–1846), who also developed the theory of instrumental errors.

Other contributors to the theory were James Stirling (1692–1770), who developed an approximation to $n!$; Marquis de Condorcet (1743–1794), who applied probability and statistics to social problems; Thomas Bayes (1702–1761), who first used probability inductively; Leonhard Euler (1707–1783), the originator of the use of the Greek letter sigma as a symbol to denote summation; and Thomas Simpson (1710–1761), who introduced the principle of continuity into the theory of mathematical probability. In his study of probability Jean Le Rond d' Alembert (1717–1783) used meteorological data; Joseph Louis Lagrange (1736–1813) applied the differential calculus; and Pierre Rémond de Montmort (1678–1719) introduced the calculus of finite differences. The Comte de Buffon (1707–1788) anticipated some aspects of modern genetics and the calculus of probabilities, and Siméon Denis Poisson (1781–1840) developed the distribution that bears his name.

Between 1835 and 1870 the Belgian scientist Lambert A. J. Quetelet (1796–1874) made great contributions to the development and use of probability and statistics. He showed that biological and anthropological measurements closely follow the normal curve. Quetelet applied statistical methods not only in biology but also in education and sociology. He displayed a tremendous breadth of interest and is credited with being the first to recognize the constancy of large numbers and one of the first to demonstrate that statistical techniques developed in one area of research are applicable in most other areas.

In Germany, Georg Friedrich Knapp (1842–1926), following up Quetelet's principles, investigated extensively the statistics of mortality, and Wilhelm Lexis (1837–1914) developed a procedure that today is called one-way analysis of variance.

During the last quarter of the nineteenth century, Sir Francis Galton (1822–1911), the founder of the School of Eugenics in England, displayed unbounded enthusiasm as he verified the principle of systematic variation in every biological variable for which he was able to accumulate adequate data. Revelation of the principle of orderliness in biological variation formed the beginning of a new era in biological research. Galton and his great successor Karl Pearson (1857–1936), using problems from genetics, developed the ideas of regression and correlation. Later, Pearson and Charles Edward Spearman (1863–1945) extended this theory and applied it to studies in the social sciences. Pearson also studied extensively the effects of errors of sampling, developed the chi-square test and introduced the terms "mean deviation" and "standard deviation" into the literature.

Early in the present century, William Sealy Gosset (1876–1937), a statistician for Guinness, an Irish brewery, writing under the pseudonym of "Student," published many papers on the interpretation of data obtained by sampling. He was the first to recognize the importance of developing methods of extracting reliable information from small samples. His methods were later popularized in England by Sir Ronald A. Fisher (1890–1962) and his colleagues, who made many contributions to science, especially to population genetics, and greatly extended the theory of experimentation, increasing the interest in statistical methods as well as their use in all areas of scientific investigation. It was Fisher who introduced the now widely-used term "null hypothesis" and developed statistical techniques for the analysis of variance.

In the twentieth century, many noteworthy statisticians, too numerous to mention, have been active in developing new theories and applications. The availability of electronic computers has greatly helped in these developments. Today the research worker considers statistics one of his most useful tools.

Everyday life is influenced more and more by decisions based on quantitative information. The scientific sequence of hypothesis, experiment, and test of hypothesis is now a familiar approach to problems in every area of activity. Today, modern statistical methods, founded on probability theory, are proving indispensable as aids in the physical and biological sciences, in economics and sociology, in psychology and education, in medicine and agriculture, and in government and industry. The astronomer predicts future positions of heavenly bodies on the basis of statistical methods; conformity to genetic segregation is ascertained statistically; life insurance premiums and annuity payments are determined from mortality tables based on statistical records; power companies cannot supply electricity efficiently without statistical data of load requirements; research workers determine significance in agricultural field trials from statistical considerations; engineers find sampling theory invaluable in controlling quality of manufactured products; and business executives and governmental analysts use statistical procedures in decision-making. Although these are widely differing fields of application, most of the statistical methods employed are the same. One aspect of statistical analysis may be stressed more in one field of application than in another, but, in general, the same statistical procedures are used in all fields.

In approaching the study of statistics, a word of caution is in order. It is important to realize that no statistical procedure can, in itself, insure against mistakes, inaccuracies, faulty reasoning, or incorrect conclusions. The original data must be accurate; the methods must be properly applied; and the results must be interpreted by one who understands not only the methods themselves but also the field to which they are applied. The statistical methods discussed in this book are to be considered as tools that, in proper hands and applied to the situation for which they are designed, can produce useful results, but that, by themselves, have no power to work miracles.

2
Organization of Data

2.1 INTRODUCTION

Frequently, the collection of information leads to large masses of data that, if they are to be understood or to be presented effectively, must be in some manner summarized. Clear and forceful presentation is an important aid to the understanding and correct interpretation of such data. Two methods of presenting quantitative data are in common use. One method involves a summarized presentation of the numbers themselves, usually in tabular form; the other consists in presenting the quantitative data in pictorial form—graphs, diagrams, or other similar representations. Representation of a mass of data by either of these methods is the part of statistical analysis that should lead to a better over-all comprehension of the data.

2.2 TABULAR AND GRAPHICAL METHODS OF PRESENTING DATA

In nearly all scientific and business publications, in government reports, in magazines and in newspapers, data of all sorts are presented by means of tables, diagrams, or pictures.

It is always possible to include quantitative data as a part of the textual material, but there are certain advantages in presenting them in tabular form. According to the *Statistical Abstract of the United States* the total deposits of all commercial banks in the United States in 1945 amounted to $149.9 billion; in 1950, $156.1 billion; in 1955, $193.2 billion; in 1960, $230.5 billion; in 1965, $333.8 billion; and in 1970, $485.5 billion. No doubt, such textual presentation of these data might be passed over by the average reader, who would fail to grasp the important aspects of the data. Certainly the presentation can be made more readable and effective by arranging the data in columnar form as a part of the text. For example, the same information can be presented as follows: Total deposits in billions of dollars of all commercial banks in the United States for the five-year intervals from 1945 to 1970 were

Year	Total deposits
1945	149.9
1950	156.1
1955	193.2
1960	230.5
1965	333.8
1970	485.5

This arrangement enables us to visualize much more easily the significant characteristics of the data. We see immediately that commercial bank deposits more than trebled in the 25-year period. Such features stand out much more clearly when the data are arranged in columnar form. However, to understand the meaning of the data in the right-hand column completely we must refer to the descriptive material in the text; this, then, must still be described as a textual presentation.

In tabular presentation the data are arranged in columns and rows, a form that retains the advantages of the columnar arrangement, and in addition includes all information necessary for a complete understanding of the table, thus eliminating reference to the text. As an example, the commercial bank-deposit data are presented in tabular form in Table 2-1, on the next page.

Tables are usually numbered for rapid and positive identification, and any textual reference to the data can be made by citing specific table numbers.

Definition 2-1.   A measurable characteristic is called a *variable*.

Definition 2-2.   Individual measurements of a variable are called *variates*.

Table 2-1
Total deposits (in billions of
dollars) of all commercial banks in
the United States from 1945 to
1970, in 5-year intervals.

Year	Total deposits
1945	149.9
1950	156.1
1955	193.2
1960	230.5
1965	333.8
1970	485.5

For the data of Table 2-1, the total bank deposit is a variable, and each deposit entry is a variate.

Definition 2-3.   Pictorial representations made for the purpose of studying changes in a single variable or comparing several similar or related variables are called *graphs*.

Although graphs do not supply any more information than can be obtained from numbers arranged in tabular form, they generally set forth the important facts more clearly and forcefully. To be effective a graph should be simple and should emphasize the significant aspects of the data.

There are many different types of pictorial representation used in delineating changes in a single variable or in comparing two or more variables. Some of the more common types will be discussed.

2.2.1   The Bar Graph

The bar graph makes comparisons by means of parallel bars placed either horizontally or vertically. From the standpoint of supplying identifying labels, horizontal bars are more convenient. Vertical bars are used chiefly for the presentation of data the magnitudes of which vary with time. Such data are known as time series, and they will be discussed in Chapter 15. In such diagrams the width and spacing of the bars are uniform, and their lengths are proportional to the variates represented. This type of graph is well adapted to the comparison of variables which are similar but which are not necessarily related to one another. Usually a space of one-half the width of the bar is left between adjacent bars in order to make easier identification of the bar by its label. Typical uses of the bar graph are illustrated in the following examples.

Example 2-1. Construct a bar graph for the data of Table 2-1.

Solution. See Figure 2-1.

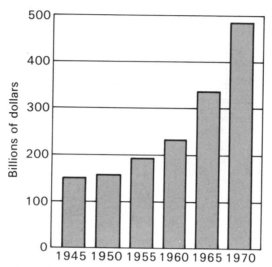

Figure 2-1

Total deposits of all commercial banks in the
United States, 1945–1970.

Example 2-2. Using the data of Table 2-2, construct a bar graph showing
the number of acres of federally owned land in seven of the western states.

Table 2-2

Total land acreage and federally owned land acreage (in millions of acres)
in seven western states, 1974.

State	Total acreage	Acreage federally owned	Acreage not federally owned
Arizona	72.69	31.93	40.76
California	100.21	45.07	55.14
Idaho	52.93	33.73	19.20
Montana	93.27	27.65	65.62
Nevada	70.26	60.83	9.43
Oregon	61.60	32.22	29.38
Washington	42.69	12.58	30.11

Solution.   See Figure 2-2.

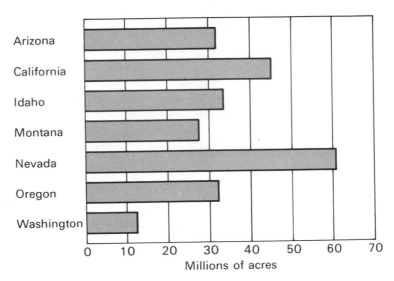

Figure 2-2
Federally owned land in seven western states, 1974.

The bar graphs illustrated in Figures 2-1 and 2-2 are composed of individual, unsegmented bars, which are the types most commonly used. Two important adaptations, the component-part bar graph and the grouped bar graph, are shown in Figures 2-3 and 2-4, respectively. The data are those shown in Table 2-2.

We see that the component-part bar graph furnishes a pictorial method for comparing changes in the whole quantity with changes in its component parts. If we wish to compare the component parts, not relating them to the whole, the grouped bar graph is the more effective device.

2.2.2 The Broken-line Graph

The broken-line graph is particularly useful in representing the relationship between two different variables and is obtained by plotting as points pairs of corresponding values of the two variables and joining consecutive points by a series of straight lines. It is used in particular to emphasize changes in some

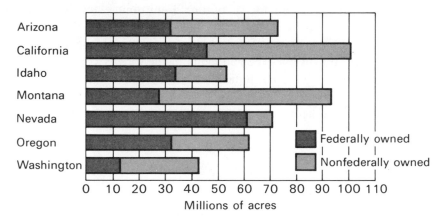

Figure 2-3
Federally and nonfederally owned land in seven western states, 1974.

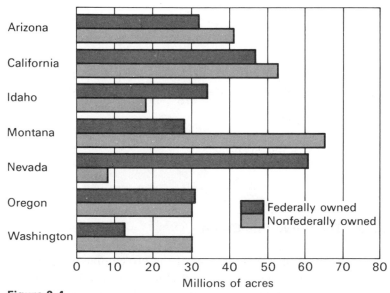

Figure 2-4
Federally and nonfederally owned land in seven western states, 1974.

variable occurring during an interval of time. For example, the data of Table 2-1 can be represented by a broken-line graph as well as by the bar graph shown in Figure 2-1. A broken-line graph can be obtained from a bar graph by joining with straight lines the midpoints of the tops of consecutive bars. A broken-line graph gives some indication of the change occuring during the interval of time represented by the width of each bar of a bar graph. In this case, therefore, a broken-line graph is more useful than a bar graph.

Example 2-3.   Draw the broken-line graph for the data of Table 2-1.

Solution.   See Figure 2-5.

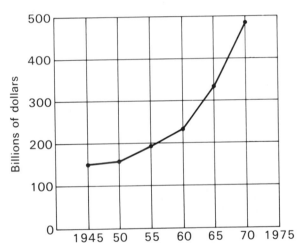

Figure 2-5
Total deposits of all commercial banks in the United States, 1945–1970.

2.2.3   The Curve-line Graph

The curve-line graph is a modification of the broken-line graph in which a smooth curve, instead of a series of straight lines, is drawn through the points. As the number of plotted points increases, the curve-line graph and the broken-line graph more nearly coincide. The curve-line graph is especially useful in showing variations in continuously changing variables such as temperature, rainfall, crop yields, bank deposits, commodity prices. Both types of graph are useful in estimating the value of one variable from a given value of the other within the plotted range of values. Estimating values outside the plotted range is frequently unreliable and should be avoided.

Although the curve-line graph may in some cases represent changes between the plotted points somewhat more realistically, there are those who prefer the broken-line graph because of its simplicity and the ease with which it can be drawn. Both are equally effective for the purpose of showing pictorially the general type of relationship existing between two variables.

Example 2-4. Draw a curve-line graph for the data of Table 2-1, and estimate the total amount of bank deposits in the continental United States in 1963.

Solution. See Figure 2-6.

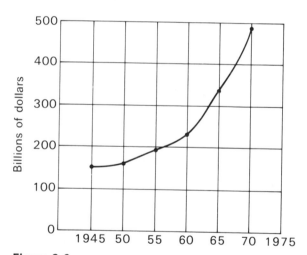

Figure 2-6

Total deposits of all commercial banks in the United States, 1945–1970.

From the graph, an estimate of the total amount of bank deposits in the continental United States in 1963 appears to be close to $300 billion.

2.2.4 The Circle Graph or Pie Diagram

This type of graph is useful in showing the apportionment of a total into its component parts. The components are expressed as percents of the whole and are represented by segments of a circle whose central angles are the corresponding percents of 360°. The pie diagram should not be subdivided into too many categories, and the labeling of the segments should be clear. Horizontal labels should be used except in the simplest cases where radial labels can be easily read.

Example 2-5. School enrollments in the United States in 1973, expressed to the nearest 100,000 students, were as follows: elementary, 31,500,000; secondary, 15,300,000; college and professional schools, 8,200,000. Represent by a circle graph the relation between enrollments in these three categories.

Solution. Of the total enrollment, 57.3% is in the elementary grades, 27.8% is in the secondary grades, and 14.9% is in the collegiate institutions. At the center of the circle the number of degrees representing the elementary school enrollment is 57.3% of 360°, or 206.3°, and so on. The resulting circle graph is shown in Figure 2-7.

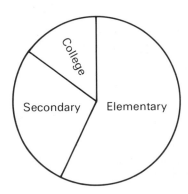

Figure 2-7
School enrollments in the United States in 1973.

2.2.5   The Ideograph or Pictorial Chart

Pictures are often used in the design of charts for popular presentation. This type of representation is common in government pamphlets and in advertisements. Such presentations register a meaningful impression in our mind almost before we think about them. They are useful in capturing the attention of persons who ordinarily would not look at a more formal chart or graph. They are most effective in displaying information which can be presented simply. The pictorial units used should be easily recognized, and differences in magnitude should be indicated by differences in the number of units, not by differences in their size. Differences in size frequently lead to difficulties in interpretation because of the danger of optical illusion. Examples of this type of pictorial representation are shown in Figures 2-8 and 2-9.

Each symbol represents 100 billion dollars

Figure 2-8

Total deposits of all commercial banks in the United States, 1945–1970.

Each symbol represents 20 million automobiles

Figure 2-9
Number of automobiles in the United States at ten-year intervals,
1940–1970.

2.3   FREQUENCY DISTRIBUTIONS AND THEIR TABULAR AND GRAPHICAL REPRESENTATIONS

Statistical data usually consist of a set of measurements of a class of objects. Such a set of data, showing for a 120-year period the total seasonal rainfall (July 1–June 30) in Sacramento, California, is shown in Table 2-3.

Table 2-3
Seasonal rainfall (in inches) at Sacramento, California, for the 120-year period
1855–1975 (read horizontally).

13.8	10.4	15.0	16.0	22.1	16.2	36.1	11.6	7.8	22.6
17.9	25.3	32.8	16.6	13.6	8.5	23.7	14.2	22.9	17.7
26.3	9.2	24.9	17.9	26.5	26.6	16.5	18.1	24.8	16.6
32.3	14.0	11.6	20.0	33.8	15.8	15.2	24.0	16.4	24.1
23.2	17.3	10.5	15.0	20.2	20.2	17.3	16.6	16.9	22.0
23.9	24.0	12.2	21.8	12.2	22.0	9.6	8.0	20.4	17.2
18.3	13.0	10.6	17.2	8.9	16.8	14.2	15.7	8.0	17.7
16.1	17.8	11.6	10.4	13.6	8.4	12.6	8.1	11.6	21.1
20.5	19.8	24.8	9.7	25.1	31.8	24.9	20.0	17.6	17.1
13.9	9.7	14.3	14.9	12.9	17.6	23.8	16.3	13.5	15.0
25.5	17.0	28.7	11.0	13.1	12.5	17.5	24.2	13.5	18.2
10.5	26.3	11.8	24.0	17.2	16.4	9.3	25.1	22.4	18.2

Definition 2-4. If a variable can assume only specific values (usually integers), it is called a *discrete* or *discontinuous* variable.

Examples of discrete variables are the number of children in a family, or the total value of all postage stamps used by a person on a given day.

Definition 2-5. If a variable can assume any values whatsoever between certain limits, it is called a *continuous* variable.

Examples of continuous variables are the seasonal rainfall in Sacramento, and any variables representing length, weight, and time.

Since no measuring device is exact, the precision of measurements of a continuous variable is subject to the accuracy of the device and its operator. All measurements should therefore be recorded only to the appropriate number of decimal places (or significant figures). In Table 2-3 the rainfall figures are recorded to one-decimal accuracy; in recording a value of 13.8 inches we assume that the true value is greater than or equal to 13.75 but less than or equal to 13.85; in doing this we adopt the usual rounding-off convention of leaving the final digit even in rounding off a number ending in a 5.

For large sets of data, as in Table 2-3, the variates must be presented in a summarized form in order for the mind to obtain any significant information from them. For concise description of the variates, some combination of them is imperative. This is achieved by constructing a frequency table which classifies the set of observations according to the number of variates falling within certain limits. A frequency table for the data of Table 2-3 is shown in Table 2-4.

Table 2-4
Frequency table for data of Table 2-3.

Class boundaries (inches)	Class midpoints (X)	Tallied frequency	Frequency (f)	Cumulative frequency
7.5–10.5	9		14	14
10.5–13.5	12		16	30
13.5–16.5	15		23	53
16.5–19.5	18		25	78
19.5–22.5	21		13	91
22.5–25.5	24		18	109
25.5–28.5	27		5	114
28.5–31.5	30		1	115
31.5–34.5	33		4	119
34.5–37.5	36		1	120

Definition 2-6.   The endpoints of the classes (7.5, 10,5, 13.5, 16.5, etc. in Table 2-4) are called *class limits* or *class boundaries*.

Definition 2-7.   The midpoint between the upper and lower class boundaries of a class is called the *class mark* or *class midpoint*.

Definition 2-8.   The difference between the largest and smallest class boundaries is called the *range* of the table. (In Table 2-4 the range is $37.5 - 7.5 = 30$.)

Definition 2-9.   The number of variates falling between the lower and upper class boundaries of a given class is called the *class frequency*, or simply, *frequency*.

Definition 2-10.   The set of frequencies with their respective classes or class midpoints is called a *frequency distribution*.

In classifying data in a frequency table, experience indicates that for most data 10–20 classes are desirable. With fewer than 10 classes too much accuracy is lost; with more than 20 classes too extensive a summary is obtained.

The class interval should therefore be about $\frac{1}{10}$ to $\frac{1}{20}$ of the difference between the largest and smallest variate, which in Table 2-3 is $36.1 - 7.8 = 28.3$. Therefore, for 10 classes some convenient number in the neighborhood of 2.83 should be selected for the class interval. The value of 3.0 was used in Table 2-4. Since the first class should contain the smallest variate, it must begin with 7.8 or less; the value of 7.5 was chosen for the smallest class boundary.

Various procedures are used in tallying a variate falling on a class boundary. We shall adopt the method of always tabulating such a variate in the higher class, which means that the variate "10.5 inches" was recorded in the class having 10.5 and 13.5 as boundaries. Another method used to avoid having variates falling on class boundaries is to express the boundaries to one-half unit beyond the accuracy of the measurement. If the data are accurate to tenths, the class limits should then be expressed to twentieths; if data are measured to the nearest quarter of an inch, the class limits should be arranged in eighths of an inch. Therefore in the example above we might choose as the first class limit 7.45 or 7.55, and the remaining class limits would be obtained by repeatedly adding 3.00 until the largest measurement is included. Some statisticians use the method of assigning, for a variate falling on a class boundary, a count of one-half to the lower class and one-half to the higher class. Often, especially for discrete data, noncontiguous boundaries are used; i.e. for scores in an examination, based on a total of 100 points, the following would be appropriate class boundaries, assuming the lowest score to be a 43 and all scores to be integers: 40–49, 50–59, etc. Clearly, a score of 50 would be placed in the second class.

Definition 2-11 The *cumulative frequency* for a given class is the sum of the frequencies of all classes up to and including the given class, assuming that the class with smallest class boundaries is written first.

Cumulative frequencies for each class in the example are listed in the last column of Table 2-4.

We have discussed a frequency table for a continuous variable. For a discrete variable, it is possible to list the distinct variates in increasing order and tally the number of times each variate occurs, thereby obtaining its frequency. As an example, consider the data of Table 2-5 representing the number of books read in the past six months by each student in a class of 25.

Table 2-5
Number of books read in the past 6
months by each student in a class of 25.

6	24	14	11	33
15	15	8	14	10
8	27	15	6	20
20	9	33	15	10
6	11	20	8	6

A frequency table for the data of Table 2-5 is shown in Table 2-6. Note the similarity with columns 2 and 4 of Table 2-4, but notice that values of X are not equally spaced. Readers may find the cumulative frequencies for Table 2-6 in the same way they were determined for Table 2-4.

Several types of graphical representations are used as aids in the study of frequency distributions. A few important ones will be discussed.

2.3.1 Histogram

A histogram is a vertical bar graph, with no space between bars. For frequency distributions with contiguous boundaries, the histogram shows class boundaries as points on the horizontal axis and frequencies as units on the vertical

Table 2-6
Frequency table for the data of
Table 2-5.

Class midpoint (X)	Frequency (f)
6	4
8	3
9	1
10	2
11	2
14	2
15	4
20	3
24	1
27	1
33	2

axis. The frequency corresponding to a class is represented by the height of a rectangle (bar) whose base is the class interval. The histogram for the data of Table 2-4 is shown in Figure 2-10. For frequency distributions with non-contiguous boundaries the vertical lines of the bars are drawn at the midpoints between the upper class boundaries of one class and the lower boundaries of the next class. Thus, in the histogram for the frequency distribution of Table 2-6, a bar for the first class, with midpoint 6, is drawn at height 4 from $X = 5.5$ to $X = 6.5$; a bar for the second class, with midpoint 7, has height 0; a bar for the third class, with midpoint 8, is drawn at height 3 from $X = 7.5$ to $X = 8.5$, and so on.

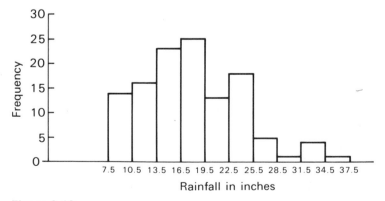

Figure 2-10
Histogram for the data of Table 2-4.

2.3.2   Frequency Polygon

A frequency polygon is a broken-line graph having class midpoints represented on the horizontal axis and frequencies on the vertical axis. The frequency corresponding to each class midpoint is marked by a point, and consecutive points are connected by straight lines. This procedure is equivalent to joining the midpoints of the tops of the rectangles of a histogram by straight lines. The frequency polygon for the data of Table 2-4 is shown in Figure 2-11.

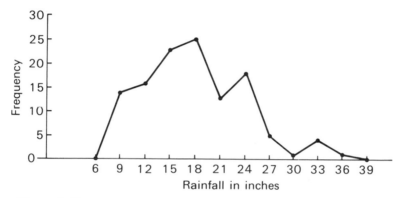

Figure 2-11
Frequency polygon for the data of Table 2-4.

2.3.3   Frequency Curve

If the class intervals are made smaller and smaller while at the same time the total number of measurements or counts is increased further and further, the histogram may approach a curve, which is called a frequency curve. It is approximated by fitting a smooth curve to the points of the frequency polygon.

2.3.4   Cumulative Frequency Polygon or Ogive

A cumulative frequency polygon or ogive is a broken-line graph having the upper class boundaries represented on the horizontal axis and the cumulative frequencies on the vertical axis. The cumulative frequency corresponding to each upper class boundary is marked by a point, and consecutive points are connected by straight lines. The cumulative frequency polygon for the data of Table 2-4 is shown in Figure 2-12.

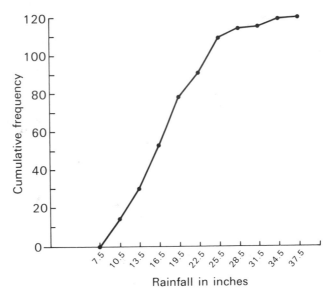

Figure 2-12
Cumulative frequency polygon for the data of Table 2-4.

EXERCISES

1. The urban and rural populations in the United States at 10-year intervals from 1910 to 1970, expressed to the nearest thousand persons, were as follows:

Year	Urban	Rural
1910	42,064,000	50,164,000
1920	54,158,000	51,553,000
1930	69,161,000	54,042,000
1940	74,424,000	57,246,000
1950	90,128,000	61,198,000
1960	125,269,000	54,054,000
1970	149,325,000	53,887,000

Construct for these data
(a) a component-part bar graph;   (c) a broken-line graph;
(b) a grouped bar graph;   (d) a curve-line graph.

2. For the urban population data of Exercise 1 construct an ideograph.

3. The numbers of subscribers to cable television in the United States for the period 1965-1973, expressed in thousands of households, were as follows:

Year	Number	Year	Number
1965	1275	1970	4500
1966	1575	1971	5300
1967	2100	1972	6000
1968	2800	1973	7300
1969	3600		

(a) Construct a bar graph for the 6-year period 1965–1970.
(b) Construct a broken-line graph for all the data.

4. Of the life insurance policy dividends paid in the United States in 1957, 21% were taken in cash, 31% were used to pay premiums, 18% were used to purchase additional paid-up life insurance, and 30% were left with the life insurance companies to earn interest. Construct a pie diagram showing these different uses of policy dividends.

5. The data for total seasonal rainfall at Davis, California, for the 70-year period 1905–1975 (read horizontally) are given below to the nearest tenth of an inch.

24.5	23.9	13.2	22.1	11.8	23.2	9.5	8.7	28.7	20.0
20.9	14.1	9.7	19.4	8.9	17.2	16.6	17.9	9.0	19.2
18.4	19.0	14.8	10.8	13.9	8.7	15.2	9.6	11.0	18.7
17.8	17.9	25.8	7.0	20.4	31.5	21.2	17.8	15.5	15.4
15.4	12.6	14.1	13.9	12.5	16.4	20.7	16.7	13.5	14.1
25.4	12.0	28.7	11.0	12.6	12.7	15.0	27.1	11.2	18.6
11.4	27.6	11.5	24.7	17.0	16.4	8.6	27.6	21.6	18.7

(a) Construct a frequency table for these data (use approximately 10 classes). For the frequency table constructed in (a) draw
(b) the histogram;
(c) the frequency polygon;
(d) the cumulative frequency polygon.

6. An industrial organization gives an aptitude test to all applicants for employment. The results in the case of 145 people taking the test are as follows:

Score	Frequency	Score	Frequency
9	5	4	21
8	10	3	15
7	19	2	11
6	23	1	6
5	35		

(a) Draw a histogram for the frequency distribution, and on the same graph draw the frequency polygon.
(b) Draw the cumulative frequency polygon.

3
Summation Notation

In statistics the sums of large numbers of terms frequently occur, and it is desirable to have a simplified notation for the summation of these terms. If a variable is denoted by X, then successive observations of that variable are written as X_1 (read "X sub one"), X_2, X_3, and so on. In general, the ith observation is written as X_i, and if there are N observations the last one is denoted by X_N. The Greek letter Σ (sigma) is used to indicate the sum of these N observations, i.e.

$$\sum_{i=1}^{N} X_i = X_1 + X_2 + X_3 + \cdots + X_N. \tag{3-1}$$

The symbol $\sum_{i=1}^{N} X_i$ is read "summation of X sub i (or sigma X sub i), where i assumes all integral values from 1 to N inclusive," or more simply "summation of X sub i, where i goes from 1 to N."

Definition 3-1. The letter below a summation sign Σ is called the *summation index*.

In the summation in Equation (3-1) the summation index is i.

Example 3-1. If $X_1 = 3$, $X_2 = 9$, $X_3 = 11$, find $\sum\limits_{i=1}^{3} X_i$.

Solution. $\sum\limits_{i=1}^{3} X_i = X_1 + X_2 + X_3 = 3 + 9 + 11 = 23$.

An explicit definition of the summation notation is as follows:

Definition 3-2. The greek letter sigma, Σ, placed before a term signifies the sum of all terms the first one of which is obtained by putting for i (or for whatever other summation index is used) in the term being summed the integer listed below the Σ-sign, the second by putting i equal to the next larger integer, and so on, until finally the last term results from putting i equal to the integer indicated above the Σ-sign.

The number of terms in the summation is one more than the difference between the integer that appears above and that that appears below the summation sign.

Example 3-2. If $X_1 = 9$, $X_2 = 6$, $X_3 = 5$, $X_4 = 8$, $X_5 = 12$, find

(a) $\sum\limits_{i=1}^{5} (X_i - 5)^2$;   (b) $\left(\sum\limits_{i=3}^{4} X_i\right)^2$;   (c) $\sum\limits_{i=1}^{3} (X_i^2 - 2X_i)$.

Solution.

(a) $\sum\limits_{i=1}^{5} (X_i - 5)^2 = (X_1 - 5)^2 + (X_2 - 5)^2 + (X_3 - 5)^2 + (X_4 - 5)^2 + (X_5 - 5)^2$

$\qquad = (9 - 5)^2 + (6 - 5)^2 + (5 - 5)^2 + (8 - 5)^2 + (12 - 5)^2$

$\qquad = 75.$

(b) $\left(\sum\limits_{i=3}^{4} X_i\right)^2 = (X_3 + X_4)^2 = (5 + 8)^2 = 13^2 = 169.$

(c) $\sum\limits_{i=1}^{3} (X_i^2 - 2X_i) = (X_1^2 - 2X_1) + (X_2^2 - 2X_2) + (X_3^2 - 2X_3)$

$\qquad = (81 - 18) + (36 - 12) + (25 - 10)$

$\qquad = 63 + 24 + 15$

$\qquad = 102.$

Example 3-3. Prove that

$$\sum_{i=1}^{N}(aX_i+b)=a\sum_{i=1}^{N}X_i+Nb.$$

Solution.

$$\sum_{i=1}^{N}(aX_i+b)=(aX_1+b)+(aX_2+b)+(aX_3+b)+\cdots+(aX_N+b)$$

$$=aX_1+aX_2+aX_3+\cdots+aX_N+\underbrace{b+b+b+\cdots+b}_{N\text{ terms}}$$

$$=a(X_1+X_2+X_3+\cdots+X_N)+Nb$$

$$=a\sum_{i=1}^{N}X_i+Nb.$$

From the preceding examples it is evident that:

1. The summation of an expression which is the sum of two or more terms is equal to the sum of the summations of the separate terms. For the case of an expression consisting of three terms this means

$$\sum_{i=1}^{N}(X_i+Y_i+Z_i)=\sum_{i=1}^{N}X_i+\sum_{i=1}^{N}Y_i+\sum_{i=1}^{N}Z_i;\qquad(3\text{-}2)$$

2. The summation of a constant multiplied by a variable is the same as the constant multiplied by the summation of the variable,

$$\sum_{i=1}^{N}aX_i=a\sum_{i=1}^{N}X_i;\qquad(3\text{-}3)$$

3. The summation of a constant is N multiplied by the constant,

$$\sum_{i=1}^{N}b=Nb.\qquad(3\text{-}4)$$

Example 3-4. Prove that

$$\sum_{i=1}^{N}(X_i-m)^2=\sum_{i=1}^{N}X_i^2-\frac{1}{N}\left(\sum_{i=1}^{N}X_i\right)^2,$$

where

$$m=\frac{\sum\limits_{i=1}^{N}X_i}{N}.$$

Solution.

$$\sum_{i=1}^{N}(X_i - m)^2 = \sum_{i=1}^{N}(X_i^2 - 2mX_i + m^2)$$

$$= \sum_{i=1}^{N}X_i^2 - 2m\sum_{i=1}^{N}X_i + Nm^2$$

$$= \sum_{i=1}^{N}X_i^2 - 2\left(\frac{\sum_{i=1}^{N}X_i}{N}\right)\sum_{i=1}^{N}X_i + N\left(\frac{\sum_{i=1}^{N}X_i}{N}\right)^2$$

$$= \sum_{i=1}^{N}X_i^2 - \frac{2}{N}\left(\sum_{i=1}^{N}X_i\right)^2 + \frac{1}{N}\left(\sum_{i=1}^{N}X_i\right)^2$$

$$= \sum_{i=1}^{N}X_i^2 - \frac{1}{N}\left(\sum_{i=1}^{N}X_i\right)^2.$$

In proving results of the type of Example 3-4 involving the quantity $m = \left(\sum_{i=1}^{N}X_i\right)/N$ it is important to carry out the procedure in the order indicated above. This means that the following sequence of steps should be used.

1. Perform the algebraic operations indicated after the summation sign and simplify (in Example 3-4 this operation consisted in squaring the binomial $X_i - m$).
2. Take the summation of the separate terms in accordance with the principle stated in Formula (3-2); at the same time place any constants before the summation signs in accordance with Formula (3-3) and replace the summation of a constant by N times the constant in accordance with Formula (3-4).
3. Replace m by $\left(\sum_{i=1}^{N}X_i\right)/N$ in the result obtained in step 2 if the final result is to be expressed in terms of $\sum_{i=1}^{N}X_i$; if the final result is to be expressed in terms of m, replace $\sum_{i=1}^{N}X_i$ by Nm.
4. Simplify the result obtained in step 3 as far as possible.

Frequently, where no confusion is possible, the summation index is omitted, and it is then understood that the summation includes all values under discussion.

Example 3-5. If $X_1 = 3$, $X_2 = 9$, $X_3 = 5$, find (a) ΣX^2; (b) $\sqrt{\Sigma 2X}$.

Solution.

(a) $\Sigma X^2 = X_1^2 + X_2^2 + X_3^2 = 9 + 81 + 25 = 115$;

(b) $\sqrt{\Sigma 2X} = \sqrt{2\Sigma X} = \sqrt{2(3 + 9 + 5)} = \sqrt{34}$.

EXERCISES

1. Write out in full the sums represented by each of the following expressions:

(a) $\displaystyle\sum_{i=2}^{6} X_i$; (d) $\displaystyle\sum_{i=1}^{3} (X_i - 2)^2$; (g) $\displaystyle\sum_{j=2}^{5} (Y_j + 7)$;

(b) $\displaystyle\sum_{i=7}^{9} X_i^2$; (e) $\displaystyle\left[\sum_{i=1}^{4} (X_i + 3)X_i\right]^2$; (h) $\displaystyle\sum_{k=1}^{4} W_k^3$;

(c) $\displaystyle\left(\sum_{i=1}^{7} X_i\right)^2$; (f) $\displaystyle\sum_{i=1}^{4} cX_i$; (i) $\displaystyle\sum_{i=1}^{4} f_i X_i^2$.

2. Write each of the following expressions, using a summation sign and appropriate limits:

(a) $X_1 + X_2 + X_3 + X_4$;
(b) $X_2^2 + X_3^2 + X_4^2$;
(c) $[(X_3 - 4) + (X_4 - 4) + (X_5 - 4)]^2$;
(d) $y_9^2 + y_{10}^2 + y_{11}^2 + y_{12}^2 + y_{13}^2$;
(e) $kn_2 + kn_3 + kn_4 + kn_5$;
(f) $(X_1 - m)^2 + (X_2 - m)^2 + (X_3 - m)^2$.

3. If $X_1 = 3$, $X_2 = 9$, $X_3 = -7$, $X_4 = -3$, find the numerical values of the following:

(a) $\displaystyle\sum_{i=2}^{4} (X_i + 1)^2$; (b) $\displaystyle\sum_{i=1}^{3} X_i(X_i - 7)$.

4. If $X_1 = 6$, $X_2 = 3$, $X_3 = 4$, $X_4 = 0$, $X_5 = 2$, find

(a) $\displaystyle\sum_{i=1}^{4} (X_i - 2)X_i$; (b) $\displaystyle\sqrt{\sum_{i=2}^{5} (X_i^2 + 5)}$.

5. If $X_1 = 7$, $X_2 = -5$, $X_3 = 4$, $X_4 = 2$, find

(a) $\displaystyle\sum_{i=1}^{4} (X_i - 10)^2$; (b) $\displaystyle\sqrt{\sum_{i=2}^{4} X_i(X_i + 4)}$.

6. If $X_1 = 8$, $X_2 = 4$, $X_3 = -4$, $X_4 = 0$, find

(a) $\left[\sum_{i=2}^{4} (X_i + 3)\right]^2$;

(b) $\sum_{i=1}^{4} (X_i - m)^2$, where $m = \dfrac{\sum_{i=1}^{4} X_i}{4}$.

7. If $X_1 = 1$, $X_2 = 4$, $X_3 = 9$, find numerical values of the following:

(a) $\left[\sum_{i=1}^{2} (X_i + 1)\right]^2$;

(b) $\sum (\sqrt{X} + 1)^2$.

8. If $X_1 = -1$, $X_2 = 3$, $X_3 = -5$, $X_4 = 6$, evaluate $\sum (X^2 + 2X - 3)$.

9. If $X_1 = 7$, $X_2 = 8$, $X_3 = 0$, $X_4 = -7$, find

(a) $\sqrt{\sum_{j=2}^{4} (X_j + 5)}$;

(b) $\sum (X + m)^2$, where $m = \dfrac{\sum X}{4}$.

10. If $X_1 = 4$, $X_2 = -5$, $X_3 = 11$, $X_4 = 0$, find

(a) $\sum (X^2 - 3)$;

(b) $\sum_{j=1}^{3} \sqrt{X_j + 5}$.

11. If $X_1 = 9$, $X_2 = 3$, $X_3 = -2$, $X_4 = 0$, $X_5 = 4$, find

(a) $\sum_{i=1}^{5} (X_i^2 - 3X_i)$;

(b) $\sqrt{\sum_{i=2}^{4} (X_i - 2)^2}$.

12. Find

(a) $\sum_{i=1}^{n} 1$;

(b) $\sum_{i=1}^{6} 3$.

In Exercises 13–31 suppose that a set of numbers $X_1, X_2, X_3, \ldots, X_N$ is given and that $m = \left(\sum_{i=1}^{N} X_i\right)/N$. Prove the relationships.

13. $\sum_{i=1}^{3} X_i^2 - \sum_{i=1}^{3} (X_i - 4m)(X_i + m) = 21m^2$.

14. $\sum_{i=1}^{5} X_i(X_i - 1) - \sum_{i=1}^{5} (X_i - 4)(X_i + 3) = 60$.

15. $\sum_{i=1}^{N} [(X_i - 3m)^2 - 2mX_i] = \sum_{i=1}^{N} X_i^2 + \dfrac{\left(\sum_{i=1}^{N} X_i\right)^2}{N}$.

16. $\sum_{i=1}^{N} (X_i - m) = 0$.

17. If b is a constant and $\sum_{i=1}^{N} (X_i - b) = 0$, then $b = m$.

18. If b is a constant, then $\sum_{i=1}^{N} (X_i - b)^2 = \sum_{i=1}^{N} (X_i - m)^2 + Nb^2$. (*Hint:* Substitute $X_i - b = (X_i - m) + (m - b)$ on the left hand side and square).

19. $\sum[(X-m)^2 + X(m-1)] = \sum X^2 - Nm.$

20. $\sum_{j=1}^{N} [X_J(X_J - m) + m^2] = \sum_{j=1}^{N} X_j^2.$

21. $\sum[X(X+m) - m^2(X+1)] = \sum X^2 - \dfrac{(\sum X)^3}{N^2}.$

22. $\sum_{i=1}^{N} [(X_i + m)(X_i - 3m) - mX_i] = \sum_{i=1}^{N} X_i^2 - 6\,\dfrac{\left(\sum\limits_{i=1}^{N} X_i\right)^2}{N}.$

23. $\sum_{i=1}^{N} [X_i(X_i - m^2) - m(1 - 2mX_i) + X_i] = \sum_{i=1}^{N} X_i^2 + \dfrac{\left(\sum\limits_{i=1}^{N} X_i\right)^3}{N^2}.$

24. $\sum_{i=1}^{N} (X_i - 1)(X_i + 2mX_i + 3) - \sum_{i=1}^{N} (X_i^2 - 3) = 2m\sum_{i=1}^{N} X_i^2 + 2mN(1-m).$

25. $\sum[(X-2m)^2 + X - m] = \sum[X(X-m) + m^2].$

26. $\sum[X(X-m) + m - X] = \sum X^2 - \dfrac{(\sum X)^2}{N}.$

27. $\sum_{i=1}^{N} [4mX_i + (X_i - 2m)(X_i + 2m)] = \sum_{i=1}^{N} X_i^2.$

28. $\sum_{j=1}^{N} (X_J - m)(X_J - 2m) + \sum_{j=1}^{N} m(X_J - 1) = \sum_{j=1}^{N} X_j^2 - \sum_{j=1}^{N} X_J.$

29. $\sum(X - 3m)(X - m) + \sum(3m^2 - mX) = \sum X^2 + \dfrac{(\sum X)^2}{N}.$

30. $\sum\left[(X-m)^2 + \dfrac{1}{N}\right] = \sum X^2 - Nm^2 + 1.$

31. $\sum_{i=1}^{N} [X_i(X_i + m) + (X_i - m)^2] = 2\sum_{i=1}^{N} X_i^2.$

4

Analysis of Data

4.1 INTRODUCTION

In order to reduce a mass of data to an understandable form that can be quickly grasped, we construct a frequency table for the data and draw the corresponding histogram or frequency curve. It is useful to simplify the presentation further by defining certain measures that describe important features of the distribution.

Definition 4-1. Any measure indicating a center of the distribution is called a *measure of central tendency*.

In studying frequency distributions, it is also important to know the extent to which the data vary or scatter about a central point; the data may be closely packed about a central point as in (*a*) of Figure 4-1, they may be uniform as in (*b*), or there may be a relatively large number of extreme cases as in (*c*).

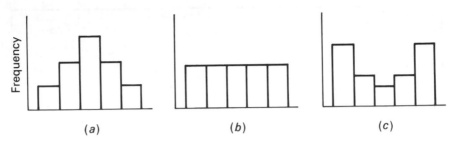

Figure 4-1
Various frequency distributions having the same center.

Definition 4-2.   A numerical value indicating the amount of scatter about a central point is called a *measure of dispersion*.

Definitions of the more important measures of central tendency and of dispersion, along with their relative merits, are set forth in this chapter.

4.2   MEASURES OF CENTRAL TENDENCY

4.2.1   The Mean (or Arithmetic Mean)

Definition 4-3.   For a set of N numbers, X_1, X_2, $\ldots$, X_N, the *mean* (or *arithmetic mean*), denoted by m, is defined as

$$m = \frac{X_1 + X_2 + \cdots + X_N}{N} = \frac{\sum\limits_{i=1}^{N} X_i}{N}. \tag{4-1}$$

Example 4-1. Find to two decimal places the mean of 11, 14, 17, 20, 16, 10.

Solution.

$$m = \frac{11 + 14 + 17 + 20 + 16 + 10}{6} = 14.67$$

It will be observed that the arithmetic mean is what is frequently called the average. However, the term "average" is also used for other measures of central tendency; consequently, whenever this term appears the reader will have to decide from the context which measure of central tendency is meant.

If there are h distinct values of X, we shall denote them by $X_1, X_2, \ldots, X_h$. Then, if X_1 occurs f_1 times, X_2 occurs f_2 times, $\ldots$, X_h occurs f_h times, collecting in the sum (4-1) the f_1 values that are X_1, the f_2 values that are X_2, and so on, we obtain

$$m = \frac{X_1 f_1 + X_2 f_2 + \cdots + X_h f_h}{f_1 + f_2 + \cdots + f_h} = \frac{\sum_{i=1}^{h} X_i f_i}{N}, \tag{4-2}$$

where

$$N = \sum_{i=1}^{h} f_i.$$

Formula (4-2) is useful when the frequency table represents the variates of a discrete variable, as in Table 2-6.

If we are dealing with a continuous variable and if the data are presented in a frequency table, we do not know the exact values of the variates but only the classes in which they lie. However, it seems reasonable to assume that the data within a class are uniformly distributed about the class midpoint, so for the purpose of calculating the mean we shall assume that all data within a class actually occur at the midpoint, which results in the following definition.

Definition 4-4.   If, in a frequency table with h classes, the class midpoints are $X_1, X_2, \ldots, X_h$ and the respective frequencies are $f_1, f_2, \ldots, f_h$, then the mean is defined as

$$m = \frac{X_1 f_1 + X_2 f_2 + \cdots + X_h f_h}{f_1 + f_2 + \cdots + f_h} = \frac{\sum_{i=1}^{h} X_i f_i}{N}, \tag{4-3}$$

where we have again denoted the total number of variates by N; that is,

$$\sum_{i=1}^{h} f_i = N.$$

Formula (4-3) is the same as Formula (4-2), but whereas (4-2) follows directly from Definition 4-3 for the case where equal values of X are collected, Formula (4-3) is the *definition* of the mean of any frequency table.

Example 4-2. Find to two decimal places the mean for the data given in Table 2-4.

Solution. In this case $X_1 = 9, f_1 = 14, X_2 = 12, f_2 = 16, \ldots, X_{10} = 36, f_{10} = 1$; consequently,

$$m = \frac{9 \cdot 14 + 12 \cdot 16 + 15 \cdot 23 + 18 \cdot 25 + 21 \cdot 13 + 24 \cdot 18 + 27 \cdot 5 + 30 \cdot 1 + 33 \cdot 4 + 36 \cdot 1}{120}$$

$$= \frac{2151}{120} = 17.92.$$

However, it is more convenient to arrange this calculation in tabular form, as shown in Table 4-1, where the fourth column (the other columns are not needed for this example) gives the values of Xf and the total of that column is equal to the numerator of m.

Table 4-1
Calculation of the mean and standard deviation for the data of Table 2-4.

1	2	3	4	5	6	7
Class boundaries	X	f	Xf	d	df	d^2f
7.5–10.5	9	14	126	−3	−42	126
10.5–13.5	12	16	192	−2	−32	64
13.5–16.5	15	23	345	−1	−23	23
16.5–19.5	18	25	450	0	0	0
19.5–22.5	21	13	273	+1	+13	13
22.5–25.5	24	18	432	+2	+36	72
25.5–28.5	27	5	135	+3	+15	45
28.5–31.5	30	1	30	+4	+ 4	16
31.5–34.5	33	4	132	+5	+20	100
34.5–37.5	36	1	36	+6	+ 6	36
		120	2151		− 3	495

If h, the number of classes, is large, the above method for calculating the mean may be lengthy. We shall therefore consider a faster method of calculating the mean from a frequency table.

Let

$A =$ any class midpoint (called the *assumed mean*),

$c =$ the difference between two consecutive class midpoints,

$d_i = \dfrac{X_i - A}{c}$; that is, d_i is the number of classes that X_i is away from the assumed mean. It is positive if X_i is greater than A and negative if X_i is less than A.

Then m can be calculated by the following formula:

$$m = A + c \frac{\sum\limits_{i=1}^{h} d_i f_i}{N}. \tag{4-4}$$

A proof of the formula will follow Example 4-3.

Example 4-3.   Find the mean of Example 4-2 by using an assumed mean.

Solution.   Table 4-1, based on an assumed mean, $A = 18$, shows the quantities required in the calculation in the fifth and sixth columns. (*Note*: The last column in Table 4-1 is not used in this example, but will be used in Example 4-14.)

Since $\sum\limits_{i=1}^{h} d_i f_i = -3$, we have $m = 18 + 3\left(\dfrac{-3}{120}\right) = 17.92$.

Proof of Formula (4-4).   Since $d_i = (X_i - A)/c$, we find $X_i = A + cd_i$, and substituting this value of X_i into the definition of m as given by (4-3), we have

$$m = \frac{\sum\limits_{i=1}^{h} X_i f_i}{N} = \frac{\sum\limits_{i=1}^{h} (A + cd_i) f_i}{N} = \frac{\sum\limits_{i=1}^{h} (A f_i + cd_i f_i)}{N} = \frac{\sum\limits_{i=1}^{h} A f_i + \sum\limits_{i=1}^{h} cd_i f_i}{N}$$

$$= \frac{A \sum\limits_{i=1}^{h} f_i + c \sum\limits_{i=1}^{h} d_i f_i}{N} = \frac{AN + c \sum\limits_{i=1}^{h} d_i f_i}{N} = A + c \frac{\sum\limits_{i=1}^{h} d_i f_i}{N}.$$

Although Formula (4-4) is valid no matter what class midpoint is assumed for A, in order to simplify calculations it is advisable to choose A near the value where the mean is expected to occur.

Theorem 4-1.   If a set of N_1 variates has mean m_1 and a set of N_2 variates has mean m_2, then the mean of the set of $N_1 + N_2$ variates obtained by combining these two sets is given by

$$m = \frac{m_1 N_1 + m_2 N_2}{N_1 + N_2}. \tag{4-5}$$

Proof.   For the set of N_1 variates let the class midpoints be $X_1, X_2, \ldots, X_h$, and the corresponding frequencies be $f_1, f_2, \ldots, f_h$; for the set of N_2 variates let the class midpoints be $X_1', X_2', \ldots, X_k'$, and the corresponding frequencies be $f_1', f_2', \ldots, f_k'$. Then

$$m = \frac{\sum\limits_{i=1}^{h} X_i f_i + \sum\limits_{i=1}^{k} X_i' f_i'}{N_1 + N_2}.$$

Since

$$m_1 = \frac{\sum\limits_{i=1}^{h} X_i f_i}{N_1} \quad \text{and} \quad m_2 = \frac{\sum\limits_{i=1}^{k} X_i' f_i'}{N_2},$$

we obtain, by solving the first equation for $\sum\limits_{i=1}^{h} X_i f_i$, the second equation for

$\sum\limits_{i=1}^{k} X_i' f_i'$,

$$\sum\limits_{i=1}^{h} X_i f_i = m_1 N_1 \quad \text{and} \quad \sum\limits_{i=1}^{k} X_i' f_i' = m_2 N_2.$$

Substituting these values into the numerator of the formula for m yields Formula (4-5).

The value of m given by Formula (4-5) is frequently called a *weighted mean*, since the means m_1 and m_2 are weighted according to the number of variates in each set. A weighted mean is only equal to the ordinary mean of m_1 and m_2 if the two sets of data contain the same number of variates.

Theorem 4-1 can be generalized to the case of combining more than two sets of data (see Exercise 16).

Example 4-4.   A mathematics class is divided into two sections, both of which are given the same test. Section 1 (41 students) has a mean score of 62; and section 2 (52 students) has a mean score of 68. Find the mean of the whole class correct to two decimal places.

Solution.   Since $N_1 = 41$, $m_1 = 62$, $N_2 = 52$, $m_2 = 68$, we find

$$m = \frac{62 \cdot 41 + 68 \cdot 52}{93} = 65.35.$$

4.2.2   The Median

The median of a set of N numbers arranged in order of ascending or descending magnitude is the middle number of the set if N is odd, and the mean of the two middle numbers if N is even.

Definition 4-5.   For a set of N numbers, $X_1, X_2, \ldots, X_N$, written in increasing order of magnitude, the median, denoted by M_d, is defined as

$$X_{\frac{N+1}{2}}$$

if N is odd, and as

$$\frac{X_{\frac{N}{2}} + X_{\frac{N}{2}+1}}{2}$$

if N is even.

Example 4-5. Find the median of 6, 7, 9, 12, 16, 20.

Solution.

$$M_d = \frac{X_3 + X_4}{2} = \frac{9 + 12}{2} = 10.5.$$

Definition 4-6. For a frequency distribution the *median class* is defined as the lowest class for which the cumulative frequency exceeds $N/2$.

To obtain a formula for calculating the median from a frequency distribution with h classes we introduce the following notation:
Let

N = total number of variates = $\displaystyle\sum_{i=1}^{h} f_i$,

c = the difference between two consecutive class midpoints,

L = the mean of the lower boundary of the median class and the upper boundary of the next lower class,

f_m = the frequency of the median class,

F_c = the cumulative frequency for the class next lower than the median class, which also equals the number of variates less than L.

Then, assuming that the data in the median class are distributed uniformly throughout the interval, by linear interpolation

$$c \cdot \frac{\frac{N}{2} - F_c}{f_m}$$

is the distance from L to the median as shown in Figure 4-2. It might appear that this expression should assume different forms depending upon whether N is even or odd; this, however, is not the case if the data are uniformly distributed throughout the median class.

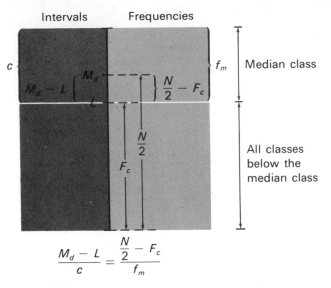

Figure 4-2
Location of median in median class.

We therefore have the following definition:

Definition 4-7. For a frequency distribution the *median* is defined as

$$M_d = L + c \cdot \frac{\frac{N}{2} - F_c}{f_m}. \tag{4-6}$$

Example 4-6. Find to two decimal places the median for the data of Table 4-1.

Solution. The median class is the 16.5–19.5 class. Therefore,

$$M_d = 16.5 + 3 \cdot \frac{\frac{120}{2} - 53}{25} = 16.5 + 3 \cdot \frac{7}{25} = 16.5 + 0.84 = 17.34.$$

Example 4-7. Find the median and median class of the data in Table 4-2, which represent the grades on an examination given to a class of 66 students.

Table 4-2
Frequency table for grades on an
examination.

Class boundaries	Frequency
30–39	6
40–49	12
50–59	15
60–69	15
70–79	8
80–89	6
90–99	4
	66

Solution.   Here the median falls between 59 and 60. Therefore, by Definition 4-6 the 60–69 class is the median class, and since $N/2 = 33$ and $F_c = 33$ we have $M_d = 59.5$, which is obvious without the use of Formula (4-6) since it is the value on both sides of which equal frequencies fall.

4.2.3   The Mode

Definition 4-8.   The class (or classes) for which the frequency is greatest is called the *modal class* (or *modal classes*).

In Table 4-1 the modal class is the 16.5–19.5 class; for the data of Table 4-2, the modal classes are the 50–59 class and 60–69 class.

Definition 4-9.   The class midpoint of a modal class is called a *crude mode*. If there are several modal classes, there will be several crude modes, as in Table 4-2.

The true mode is easy to define but very difficult to determine.

Definition 4-10.   If a frequency curve has been fitted to a set of data, a *true mode* is the value on the horizontal axis for which the frequency curve reaches a maximum value.

4.2.4   The Geometric Mean

The geometric mean is less frequently used than the measures of central tendency discussed so far.

Definition 4-11.   For a set of N nonnegative numbers, $X_1, X_2, \ldots, X_N$, the *geometric mean*, denoted by M_g, is defined as

$$M_g = \sqrt[N]{X_1 \cdot X_2 \cdot X_3 \cdots X_N}. \tag{4-7}$$

It is convenient to express this in logarithmic form; thus,

$$\log M_g = \frac{\log X_1 + \log X_2 + \cdots + \log X_N}{N} = \frac{\displaystyle\sum_{i=1}^{N} \log X_i}{N}. \tag{4-8}$$

The geometric mean is applied to data for which the ratio of any two consecutive numbers is either constant (in which case the data form a geometric progression) or nearly constant. This occurs, for example, in data representing the size of a population at consecutive time intervals or the value of a sum of money which is increasing at compound interest.

Example 4-8.   Table 4-3 shows the percent increases in pay received by California State University and College faculty members for each of the academic years from 1966–1967 to 1969–1970. Find the average percent increase for this period.

Table 4-3
Percent increases in pay for California
State University and College faculty.

Year	Percent increase
1966–1967	6.1
1967–1968	8.5
1968–1969	10.5
1969–1970	5.2

Solution.   We need to find an average with the property that, if the faculty member had received this average increase in each of the four academic years, this person's final salary when computed from the annual average increases would equal the actual final salary for 1969–1970. Denote the salaries for each of the five academic years from 1965–1966 to 1969–1970 by $s_0, s_1, \ldots, s_4$ and let $r_1 = s_1/s_0$, $r_2 = s_2/s_1$, and so forth. Thus, since the first increase was 6.1%, we have $r_1 = 1.061$. From Table 4-3 we then obtain the data given in Table 4-4.

Table 4.4

Percent increases in pay for California State University and College faculty.

Year	Percent increase	r_i	$\log r_i$
1966–1967	6.1	1.061	0.02572
1967–1968	8.5	1.085	0.03543
1968–1969	10.5	1.105	0.04336
1969–1970	5.2	1.052	0.02202
			0.12653

The faculty member's final salary is

$$S_4 = S_0 \cdot \frac{S_1}{S_0} \cdot \frac{S_2}{S_1} \cdot \frac{S_3}{S_2} \cdot \frac{S_4}{S_3} = S_0 \, r_1 r_2 r_3 r_4 \,.$$

If that person had received the same percent increase each year, his or her final salary would be

$$S_0 \, rrrr = S_0 \, r^4.$$

Therefore we need to find r such that

$$S_0 \, r^4 = S_0 \, r_1 r_2 r_3 r_4 \,,$$

which means

$$r^4 = r_1 r_2 r_3 r_4$$

or

$$r = \sqrt[4]{r_1 r_2 r_3 r_4} \,.$$

This shows that the geometric mean of r_1, r_2, r_3, and r_4 is needed; that is, $r = M_g$. In this case, we find

$$M_g = \sqrt[4]{r_1 r_2 r_3 r_4}$$

and

$$\log M_g = \frac{\sum_{i=1}^{4} \log r_i}{4} = \frac{0.12653}{4} = 0.03163,$$

from which we obtain

$$M_g = 1.07555,$$

so that the average percent increase for this period was 7.555%.

The arithmetic mean for Example 4-8 is 7.575%. The geometric mean is never greater than the arithmetic mean.

For data presented in a frequency table the following definition applies.

Definition 4-12.   If, in a frequency table, the class midpoints are nonnegative and are denoted by X_1, X_2, ..., X_h, and the respective frequencies are denoted by $f_1, f_2, \ldots, f_h$, then the geometric mean is defined as

$$M_g = \sqrt[N]{X_1^{f_1} \cdot X_2^{f_2} \cdots X_h^{f_h}}, \qquad (4\text{-}9)$$

where

$$N = \sum_{i=1}^{h} f_i.$$

4.2.5   The Harmonic Mean

Another measure of central tendency, which is only occasionally used, is the *harmonic mean*.

Definition 4-13.   For a set of N numbers, X_1, X_2, ..., X_N, the *harmonic mean*, denoted by M_h, is defined as

$$M_h = \frac{N}{\dfrac{1}{X_1} + \dfrac{1}{X_2} + \dfrac{1}{X_3} + \cdots + \dfrac{1}{X_N}} = \frac{N}{\sum\limits_{i=1}^{N} \dfrac{1}{X_i}}. \qquad (4\text{-}10)$$

The harmonic mean is used most frequently in averaging speeds where the distances for which each speed is applicable are the same. Let us assume that a car travels first d miles at speed r_1, then d miles at speed r_2, and finally d miles at speed r_3. Then the average speed is equal to the total distance traveled divided by the total time, or

$$\text{average speed} = \frac{d + d + d}{t_1 + t_2 + t_3}.$$

Now since t_1, the time at which the car traveled at speed r_1, equals

$$t_1 = \frac{d}{r_1}$$

and t_2 and t_3 are obtained similarly, we find

$$\text{average speed} = \frac{3d}{\dfrac{d}{r_1} + \dfrac{d}{r_2} + \dfrac{d}{r_3}} = \frac{3d}{d\left(\dfrac{1}{r_1} + \dfrac{1}{r_2} + \dfrac{1}{r_3}\right)} = \frac{3}{\dfrac{1}{r_1} + \dfrac{1}{r_2} + \dfrac{1}{r_3}},$$

which is the harmonic mean of the three speeds.

Thus, for example, if a car travels a certain distance with an average speed of 20 miles per hour and returns over the same route with an average speed of 30 miles per hour, the average speed for the round trip is given by

$$\frac{2}{\dfrac{1}{20} + \dfrac{1}{30}} = \frac{2}{\dfrac{30 + 20}{600}} = \frac{1200}{50} = 24 \text{ miles per hour,}$$

and is not equal to 25 miles per hour, as might have been guessed.

Example 4-9.   An airplane travels a distance of 900 miles. If it covers the first third and last third distance of the trip at 250 miles per hour and the middle third at 300 miles per hour, find its average speed.

Solution.   Since the distances traveled at each speed are the same, we need to determine the harmonic mean. Using Formula (4-10) we find

$$M_h = \frac{3}{\dfrac{1}{250} + \dfrac{1}{300} + \dfrac{1}{250}} = \frac{3}{\dfrac{6 + 5 + 6}{1500}} = \frac{4500}{17} = 264.71 \text{ miles per hour.}$$

However, if an average speed is to be calculated where the *times* (but not the distances) for which each speed is applicable are the same, the appropriate average is the arithmetic mean.

4.2.6   Relative Merits of Mean, Median, and Mode

The most frequently used measure of central tendency is the arithmetic mean. It is easy to compute, easy to define, takes all measurements into consideration, and is well designed for algebraic manipulation. From the arithmetic means of different sets of data the arithmetic mean of the combined group may easily be calculated, as shown in Theorem 4-1. One of the chief advantages of the mean is its reliability in sampling. If we regard a set of data as being a sample taken from a certain population and then take another sample from the same population, the means of the two samples will in general show closer agreement than will the two medians, i.e. generally speaking, the mean is more stable than the median. For these reasons the mean is the most widely used measure of central tendency in statistics.

The median, although easy to calculate and easy to define, is not influenced by extreme measurements, which for certain situations may prove to be advantageous. In economic statistics it is often desirable to disregard extreme variates, which may be due to unusual circumstances. The median is particularly useful when the magnitude of the extreme data is not given but when their number is known. In this case, the mean cannot be calculated.

The mode is even less important than the median because of its ambiguity. It is possible that two nonadjacent classes will have the same maximum frequency, thereby yielding two modal values, which may be purely accidental because of the particular choice of the class interval. On the other hand, the mode, like the median, is easy to understand and is not greatly influenced by extreme cases.

4.3   MEASURES OF DISPERSION

Although the arithmetic mean is a satisfactory measure of central tendency, it does not in itself give a clear picture of a set of variates or their distribution. Figure 4-1 shows very different frequency distributions having identical means, and Table 4-5 shows three sets of variates having the same mean values. We will explain the other terms used in the table later; for the moment only note that the means are the same.

Obviously, the mean value, $m = 4$, does not describe the three sets in Table 4-5 equally well. It represents set (2) perfectly and set (3) the least well. We need, therefore, some measure of the spread or dispersion of a set of variates about their mean. The first measure that might occur to us is the difference between the largest and smallest values. This is known as the *range*.

4.3.1   The Range

Definition 4-14.   For a set of numbers the *range* is the difference between the largest and smallest numbers.

Definition 4-15.   For a frequency table the *range* is the difference between the upper boundary of the highest class and the lower boundary of the lowest class.

Although the range does give some indication of the spread of the data about the mean, it depends solely on extreme values, which may be purely accidental, and tells us nothing about the distribution of the variates between these two extremes or the concentration of the variates about the center.

Table 4-5
Three sets of variates having identical means.

	1				2				3		
X	X−m	|X−m|	(X−m)²	Y	Y−m	|Y−m|	(Y−m)²	Z	Z−m	|Z−m|	(Z−m)²
3	−1	1	1	4	0	0	0	10	6	6	36
5	1	1	1	4	0	0	0	1	−3	3	9
6	2	2	4	4	0	0	0	0	−4	4	16
3	−1	1	1	4	0	0	0	0	−4	4	16
3	−1	1	1	4	0	0	0	9	5	5	25
20	0	6	8	20	0	0	0	20	0	22	102

Mean	4	4	4
Range	3	0	10
Mean Deviation	1.2	0	4.4
Variance	1.6	0	20.4
Standard Deviation	1.26	0	4.52

4.3.2 The Mean Deviation

A second measure that immediately suggests itself is the sum of the deviations of a set of variates from their mean. From Table 4-5 it appears that this expression, $\Sigma(X - m)$, may always be zero. This is easily shown to be the case for any set of N variates.

Theorem 4-2. The sum of the deviations of a set of variates from their mean is equal to zero.

Proof.

$$\sum_{i=1}^{N}(X_i - m) = \sum_{i=1}^{N} X_i - Nm = \sum_{i=1}^{N} X_i - N\left(\frac{\sum_{i=1}^{N} X_i}{N}\right) = \sum_{i=1}^{N} X_i - \sum_{i=1}^{N} X_i = 0.$$

Similarly, for a frequency distribution $\sum_{i=1}^{h} f_i(X_i - m) = 0.$ (See Exercise 29.) Although, as follows from Theorem 4-2, this expression is of no value in measuring the dispersion, it furnishes a useful device for checking the accuracy of a calculated mean.

Since for purposes of describing the dispersion it is immaterial whether a variate is at a certain distance above or below the mean, the sign of the deviation can be ignored. The average of the deviations, all taken as positive, therefore measures the dispersion of the data about the mean, and is called the *mean deviation*. Its definition involves the concept of *absolute value*, which for a number a is denoted by $|a|$ and is defined as

$$|a| = \begin{cases} a \text{ if } a \geq 0 \\ -a \text{ if } a < 0 \end{cases}.$$

This means that $|a|$ is always nonnegative. Thus, for example, $|-7| = 7$, $|6| = 6$, $|0| = 0$.

Definition 4-16. For a set of N numbers, $X_1, X_2, X_3, \ldots, X_N$, whose mean is m, the *mean deviation*, denoted by $M.D.$ is defined as

$$M.D. = \frac{\sum_{i=1}^{N} |X_i - m|}{N}. \tag{4-11}$$

More accurately, this measure should be called the mean absolute deviation, since the absolute values of the deviations from the mean are used in its calculation. In accordance with common usage we shall refer to it hereafter merely as the mean deviation.

Example 4-10. Find the mean deviation of 11, 14, 17, 20, 16, 10.

Solution. Since $m = 14.67$,

$$M.D. =$$
$$\frac{|11 - 14.67| + |14 - 14.67| + |17 - 14.67| + |20 - 14.67| + |16 - 14.67| + |10 - 14.67|}{6}$$

$$= \frac{3.67 + 0.67 + 2.33 + 5.33 + 1.33 + 4.67}{6} = \frac{18.00}{6} = 3.00.$$

In this case, therefore, we can say that, on the average, the variates differ by 3.00 from the mean.

It is customary to denote the difference $X_i - m$ by x_i; that is,

$$x_i = X_i - m. \tag{4-12}$$

The mean deviation may then be written as

$$M.D. = \frac{\sum\limits_{i=1}^{N} |x_i|}{N}. \tag{4-13}$$

Definition 4-17. If, in a frequency table, the class midpoints are X_1, $X_2, \ldots, X_h$, and the respective frequencies are $f_1, f_2, \ldots, f_h$, then the mean deviation is defined as

$$M.D. = \frac{\sum\limits_{i=1}^{h} |X_i - m| f_i}{N} = \frac{\sum\limits_{i=1}^{h} |x_i| f_i}{N}, \tag{4-14}$$

where

$$N = \sum\limits_{i=1}^{h} f_i.$$

Example 4-11. Find the mean deviation for the data of Table 4-1.

Solution. Since $m = 17.92$, we find the absolute values of the deviations as listed in the fourth column of Table 4-6, and the products of these values with their respective frequencies in the fifth column. (The last two columns of Table 4-6 are not needed in this example.)

Table 4-6
Calculation of mean deviation and standard deviation for the data of Table 4-1.

1	2	3	4	5	6	7
Class boundaries	X	f	$\lvert X-m \rvert$	$\lvert X-m \rvert f$	$(X-m)^2$	$(X-m)^2 f$
7.5–10.5	9	14	8.92	124.88	79.5664	1113.9296
10.5–13.5	12	16	5.92	94.72	35.0464	560.7424
13.5–16.5	15	23	2.92	67.16	8.5264	196.1072
16.5–19.5	18	25	0.08	2.00	0.0064	0.1600
19.5–22.5	21	13	3.08	40.04	9.4864	123.3232
22.5–25.5	24	18	6.08	109.44	36.9664	665.3952
25.5–28.5	27	5	9.08	45.40	82.4464	412.2320
28.5–31.5	30	1	12.08	12.08	145.9264	145.9264
31.5–34.5	33	4	15.08	60.32	227.4064	909.6256
34.5–37.5	36	1	18.08	18.08	326.8864	326.8864
				574.12		4454.3280

Consequently,

$$M.D. = \frac{574.12}{120} = 4.78.$$

The mean deviation includes all the variates, is easy to define, and is easy to compute. It gives due weight to extreme cases. Its disadvantage is its unsuitability for algebraic manipulations, since signs must be adjusted in its definition. For example, there is no shorter method of calculating the mean deviation, as was the case for the mean. Also, if the means and mean deviations of two sets of variates are known, there is no formula allowing the calculation of the mean deviation of the combined set without going back to the original data.

4.3.3   The Variance and Standard Deviation

A method of converting positive and negative quantities into quantities that are all positive is the process of squaring. The average of the squared deviations from the mean, called the *variance*, is used as a measure of dispersion.

Definition 4-18.   For a set of N numbers, X_1, X_2, X_3, ..., X_N, whose mean is m, the *variance*, denoted by σ^2, is defined as

$$\sigma^2 = \frac{\sum\limits_{i=1}^{N}(X_i - m)^2}{N} = \frac{\sum\limits_{i=1}^{N}x_i^2}{N}, \qquad (4\text{-}15)$$

where

$$x_i = X_i - m.$$

The variance is a widely used measure of dispersion in statistical analysis. It is expressed in units that are the squares of the original units. For many purposes it is desirable that a measure of dispersion be expressed in the same units as the original variates and their mean. Such a measure of dispersion is the standard deviation, which is obtained by taking the square root of the variance.

Definition 4-19.   The square root of the variance is called the *standard deviation* and is defined as

$$\sigma = \sqrt{\frac{\sum\limits_{i=1}^{N}(X_i - m)^2}{N}} = \sqrt{\frac{\sum\limits_{i=1}^{N}x_i^2}{N}}. \qquad (4\text{-}16)$$

Example 4-12.   If $X_1 = 3$, $X_2 = 6$, $X_3 = 12$, find the standard deviation.

Solution.   Since $m = 7$,

$$\sigma = \sqrt{\frac{(3 - 7)^2 + (6 - 7)^2 + (12 - 7)^2}{3}} = \sqrt{\frac{42}{3}} = \sqrt{14} = 3.74.$$

Definition 4-20.   If in a frequency table the class midpoints are $X_1, X_2, \ldots, X_h$, and the respective frequencies are $f_1, f_2, \ldots, f_h$, then the standard deviation is defined as

$$\sigma = \sqrt{\frac{\sum\limits_{i=1}^{h}(X_i - m)^2 f_i}{N}} = \sqrt{\frac{\sum\limits_{i=1}^{h}x_i^2 f_i}{N}}, \qquad (4\text{-}17)$$

where

$$N = \sum_{i=1}^{h} f_i.$$

Example 4-13.   Find the standard deviation for the data of Table 4-1.

Solution.   The values of $(X - m)^2$ and $(X - m)^2 f$ needed for the calculation of σ are listed in the last two columns of Table 4-6; we find

$$\sigma = \sqrt{\frac{4454.3280}{120}} = \sqrt{37.12} = 6.09.$$

It is not as easy to grasp the meaning of a standard deviation as it is to interpret a mean deviation, since the standard deviation is not the mean of a set of deviations. On the other hand, the standard deviation has the very remarkable property that in many frequency distributions (in particular the so-called normal distribution, which is discussed in Chapter 7) approximately two-thirds of the variates fall between $m - \sigma$ and $m + \sigma$. Thus, for example, if this property is true for the data of Table 2-4, approximately 80 (which is two-thirds of 120) variates should fall between $17.92 - 6.09 = 11.83$ and $17.92 + 6.09 = 24.01$. It can easily be verified from Table 2-3 that 77 variates actually fall within this range.

For many purposes it is convenient to write the standard deviation in the alternate forms

$$\sigma = \sqrt{\frac{\sum_{i=1}^{h} X_i^2 f_i - \dfrac{\left(\sum_{i=1}^{h} X_i f_i\right)^2}{N}}{N}}, \tag{4-18}$$

or

$$\sigma = \frac{1}{N}\sqrt{N\sum_{i=1}^{h} X_i^2 f_i - \left(\sum_{i=1}^{h} X_i f_i\right)^2}. \tag{4-19}$$

Proof of Formulae (4-18) and (4-19).

$$\sigma^2 = \frac{1}{N}\sum_{i=1}^{h}(X_i - m)^2 f_i = \frac{1}{N}\sum_{i=1}^{h}(X_i^2 f_i - 2mX_i f_i + m^2 f_i)$$

$$= \frac{1}{N}\left(\sum_{i=1}^{h} X_i^2 f_i - 2m\sum_{i=1}^{h} X_i f_i + Nm^2\right)$$

$$= \frac{1}{N}\left[\sum_{i=1}^{h} X_i^2 f_i - 2\left(\frac{\sum_{i=1}^{h} X_i f_i}{N}\right)\left(\sum_{i=1}^{h} X_i f_i\right) + N\left(\frac{\sum_{i=1}^{h} X_i f_i}{N}\right)^2\right]$$

$$= \frac{1}{N} \left[\sum_{i=1}^{h} X_i^2 f_i - 2 \frac{\left(\sum_{i=1}^{h} X_i f_i \right)^2}{N} + \frac{\left(\sum_{i=1}^{h} X_i f_i \right)^2}{N} \right]$$

$$= \frac{1}{N} \left[\sum_{i=1}^{h} X_i^2 f_i - \frac{\left(\sum_{i=1}^{h} X_i f_i \right)^2}{N} \right],$$

from which (4-18) is obtained by taking the square root of both sides. Formula (4-19) follows from (4-18) by multiplying numerator and denominator of the fraction under the radical by N.

Formulae (4-18) and (4-19) are particularly suitable for machine calculation of the standard deviation, since these formulae require only one subtraction whereas Formula (4-17) requires h subtractions. They are, therefore, sometimes referred to as the *computing forms* of the standard deviation.

Since the calculation of σ from its definition may be tedious if the number of classes is large, a faster method of calculation, employing an assumed mean, will be given. Let A represent a class midpoint, to be used as assumed mean. Then σ may be calculated by

$$\sigma = c \sqrt{\frac{\sum_{i=1}^{h} d_i^2 f_i - \frac{\left(\sum_{i=1}^{h} d_i f_i \right)^2}{N}}{N}}, \tag{4-20}$$

or by

$$\sigma = \frac{c}{N} \sqrt{N \sum_{i=h}^{h} d_i^2 f_i - \left(\sum_{i=1}^{h} d_i f_i \right)^2}, \tag{4-21}$$

where c and d_i have the same meanings as those given in the calculation of the mean by Formula (4-4). A proof of Formula (4-21) follows Example 4-14.

Example 4-14. Solve Example 4-13, using an assumed mean.

Solution. Using an assumed mean, $A = 18$, refer to the last column of Table 4-1 for the values of $d^2 f$. Substituting the total of this column and that of the df-column into Formula (4-21) we obtain

$$\sigma = \frac{3}{120} \sqrt{120 \cdot 495 - 9} = \frac{1}{40} \sqrt{59,391} = 6.09.$$

The entries in the $d^2 f$-column are most easily obtained by multiplying the respective numbers in the d- and df-columns together; as an example, $d_1^2 f_1 = d_1(d_1 f_1) = (-3)(-42) = 126$.

Proof of Formula (4-21). As in the proof of Formula (4-4), $X_i = A + cd_i$.
Substituting this into (4-19) we obtain

$$\sigma^2 = \frac{1}{N^2}\left[N\sum_{i=1}^{h} X_i^2 f_i - \left(\sum_{i=1}^{h} X_i f_i\right)^2 \right]$$

$$= \frac{1}{N^2}\left\{ N\sum_{i=1}^{h} (A + cd_i)^2 f_i - \left[\sum_{i=1}^{h} (A + cd_i) f_i\right]^2 \right\}$$

$$= \frac{1}{N^2}\left\{ NA^2 \sum_{i=1}^{h} f_i + 2NAc \sum_{i=1}^{h} d_i f_i + Nc^2 \sum_{i=1}^{h} d_i^2 f_i - \left[A\sum_{i=1}^{h} f_i + c\sum_{i=1}^{h} d_i f_i\right]^2 \right\}$$

$$= \frac{1}{N^2}\left[N^2 A^2 + 2NAc \sum_{i=1}^{h} d_i f_i + Nc^2 \sum_{i=1}^{h} d_i^2 f_i \right.$$

$$\left. - N^2 A^2 - 2NAc \sum_{i=1}^{h} d_i f_i - c^2 \left(\sum_{i=1}^{h} d_i f_i\right)^2 \right]$$

$$= \frac{c^2}{N^2}\left[N\sum_{i=1}^{h} d_i^2 f_i - \left(\sum_{i=1}^{h} d_i f_i\right)^2 \right],$$

from which (4-21) is obtained by taking the square root of both sides.

The variance and standard deviation have all the advantages of the mean
deviation, and in addition are suitable for algebraic work. This, for example,
explains the existence of a short method for calculating the standard deviation. It also makes possible the derivation of a formula (see Exercise 34) that
allows the calculation of the standard deviation of a set of variates, obtained
by combining two sets of variates, whose respective means and standard
deviations (or variances) are known. The variance and standard deviation are
therefore by far the most frequently used measures of dispersion.

EXERCISES

1. For the set of numbers 6.5, 8.5, 4.7, 9.4, 11.3, 8.5, 9.7, 9.4, calculate to two
   decimal places

   (a) the mean;   (c) the mean deviation;
   (b) the median;   (d) the standard deviation.

2. Answer the questions of Exercise 1 for the set of numbers 13.7, 14.6, 15.9, 18.2,
   15.6, 14.3, 20.2.

3. Given the following frequency distribution:

Class boundaries	Frequencies
6–12	3
12–18	5
18–24	10
24–30	12
30–36	7
36–42	3

Find to two decimal places

(a) the mean, using the definition;
(b) the mean, using an assumed mean;
(c) the median;
(d) the modal class;
(e) the mean deviation;
(f) the standard deviation, using the definition;
(g) the standard deviation, using an assumed mean.

4. Measurements of the lengths, in feet, of 50 iron rods are distributed as follows:

Class boundaries	Frequencies
2.35–2.45	1
2.45–2.55	4
2.55–2.65	7
2.65–2.75	15
2.75–2.85	11
2.85–2.95	10
2.95–3.05	2

Answer the questions given for Exercise 3.

5. A set of 6 cards is shaken in a box and dropped on a table 100 times. The number of cards (X) falling face up and the corresponding frequencies (f) are as follows:

X:	0	1	2	3	4	5	6
f:	2	10	24	35	22	6	1

Find (a) the mean, and (b) the standard deviation, for this distribution.

6. Given the following frequency distribution:

Class boundaries	Frequencies
4– 8	3
8–12	8
12–16	11
16–20	15
20–24	18
24–28	3
28–32	2

Answer the questions given for Exercise 3.

7. For the frequency table obtained in Exercise 5 of Chapter 2, answer the questions given for Exercise 3.

8. Given the frequency distribution:

Class boundaries	Frequencies	Class boundaries	Frequencies
0.5– 3.5	2	12.5–15.5	1
3.5– 6.5	3	15.5–18.5	0
6.5– 9.5	6	18.5–21.5	1
9.5–12.5	7		

Find

(a) the mean, without using an assumed mean;
(b) the mean, using an assumed mean;
(c) the median class;
(d) the modal class;
(e) the mean deviation;
(f) the variance, without using an assumed mean;
(g) the variance, using an assumed mean.

9. In two sections of a mathematics course the following distributions of grades are obtained:

Section 1		Section 2	
Range of grades	Frequencies	Range of grades	Frequencies
80–89	4	90–99	2
70–79	10	80–89	6
60–69	18	70–79	12
50–59	9	60–69	20
40–49	5	50–59	10
30–39	2	40–49	2

Note: Any class not listed indicates that there are no scores in that class.

Find

(a) the mean of Section 1;
(b) the mean of Section 2;
(c) the mean of the whole group, using Formula (4-5).

10. Make up a frequency table for the whole group consisting of the two sections in Exercise 9, and from this table find for the whole group
(a) the mean;
(b) the mean deviation;
(c) the standard deviation.

11. If the mean of 75 items is 52.6 gallons and the mean of 25 items is 48.4 gallons, find the mean of the 100 items.

12. Of 500 freshmen, whose mean height is 67.8 inches, 150 are girls. If the mean height of the 150 girls is 62.0 inches, what is the mean height of the boys?

13. A group of 200 students whose mean height is 60.96 inches is divided into two groups, one having a mean height of 63.4 inches, the other having a mean height of 57.3 inches. How many students are there in each of the two groups?

14. In a test given to two sections of a mathematics course the average grade is 60.98. Section 1 has a mean of 57.30, section 2 a mean of 65.30. If there are 27 students in section 1, how many students are there in section 2?

15. In an athletic competition in which 45 participants are entered, 20 compete on the first day, the others on the second day. If the group of 20 participants on the first day scores an average of 48.4 points, what is the smallest average number of points the remaining group of participants has to score on the second day in order for the average score of the 45 participants to be at least equal to 55.0?

16. Three sets of data having N_1, N_2, N_3 variates, respectively, and means m_1, m_2, m_3, respectively, are combined into a single set. Prove that the mean m of this set is

$$m = \frac{m_1 N_1 + m_2 N_2 + m_3 N_3}{N_1 + N_2 + N_3}.$$

17. In a certain college 200 people, consisting of 40 faculty members, 60 graduate students, and 100 undergraduates, were found to have an average weight of 159.5 pounds. If the faculty averaged 10 pounds more per person than the graduate students, and the graduate students 5 pounds more than the undergraduates, what was the average weight of each of the three groups?

18. A transatlantic plane carries 140 passengers. It is known that the average weight of all passengers on that plane is 147.14 pounds, that the men on that plane weigh on the average 170 pounds, and women 130 pounds. How many men and how many women were on the plane?

19. In a final examination given to three sections of a mathematics class totaling 91 students the average grade of all students was 69.3. The averages for sections 1 and 2 were 70.4 and 64.2, respectively. Records giving the number of students in each section and the average grade of section 3 were lost, but the instructors of sections 1 and 2 remember that they had exactly the same number of students while the instructor of section 3 recalls that he had 5 students less than section 1. What was the average grade for section 3?

20. The records about people who died in a particular city during a certain year show that the average age of death was 69.4 years. If the average age of death for the men in this group was 66.4 years and that of the women 72.0 years, what is the ratio of the number of men to the number of women in the records examined?

21. In an examination given to a certain class the average grade was 70.6. The mean grade for all girls in the class was 75.0, and the one for all boys 65.0. What percent of the class were girls and what percent boys?

22. The population of the United States from 1910 to 1950 is given in the following table. Calculate the geometric mean of these data.

Year	Population (in millions)
1910	92.0
1920	105.7
1930	122.8
1940	131.7
1950	150.7

23. The amount of $2000 is invested at 5% on January 1, 1976. If interest is added annually on January 1 to this amount, calculate the average amount invested between January 1, 1976, and December 31, 1980.

24. On July 1, 1976, $5000 is placed into a savings account at 4% interest, compounded semiannually. Find the average amount deposited in the savings account between July 1, 1976, and December 31, 1978, if no withdrawal is made from the account in this period.

25. A man makes an automobile trip lasting 4 days. Each day he travels 8 hours. If the first day he drives at 50 miles per hour, the second day at 52, the third day at 58, and the last day at 60 miles per hour, what is his average speed for the trip?

26. A salesman makes a trip by car lasting 4 days. If he travels 200 miles each day, but drives the first and last days at 50 miles per hour, the second at 55, and the third at 60 miles per hour, what is his average speed on the trip?

For a set of N numbers, X_1, X_2, X_3, ..., X_N, whose mean is m, prove in Exercises 27 and 28 that

27. $\Sigma(X + 2m)^2 - 8Nm^2 = \Sigma X^2$.

28. $\Sigma X(X - m)^2 - m\Sigma(m - 2X)X = \Sigma X^3$.

29. In a frequency distribution, X_1 occurs with frequency f_1, X_2 occurs with frequency f_2, and so on. Prove that

$$\sum_{i=1}^{h}(X_i - m)f_i = 0,$$

where m is the mean and h is the number of classes.

30. If for a set of N values of X it is known that $\Sigma X = 10$, $\Sigma X^2 = 260$, and $\sigma^2 = 25$, find N. (*Hint*: See Example 3-4.)

31. The mean of a set of 10 variates is 3 and the sum of their squares is 100. Find the standard deviation of the set.

32. For a set of N values of X, it is known that $\Sigma X^2 = 1360$, $\Sigma X = 40$, $N\sigma^2 = 1280$; find N.

33. Prove that the geometric mean of two nonnegative numbers is never greater than their arithmetic mean.

34. A set of N_1 variates has mean m_1 and standard deviation σ_1, while another set of N_2 variates has mean m_2 and standard deviation σ_2. If the two sets are combined into a single set, prove that the variance of this set, which we shall denote by σ^2, is given by

$$\sigma^2 = \frac{N_1\sigma_1^2 + N_2\sigma_2^2 + N_1(m_1 - m)^2 + N_2(m_2 - m)^2}{N_1 + N_2},$$

where m is the mean of the combined set.

5

Elementary Probability, Permutations, and Combinations

5.1 DEFINITION OF PROBABILITY

Experience resulting from repeated experiments or from recurrent observations is frequently used to predict the outcome of future events. If we toss a properly balanced coin we can predict with certainty that, excluding the possibility of the coin-on-edge event, it will fall with either a head or a tail showing. We say that the chance of its showing heads for a single toss is 50 : 50, or that it will probably show heads about one-half of the time. From past and present knowledge of weather conditions in a particular locality we may say that it will probably be warm tomorrow or that there is a probability of rain. In this sense we are using the word probability to denote a belief founded on a certain amount of evidence and in many cases, no doubt, influenced by considerable wishful thinking. In mathematical usage the meaning of the word probability is established by definition and is not connected with beliefs or any form of wishful thinking.

Statistics and probability are so fundamentally interrelated that it is impossible to discuss statistics without an understanding of the meaning of probability. A knowledge of probability theory makes it possible to interpret statistical results, since many statistical procedures involve conclusions based on samples which are always affected by random variation, and it is by means of probability theory that we can express numerically the inevitable uncertainties in the resulting conclusions.

Before defining probability let us consider an example. In studying the effect of seed treatments on emergence of cotton seedlings it is necessary to know the percent of emergence or viability of untreated seed. To find this we might start by planting 100 untreated cotton seeds and counting the seedlings that emerge. If 49 seeds germinate, that is, if there are 49 successes (by *success* in statistics we mean the occurrence of the event under discussion) in 100 trials, we say that the relative frequency of success (number of successes divided by number of trials) is 0.49. Next let us plant 200 seeds, and suppose that out of these 103 germinate, resulting in a relative frequency of $103 \div 200 = 0.515$. If we continue by planting more and more seeds, a whole sequence of values for the respective relative frequencies is obtained. If these relative frequencies approach a limiting value, we call this limit the probability of success in a single trial. From the data of Table 5-1 it appears that the

Table 5-1
Relative frequency of emergence of cotton seedlings from untreated seed.

Number of trials (n)	Number of successes (s)	Relative frequency (s/n)
100	49	0.49
200	103	0.515
500	262	0.524
1000	517	0.517
10,000	5193	0.5193

relative frequencies are approaching the value 0.52, which we would then call the probability of a cotton seedling emerging from an untreated seed.

We now can formulate a definition of probability.

Definition 5-1. If the number of successes in n trials is denoted by s, and if the sequence of relative frequencies s/n obtained for larger and larger values of n approaches a limit, then this limit is defined as the *probability of success in a single trial.*

In actual practice a probability is frequently taken as the relative frequency for the largest number of trials for which the experiment has been performed. Thus, for the data of Table 5-1, if the experiment was performed for 10,000 untreated cotton seeds, but not for any larger number, we might take the corresponding relative frequency 0.5193 as the probability of a cotton seedling emerging from an untreated seed. We will also find it convenient for other purposes—for example, for the understanding and proving of theorems on probability—to think of probability as the relative frequency applicable to a large, but finite, number of trials. While this is not strictly accurate, it will never lead us to a wrong conclusion, and will enable us to avoid a discussion of limits, which is beyond the scope of this text.

In many cases, notably in games of chance, the probability may be stated without collecting data on frequencies. Without making trials we can say that the probability of getting a head in tossing a properly balanced coin is $\frac{1}{2}$. We make this statement since, in tossing a properly balanced coin, only two events are possible—namely head or tail—and both these events are equally likely to occur. Similarly, even before throwing an honest die, we can say that the probability of any particular face being on top is $\frac{1}{6}$, since, if we throw one, there are six equally likely events which can occur. For any such cases, in which successes or failures constitute a finite number of equally likely events, the following definition of probability is appropriate.

Definition 5-2.   If an event can happen in s ways and fail to happen in f ways, and if each of these $s + f$ ways is equally likely to occur, the probability of success (the event happening) in a single trial is

$$p = \frac{s}{s+f},$$

and the probability of failure is

$$q = \frac{f}{s+f}.$$

It should be pointed out that this definition is in a sense circular in nature, since the expression "equally likely to occur" itself involves the concept of probability. However, since this term is generally intuitively understood, the concept of "equally likely" will be left undefined. A precise definition of probability without this deficiency can be given in terms of elementary notions concerning sets. Since these notions may, however, not be familiar to the reader, this definition is omitted here.

Definition 5-2 clearly is a special case of Definition 5-1. Thus, the probability of getting a head in the toss of a well-balanced coin can be obtained from either of these definitions. Obviously it is much easier to obtain the probability from the second definition, since in a toss of a coin a head can occur only in $s = 1$ way and a tail in $f = 1$ way, and both ways are equally likely to occur; thus

$$p = \frac{1}{1+1} = \frac{1}{2}.$$

If we use Definition 5-1 we would have had to toss the coin many times and record the number of heads. Clearly, the limit of the ratio of the number of heads to the number of tosses would approach the value $\frac{1}{2}$ for a larger and larger number of trials.

From Definition 5-2 we see that

$$p + q = \frac{s}{s+f} + \frac{f}{s+f} = \frac{s+f}{s+f} = 1.$$

Since an event either happens or fails to happen, certainty must be expressed by unity. From either of the definitions it should be observed that a probability is a number never greater than 1 nor less than 0. If $p = 0$, the event cannot happen at all.

Example 5-1.   From a bag containing 7 black balls and 4 white balls a ball is drawn at random. What is the probability that it is white?

Solution.   A white ball may be drawn in four different ways, and each of the 11 balls has an equal chance of being drawn; therefore

$$s = 4, \qquad f = 7, \qquad \text{and } p = \frac{4}{11}.$$

Hence, the probability of drawing a white ball is $\frac{4}{11}$.

Example 5-2.   In a single throw of two dice what is the probability of getting (a) a total of 9;   (b) a total different from 9?

Solution.   (a) A total of 9 can be obtained as follows: $6+3$, $3+6$, $5+4$ $4+5$, where the first number is the face showing on the first die and the second number is that on the second die. Therefore, there are four possible ways to get a total of 9 out of 36 equally likely ways in which two dice can fall, and the probability of getting a total of 9 is

$$p = \frac{4}{36} = \frac{1}{9}.$$

(b) It is certain that the faces of the two dice total 9 or a number different from 9, and therefore the probability of getting a total other than 9 is

$$q = 1 - \frac{1}{9} = \frac{8}{9}.$$

This result could have been obtained by enumeration. If there are four ways out of 36 of getting a total of 9, there must be 32 ways of getting a total different from 9, and therefore the probability of getting a total different from 9 is

$$q = \frac{32}{36} = \frac{8}{9}.$$

Although students of mathematics should always be encouraged to guess the answer to a problem before solving it, it should be pointed out that guessing answers to a probability problem is hazardous, since answers to such problems frequently are completely different from what intuition might indicate. The following example and Example 5-10 serve to illustrate this point. In both, the reader is advised not to look at the solution until he has made a guess or, to lend added emphasis to the point, let some of his friends make a guess.

Example 5-3.   A hat-check girl checks the hats of 100 men going to a theater. Accidentally she mixes these hats up completely. Not willing to admit her error, she hands back these 100 hats to the men as they leave the theater, completely at random, hoping that at least some will get their own hats back. What is the probability that at least one of these men gets his own hat back?

Solution.   In order to understand this problem better, let us first solve it for the case in which 100 is replaced by 2 everywhere in the example, that is, where only 2 men go to the theater. To list successes and failures, let us denote by A the first man leaving the theater, by B the second man leaving the theater, by H_A the hat belonging to A, and by H_B the hat belonging to B. Then there are two equally likely possibilities when the hat-check girl returns the hats:

$$H_A H_B$$
$$H_B H_A.$$

In the first, since A leaves the theater first, both A and B get their own hats back, in the second neither, so that $s = 1, f = 1$, and $p = \frac{1}{2}$.

Now let us look at the problem for the case in which 100 is replaced by 3 everywhere in the example. Again, let us denote by A the first man leaving the theater, by B the second, and by C the third. Then there are the following equally likely possibilities when the girl returns the hats:

$$H_A H_B H_C$$
$$H_A H_C H_B$$
$$H_B H_A H_C$$
$$H_B H_C H_A$$
$$H_C H_A H_B$$
$$H_C H_B H_A.$$

Of course, the first, second, third, and sixth are successes, since in each of these at least one of the men receives his own hat back; for example, in the sixth, because B leaves the theater second he receives his own hat back. Thus, there are four successes and two failures and

$$p = \frac{4}{6} = \frac{2}{3}.$$

At this stage, the reader is undoubtedly amazed at the high probabilities that were obtained and is prepared to guess that, since the probability has increased from $\frac{1}{2}$ to $\frac{2}{3}$ as we increased the number of men entering the theater from 2 to 3, it will increase further as we consider the case in which 4 men enter the theater. Not so, however. As the reader may verify himself (see Exercise 9), the probability for this case turns out to be lower than $\frac{2}{3}$. The following table shows the probability of at least one man getting his own hat back for increasing numbers of men entering the theater.

Number of men	Probability of at least one man getting his own hat back
2	0.5000
3	0.6667
4	0.6250
5	0.6333
6	0.6319
7	0.6321
8	0.6321

As this table indicates, the probability alternately increases and decreases and stabilizes at a value of 0.6321, if we consider four-decimal accuracy. If these values were indicated to a higher degree of accuracy, it would be seen that the feature of alternating increases and decreases continues indefinitely.

It is clear that the solution of this problem for increasing numbers of men by the indicated procedure of listing all possible successes and failures becomes increasingly tedious. Methods that allow a much more rapid calculation of the solution are available.

5.2   EXPECTATION

Definition 5-3.   If p is the probability of the occurrence of an event in a single trial, then the *expected number of occurrences* or *expectation* of that event in n trials is defined as np.

Example 5-4.   In 900 trials of a throw of two dice, what is the expected number of times that the sum will be less than 5?

Solution.   To find the probability p of obtaining a total less than 5 in a single throw of two dice, we use Definition 5-2 and find that $s = 6$, $s + f = 36$, so $p = \frac{6}{36} = \frac{1}{6}$, and the expected number of times the total will be less than 5 in 900 trials is $\frac{1}{6} \cdot 900 = 150$.

The expectation is easily seen to be the most likely number of successes to occur in n trials. Thus, in Example 5-4 the most likely event is that there will be 150 throws with sum less than 5. This, however, does not mean that this event must happen when two dice are thrown 900 times; in fact, it does not even mean that this event is very likely to happen. It does mean that this event (150 throws with sum less than 5) is more likely to occur than other events. If the probability p has been determined as the relative frequency in n trials, then the expectation in these n trials is easily seen to equal the actual number of successes in these n trials. Thus, if the probability of emergence of cotton seedlings from untreated seed, as taken from the last line of Table 5-1, is 0.5193, then the expectation in 10,000 trials is $0.5193 \cdot 10.000 = 5193$, which equals the actual number of seedlings that emerged.

Definition 5-4.   If p is the probability that a person will win m dollars, his *expectation* is defined as mp.

Example 5-5.   What is the expectation of a person who is to receive $80 if he obtains 3 heads in a single toss of 3 coins?

Solution.   In a single trial, if H stands for head and T for tail, the following possibilities exist: *HHH, HHT, HTH, THH, HTT, THT, TTH, TTT*. Therefore, $s = 1$, $s + f = 8$, and $p = \frac{1}{8}$. Then the expectation is $\frac{1}{8} \cdot 80 = \$10$.

We may say that a player's expectation mp is the fair price to pay per game for the privilege of playing. When the player pays more per game than mp he is sure to lose if he plays long enough, and if he pays less per game he is certain to come out ahead in the long run. The analysis of all games of chance requires the evaluation of the player's expectation.

5.3   THREE ELEMENTARY PROBABILITY LAWS

Probability laws are useful for faster calculation of probabilities. The first of these laws requires the following definition.

Definition 5-5.   If two or more events are such that not more than one of them can occur in a single trial, they are said to be *mutually exclusive*.

In a single draw from a deck of playing cards the two events of obtaining an ace and obtaining a king are mutually exclusive, whereas the two events of drawing an ace and drawing a spade are *not* mutually exclusive since in a particular draw a card may be both an ace and a spade.

Theorem 5-1.   If p_1, p_2, p_3, ..., p_r are the separate probabilities of the occurrence of r mutually exclusive events, the probability P that some *one* of these events will occur in a single trial is

$$P = p_1 + p_2 + p_3 + \cdots + p_r. \tag{5-1}$$

A proof of the theorem follows Example 5-6.

Example 5-6.   Find the probability of drawing an ace, king, or queen from a regular 4-suit, 52-card deck of cards.

Solution.   The probability p_1 of drawing an ace in a single trial is $\frac{4}{52} = \frac{1}{13}$; the probability p_2 of drawing a king is $\frac{1}{13}$; and the probability of drawing a queen is $\frac{1}{13}$. Since these events are mutually exclusive, the probability of drawing an ace or a king or a queen is

$$P = \frac{1}{13} + \frac{1}{13} + \frac{1}{13} = \frac{3}{13},$$

which could also have been found directly from the definition of probability.

Proof of Theorem 5-1.   Consider a large number of trials n. For the purposes of this proof it is most convenient to think of n as the number of trials from which the probabilities $p_1, p_2, \ldots, p_r$ have been determined as relative frequencies so that the expected numbers of occurrences (see Definition 5-3) of each of these r events is equal to its actual number of occurrences. To determine the total number of successes we first indicate in the n boxes of Figure 5-1 the results of the n trials. The order of the trials is indicated by the numbers in

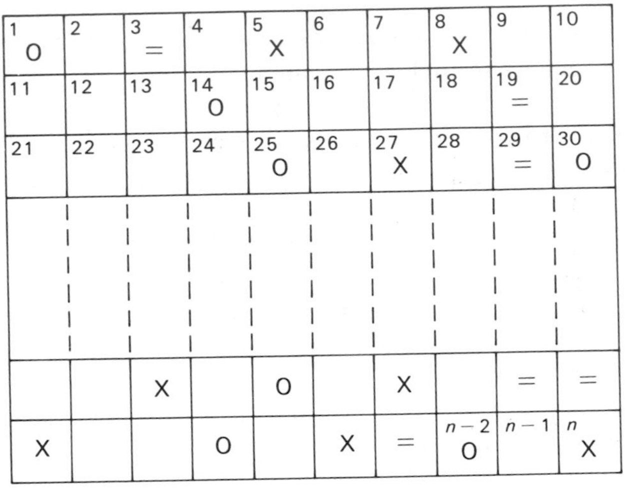

1 $\circ$	2	3 $=$	4	5 $\times$	6	7	8 $\times$	9	10
11	12	13	14 $\circ$	15	16	17	18	19 $=$	20
21	22	23	24	25 $\circ$	26	27 $\times$	28	29 $=$	30 $\circ$
		$\times$		$\circ$		$\times$		$=$	$=$
$\times$			$\circ$		$\times$	$=$	$n-2$ $\circ$	$n-1$	n $\times$

Figure 5-1
Number of occurrences of r mutually exclusive events in n trials.

the boxes; thus, the result of the first trial is marked in box number 1. A cross ($\times$) in a box indicates the occurrence of the first event, a circle ($\circ$) the occurrence of the second event, two horizontal lines ($=$) the occurrence of the third event, and so on.

Then the expected (or, in this case, also actual) number of occurrences for the first event, whose probability is p_1, is equal to $p_1 n$ or the number of boxes marked with a cross; similarly, the expected number of occurrences for the second event is $p_2 n$ or the number of boxes marked with a circle; and the expected number of occurrences for the third event is $p_3 n$ or the number of boxes marked with two horizontal lines, and so on. Note that no box will have more than one mark, i.e. a box will contain either one mark or no mark at all; this is guaranteed by the fact that the events are mutually exclusive.

Now the total number of successes (success in this case happens if some one of the events occurs) is equal to the total number of marked boxes; thus $s = p_1 n + p_2 n + \cdots + p_r n$; consequently, in accordance with the definition of probability,

$$P = \frac{s}{n} = \frac{p_1 n + p_2 n + \cdots + p_r n}{n} = p_1 + p_2 + \cdots + p_r.$$

This derivation makes it clear why Theorem 5-1 is not applicable if the events are not mutually exclusive. It also suggests how to modify this theorem for that case. Consider two events which are not mutually exclusive; let p_1 and p_2 be their separate probabilities and $p_{1,2}$ the probability that both events will occur together in a single trial. Then from a figure corresponding to Figure 5-1 the expected number of occurrences of either the first or the second event, or of both events, is given by the number of boxes marked with a cross plus the number of boxes marked with a circle minus the number of those which have been counted twice (those marked with a cross and a circle), or $s = p_1 n + p_2 n - p_{1,2} n$. Consequently, in this case

$$P = \frac{s}{n} = \frac{p_1 n + p_2 n - p_{1,2} n}{n} = p_1 + p_2 - p_{1,2}. \qquad (5\text{-}2)$$

Since we shall be primarily interested in events which are mutually exclusive, we shall make no further use of (5-2).

Before stating the second probability law we require the following definition.

Definition 5-6.   Two or more events are said to be *independent* if the probability of occurrence of any of them is not influenced by the occurrence of any other. Otherwise the events are *dependent*.

If a die is tossed twice, the occurrence of a 3 on the first throw and the occurrence of a 5 on the second throw are independent events. If, however, two cards are drawn in succession from a deck of cards without replacement, the appearance of an ace on the first draw and the appearance of a king on the second are not independent events, since the probability of getting a king on the second draw depends on what happened on the first draw. If after the first draw the card is replaced, then the outcome of the second draw is independent of the result of the first draw.

Theorem 5-2.   If p_1, p_2, p_3, ..., p_r are the separate probabilities of the occurrence of r independent events, the probability P that *all* of these events will occur in a single trial is

$$P = p_1 p_2 p_3 \cdots p_r. \qquad (5\text{-}3)$$

A proof will be given following Example 5-7.

Example 5-7. What is the probability of obtaining three heads in a single toss of three coins?

Solution. The probability of obtaining a head with the first coin is $p_1 = \frac{1}{2}$; similarly, $p_2 = \frac{1}{2}$, and $p_3 = \frac{1}{2}$. These events are independent, and the probability of tossing three heads with three coins is

$$P = \frac{1}{2} \cdot \frac{1}{2} \cdot \frac{1}{2} = \frac{1}{8}.$$

Compare the result with that in Example 5-5.

Proof of Theorem 5-2. Again let n be a large number of trials, most conveniently thought of as the number of trials from which the probabilities $p_1, p_2, \ldots, p_r$ have been determined as relative frequencies. As in the proof of Theorem 5-1, we first indictate in the n boxes of Figure 5-2 the results of the n trials, again using crosses, circles, or two horizontal lines to mark the boxes for which the first, second, third, and so on, event occurred.

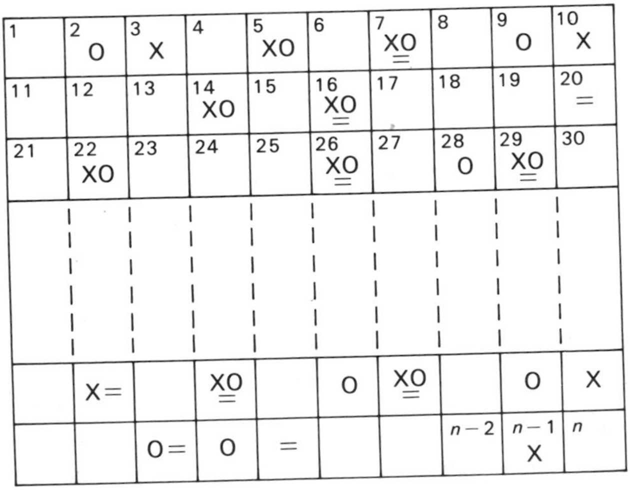

Figure 5-2
Number of occurrences of r independent events in n trials.

The expected (or actual) number of occurrences of the first event is again equal to p_1n. Since success in this case requires that all events occur, we can exclude from further consideration all boxes except those marked with a cross, since for all boxes without a cross the first event did not occur and therefore cannot be included among the successes. Now among the p_1n boxes marked with a cross we must single out those for which also the second event occurred, i.e. those which also are marked with a circle. Since the probability of occurrence of the second event, p_2, is the same for boxes marked with a cross and for boxes without a cross (here we use the fact that the events are independent), we find the expected number of occurrences of the second event among the p_1n boxes to be $p_2(p_1n) = p_1p_2n$, which equals the expected number of occurrences of the first two events. Clearly, the expected number of occurrences of the third event among these p_1p_2n boxes is $p_3(p_1p_2n) = p_1p_2p_3n$. Continuing this process, we see that the expected number of occurrences of all r events is $s = p_1p_2 \cdots p_r n$; thus, in accordance with the definition of probability,

$$P = \frac{s}{n} = \frac{p_1p_2 \cdots p_r n}{n} = p_1p_2 \cdots p_r.$$

Note that the main distinction between Theorems 5-1 and 5-2 is that in the former we are interested in the occurrence of the first *or* the second *or* the third, and so on, event, whereas in the latter we desire the occurrence of the first *and* the second *and* the third, and so on, event. Generally, therefore, a probability problem involving *or* requires the addition of the probabilities of the events (provided of course that they are mutually exclusive); a problem involving *and*, on the other hand, requires the multiplication of the probabilities of the events (provided that they are independent).

Theorem 5-3. In the case of r dependent events, if the probability of the occurrence of a first event is p_1, and if, after this event has occurred, the probability of the occurrence of a second event is p_2, and if, after the first and second events have occurred, the probability of the occurrence of a third event is p_3, and so on, the probability P that all events will occur in the specified order is

$$P = p_1p_2p_3 \cdots p_r. \tag{5-4}$$

A proof of the theorem follows Example 5-8.

Example 5-8. From a deck of 52 cards a card is withdrawn at random and not replaced. A second card is then drawn. What is the probability that the first card is an ace and the second a king?

Solution. The probability of getting an ace on the first draw is $p_1 = \frac{1}{13}$ and of getting a king on the second draw, after an ace has been withdrawn, is $p_2 = \frac{4}{51}$. Therefore, the desired probability is

$$P = \frac{1}{13} \cdot \frac{4}{51} = \frac{4}{663}.$$

This means that in a long run of drawing two cards in a row, when the first card drawn is not replaced, an ace followed by a king will occur only 4 in 663 times or about once in 166 trials.

Proof of Theorem 5-3. The proof of this theorem is almost identical to that of Theorem 5-2. Again using Figure 5-2, we consider the $p_1 n$ boxes marked with a cross, i.e. those for which the first event occurred. As before, all boxes not marked with a cross can be ruled out from further consideration. Restricting ourselves therefore to the $p_1 n$ boxes with a cross, we need to determine the expected number among them for which the second event occurred. Since p_2 in this case was defined exactly for boxes for which the first event had occurred, we find the expected number of occurrences of the second event among these $p_1 n$ boxes to be $p_2(p_1 n) = p_1 p_2 n$. The completion of this proof is now evident.

Frequently Theorem 5-2 is considered a special case of Theorem 5-3. Obviously, Theorem 5-3 is applicable also to independent events, since in that case p_2, for example, is the same whether or not the first event has occurred, and so, for independent events, Theorem 5-3 becomes Theorem 5-2. In general, however, it is more convenient to think of these two theorems as two separate probability laws.

In solving probability problems the work will be greatly simplified if we proceed as follows:

1. List all mutually exclusive events that are successes for the problem.
2. For each of the events listed in step 1 determine the probability.
3. Add all of the probabilities obtained in step 2.
4. If desirable, check the result by listing the failures for the problem.

This procedure is illustrated in the following three examples.

Example 5-9. The probability that a certain beginner at golf gets a good shot if he uses the correct club is $\frac{1}{3}$, and the probability of a good shot with an incorrect club is $\frac{1}{4}$. In his bag are 5 different clubs, only one of which is correct for the shot in question. If he chooses a club at random and takes a stroke, what is the probability that he gets a good shot?

Solution. In accordance with step 1 of the procedure stated above, we list the two events that are successes; next we list on the same lines their corresponding probabilities.

(a) Right club, good shot: $p_1 = \dfrac{1}{5} \cdot \dfrac{1}{3} = \dfrac{1}{15} = \dfrac{1}{15}$

(b) Wrong club, good shot: $p_2 = \dfrac{4}{5} \cdot \dfrac{1}{4} = \dfrac{1}{5} = \dfrac{3}{15}$.

$$\dfrac{4}{15}$$

Since either of these cases will result in a good shot, i.e. success, we need to add the probabilities of the two events:

$$P = \frac{1}{15} + \frac{3}{15} = \frac{4}{15}.$$

We can now check this result by listing the cases that are failures together with their respective probabilities.

(c) Right club, poor shot: $p_3 = \dfrac{1}{5} \cdot \dfrac{2}{3} = \dfrac{2}{15} = \dfrac{2}{15}$

(d) Wrong club, poor shot: $p_4 = \dfrac{4}{5} \cdot \dfrac{3}{4} = \dfrac{3}{5} = \dfrac{9}{15}$.

$$\dfrac{11}{15}$$

Since the probability of failure, $\frac{11}{15}$, and the probability of success, $\frac{4}{15}$, total 1, our result checks.

In some problems the number of situations that are failures may be considerably less than the number of situations that are successes. In such cases it is more convenient to determine the probability of failure and subtract it from 1 to obtain the probability of success. The following is a typical example.

Example 5-10. What is the probability that among 23 randomly selected people at least two have birthdays falling on the same day, that is, on the same month and day (not necessarily the same year)?

Solution. Here failure occurs only if all 23 people have birthdays falling on different days of the year. Let us, therefore, calculate this probability. We shall assume that each year has 365 days and that the probability of a person having a birthday on any of these days is the same. Now all 23 people will have different birthdays if all of the following 23 dependent events occur: the first person has a birthday on any day of the year, which occurs, of course, with probability 1; the second person has a birthday on any day of the year except the one on which the first person has his birthday, for which the probability is $\frac{364}{365}$; the third person has a birthday on any day of the year except those of the first two, for which the probability is $\frac{363}{365}$, and so on. Therefore the probability Q that all 23 people have different birthdays, in accordance with Theorem 5-3, is found to be

$$Q = 1 \cdot \frac{364}{365} \cdot \frac{363}{365} \cdot \frac{362}{365} \cdots \frac{343}{365} = 0.493,$$

so that the desired probability P that at least two people have birthdays falling on the same day is given by

$$P = 1 - 0.493 = 0.507.$$

This means that the chance is better than even that among 23 people in a room at least two have the same birthday, a result few people would guess. Even more surprising is the fact that this probability increases very rapidly with increasing numbers of people in the room, as shown in the following table.

Number of people	Probability of at least 2 people having the same birthday
20	0.411
21	0.444
22	0.476
23	0.507
24	0.538
30	0.706
40	0.891
50	0.970
60	0.994

These probabilities were all calculated on the assumption that the probability of a person's birthday falling on any one day of the year is equal to that for any other day. In reality, this is not true, and the actual probabilities involved are even greater than those given in the table.

There are also problems in which there are infinitely many situations that are successes, such as the following.

Example 5-11. *A, B,* and *C* in order toss a coin. The first one to throw a head wins. What are their respective chances of winning?

Solution. Let us first determine the probability that *A* will win; we shall denote this probability by P_A. If *H* stands for head and *T* for tail, then the following situations will mean success for *A* (note that *TTTH* means that *A, B,* and *C* all obtained tails on their first throws, but *A* obtained a head on the second round):

$$H \qquad p_1 = \frac{1}{2}$$

$$TTTH \qquad p_2 = \frac{1}{2} \cdot \frac{1}{2} \cdot \frac{1}{2} \cdot \frac{1}{2} = \frac{1}{16}$$

$$TTTTTTH \qquad p_3 = \frac{1}{2} \cdot \frac{1}{2} \cdot \frac{1}{2} \cdot \frac{1}{2} \cdot \frac{1}{2} \cdot \frac{1}{2} \cdot \frac{1}{2} = \frac{1}{128}$$

$$\begin{matrix} \cdot \\ \cdot \\ \cdot \end{matrix} \qquad\qquad \begin{matrix} \cdot \\ \cdot \\ \cdot \end{matrix}$$

To obtain the desired probability we need to add the probabilities of these mutually exclusive events, i.e.

$$P_A = \frac{1}{2} + \frac{1}{16} + \frac{1}{128} + \cdots .$$

This is an infinite geometric series and we recall that for such a series

$$a + ar + ar^2 + ar^3 + \cdots = \frac{a}{1-r},$$

where *r*, the ratio between any two consecutive numbers, is between -1 and 1. In this case, therefore, since $a = \frac{1}{2}, r = \frac{1}{8}$,

$$P_A = \frac{\frac{1}{2}}{1 - \frac{1}{8}} = \frac{\frac{1}{2}}{\frac{7}{8}} = \frac{1}{2} \cdot \frac{8}{7} = \frac{4}{7}.$$

The calculation of P_B and P_C is left as an exercise (see Exercise 111).

5.4   CONDITIONAL PROBABILITY

We have seen how Theorem 5-3 can be applied to calculate the probability P of the occurrence of two dependent events, say A and B, if we know the probability p_1 of the occurrence of the event A, and the probability p_2 of the occurrence of the event B after the event A has occurred. Thus, in Example 5-8, we found the probability of obtaining an ace and a king when withdrawing two cards from a deck of 52 cards to be

$$P = p_1 p_2 = \frac{1}{13} \cdot \frac{4}{51} = \frac{4}{663}.$$

Now consider the problem in which we are given the fact that the event A has occurred and we wish to find the probability p_2 of the occurrence of the event B. For example, let us find the probability p_2 of withdrawing a king after an ace has been withdrawn. This problem can be solved in two ways: either by noting that after an ace has been withdrawn, there are 4 kings left among the remaining 51 cards, so that $p_2 = \frac{4}{51}$, or by solving the equation

$$P = p_1 p_2$$

for p_2, so that

$$p_2 = \frac{P}{p_1}, \tag{5-5}$$

from which we obtain

$$p_2 = \frac{\dfrac{4}{663}}{\dfrac{1}{13}} = \frac{4}{51}.$$

Definition 5-7.   The probability of the occurrence of an event B, given that an event A has occurred, is called the *conditional probability* of the occurrence of the event B.

To indicate the events involved we shall now use $P(B|A)$, instead of p_2, as notation for the probability of the occurrence of an event B, given that an event A has occurred. To indicate the probability of the occurrence of an event B, when we do not know whether another event has occurred, we shall simply write $P(B)$.

The probability $P(B)$, therefore, is the probability of the occurrence of an event B before any experiments have taken place; accordingly it is called a *prior* or an *a priori* probability. On the other hand, the probability $P(B|A)$

represents the probability of the occurrence of an event B after some experimentation has taken place—namely, after it has become known that the event A has occurred—accordingly, it is called a *posterior* or an *a posteriori* probability.

In terms of the notation just introduced, Formula (5-5) can be written as

$$P(B|A) = \frac{P(AB)}{P(A)}, \tag{5-6}$$

where $P(AB)$ denotes the probability of the occurrence of both events A and B.

In many problems, it is of interest to calculate an *a posteriori* probability from given *a priori* probabilities. Such a formula was first stated by the English clergyman Thomas Bayes, who died in 1761, but whose now famous formula was not published until 1763. It is stated in the following theorem.

Theorem 5-4. If $B_1, B_2, \ldots, B_n$ are n mutually exclusive events, of which some one must occur in a given trial, that is, $P(B_1) + P(B_2) + \cdots + P(B_n) = 1$, and A is any event for which $P(A) \neq 0$, then the conditional probability $P(B_i|A)$ for any one of the events B_i, given that the event A has occurred, is given by

$$P(B_i|A) = \frac{P(B_i)P(A|B_i)}{P(B_1)P(A|B_1) + P(B_2)P(A|B_2) + \cdots + P(B_n)P(A|B_n)}.$$

A proof of Theorem 5-4 follows Example 5-12. The formula is known as *Bayes' Theorem, Bayes' Law,* or *Bayes' Formula.* It is a useful formula, but its applicability is limited by the fact that in many cases the *a priori* probabilities involved in the formula are unknown.

Example 5-12. In Example 5-9, if it is known that a good shot has resulted, what is the probability that the right club was used?

Solution. Let A represent the event of obtaining a good shot, B_1 the event of selecting the right club, and B_2 that of selecting the wrong club. Then $P(B_1) = \frac{1}{5}, P(B_2) = \frac{4}{5}, P(A|B_1) = \frac{1}{3}, P(A|B_2) = \frac{1}{4}$ and, using Bayes' Formula,

$$P(B_1|A) = \frac{P(B_1)P(A|B_1)}{P(B_1)P(A|B_1) + P(B_2)P(A|B_2)} = \frac{\frac{1}{5} \cdot \frac{1}{3}}{\frac{1}{5} \cdot \frac{1}{3} + \frac{4}{5} \cdot \frac{1}{4}} = \frac{\frac{1}{15}}{\frac{4}{15}} = \frac{1}{4}.$$

One can conclude that only one out of four good shots can be attributed solely to use of the right club.

This example might have been solved, without use of Bayes' Formula, as follows. In n trials, the expected number of good shots with the right club is $(\frac{1}{5} \cdot \frac{1}{3})n = \frac{1}{15}n$, and the expected number of all good shots is $(\frac{1}{5} \cdot \frac{1}{3} + \frac{4}{5} \cdot \frac{1}{4})n = \frac{4}{15}n$. Therefore, the probability that the right club was used is that proportion of the expected number of good shots in which the right club was used, namely

$$P(B_1 \mid A) = \frac{\dfrac{1}{15}n}{\dfrac{4}{15}n} = \frac{1}{4}.$$

In this case, knowledge that the shot was good has increased the probability from $P(B_1) = \frac{1}{5}$ to $P(B_1 \mid A) = \frac{1}{4}$. In other problems, the added information may result in a decrease in the probability obtained.

Proof of Theorem 5-4. The probability of the joint occurrence of two events A and B_i, where B_i is any one of the events $B_1, B_2, \ldots, B_n$, in accordance with Formula (5-6), may be written as

$$P(AB_i) = P(A)P(B_i \mid A) = P(B_i)P(A \mid B_i),$$

from which it follows, if $P(A) \neq 0$, that

$$P(B_i \mid A) = \frac{P(B_i)P(A \mid B_i)}{P(A)}. \tag{5-7}$$

Now, since $B_1, B_2, \ldots, B_n$ are n mutually exclusive events of which some one must occur in a given trial, we have

$$P(A) = P(AB_1) + P(AB_2) + \cdots + P(AB_n)$$
$$= P(B_1)P(A \mid B_1) + P(B_2)P(A \mid B_2) + \cdots + P(B_n)P(A \mid B_n). \tag{5-8}$$

Substituting this value of $P(A)$ in the denominator of Formula (5-7) yields Bayes' Theorem.

Formula (5-7) is, of course, an alternative version of Bayes' Formula and one which is easier to remember. Indeed, it is usually best to calculate $P(A)$ separately first by use of Formula (5-8) and then substitute the result into Formula (5-7). This is illustrated in Example 5-13.

Example 5-13. Three automatic machines produce similar automotive parts. Machine A produces 40% of the total, machine B, 25% and machine C, 35%. On the average, 10% of the parts turned out by machine A do not conform to the specification, and for machines B and C the corresponding percents are 5% and 1%, respectively. If one part is selected at random from the combined output and is found not to conform to the specifications, what is the probability that it was produced by machine A?

Solution. Let D represent the event of selecting a defective part. We need to determine the probability, $P(A|D)$, of the part having been selected from machine A, assuming that the part was defective, which, according to Formula (5-7), is given by

$$P(A|D) = \frac{P(A)P(D|A)}{P(D)}.$$

Now, $P(D)$, the probability of obtaining a defective part, is given—just as was shown in Section 5.3—as the sum of the following probabilities:

$$P(DA) = P(A)\,P(D|A) = (0.40)(0.10) = 0.0400$$
$$P(DB) = P(B)\,P(D|B) = (0.25)(0.05) = 0.0125$$
$$P(DC) = P(C)\,P(D|C) = (0.35)(0.01) = 0.0035$$
$$\overline{0.0560.}$$

Therefore,

$$P(A|D) = \frac{(0.40)(0.10)}{0.0560} = 0.714,$$

which shows that on the average machine A produces 71.4% of the defective parts.

5.5 COMBINATIONS AND PERMUTATIONS

Combinations and permutations deal with the grouping and arrangement of objects, and in calculating probabilities they are useful in determining the number of favorable cases as well as the total number of possible cases.

Definition 5-8. Each of the sets that can be made by using all or part of a given collection of objects, without regard to order of the objects in the set, is called a *combination*.

Definition 5-9. Each different ordering or arrangement of all or part of a set of objects is called a *permutation*.

Example 5-14. Find
(a) the number of combinations,
(b) the number of permutations,
of the letters A, B, C, D in sets of 3.

Solution. (a) The letters A, B, C, D can be taken in sets of 3, without regard to order, in the following ways: ABC, ABD, ACD, and BCD. Therefore, there are four such combinations; i.e. there are four combinations of four objects taken 3 at a time.
(b) If ordering is considered also, there are the following permutations of the letters A, B, C, D in sets of 3: ABC, ACB, BAC, BCA, CAB, CBA, ABD, ADB, BAD, BDA, DAB, DBA, ACD, ADC, CAD, CDA, DAC, DCA, BCD, BDC, CBD, CDB, DBC, DCB. Therefore, there are 24 such permutations; i.e. there are 24 permutations of four objects taken 3 at a time.

Example 5-15. Find
(a) the number of combinations,
(b) the number of permutations
of four objects taken four at a time.

Solution. (a) Since all 4 objects have to be used in the set, there is only one such combination in this case.
(b) The number of orderings of the 4 letters A, B, C, D is easily found to be 24, thus there are 24 permutations of 4 objects taken four at a time.

To find simple formulae for the number of combinations and permutations, we shall first consider the special case of finding the number of permutations of n objects (say letters) in sets of n items.

To write all such permutations we observe that we have n choices for the first letter; to each of these corresponds one of the "trees" of Figure 5-3, which represents the case $n = 4$. After having chosen the first letter (say, for example, A) we have $n - 1$ choices left for the second letter; thus, for the first two letters there are $n(n - 1)$ possible choices—as many choices as there are branches emanating from the left tips of the trees of Figure 5-3. Now, after the first two letters have been chosen there remain $n - 2$ choices for the third letter, so there are $n(n - 1)(n - 2)$ choices for the first three letters. Continuing this process, we see that there is only one choice left for the nth letter; consequently the n letters can be arranged in $n(n - 1)(n - 2) \cdots 2 \cdot 1$ ways.

If the symbol $n!$ (read "n factorial") denotes the product of the first n positive integers, that is,

$$n! = n(n - 1)(n - 2) \cdots 2 \cdot 1, \tag{5-9}$$

and $P_{n,n}$ denotes the number of permutations of n objects arranged in sets of n, we have shown that

$$P_{n,n} = n! \qquad (5\text{-}10)$$

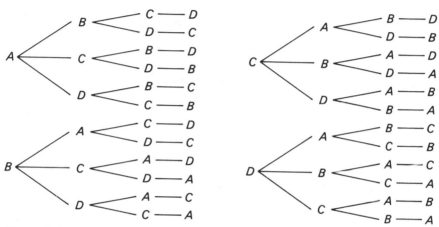

Figure 5-3
The permutations of A, B, C, D.

Example 5-16.   Solve part (b) of Example 5-15 by use of Formula (5-10).

Solution.   In this case we need the number of permutations of 4 objects in sets of 4 (or $P_{4,4}$), and we find

$$P_{4,4} = 4! = 4 \cdot 3 \cdot 2 \cdot 1 = 24,$$

which agrees with the result found directly from the definition of permutations in the solution to Example 5-15 (b).

To develop a formula for the number of permutations of n things taken r at a time, we consider the special case of 4 objects arranged in sets of 3. In this case there are 4 choices for the first letter (again see Figure 5-3), and after the first letter has been chosen there are 3 choices left for the second letter; thus, for the first 2 letters there are $4 \cdot 3 = 12$ possible choices. Now, for the third letter there remain 2 choices, so there are $4 \cdot 3 \cdot 2 = 24$ choices of 3 letters out of a total of 4. This shows that, in general, if we denote the number of permutations of n objects taken r at a time by $P_{n,r}$, we need to take the product of the r numbers $n, n - 1, n - 2, \ldots$, or

$$\overbrace{P_{n,r} = n(n - 1)(n - 2) \cdots (n - r + 1)}^{r \text{ factors}}. \tag{5-11}$$

If we multiply and divide the right-hand side of (5-11) by

$$(n - r)(n - r - 1) \cdots 2 \cdot 1,$$

the numerator becomes $n!$; thus we can write

$$P_{n,r} = \frac{n!}{(n - r)!}. \tag{5-12}$$

Example 5-17.   Solve part (b) of Example 5-14 by use of Formula (5-11).

Solution.   Here $n = 4$, $r = 3$, so

$$P_{4,3} = \frac{4!}{1!} = 4 \cdot 3 \cdot 2 = 24,$$

which agrees with the result found directly from the definition in the solution to Example 5-14 (b).

To develop a formula for the number of combinations of n things taken r at a time, we note that for every set of r letters there is just one combination if all r letters are to be used, whereas there are $r!$ permutations of these r letters. Thus, there are $r!$ times as many permutations of n objects taken r at a time as there are combinations of n objects taken r at a time. This shows that, if we denote the number of combinations of n objects taken r at a time by $C_{n,r}$, we have

$$C_{n,r} = \frac{n!}{r!(n - r)!}, \tag{5-13}$$

or, if we divide numerator and denominator by $(n - r)!$,

$$C_{n,r} = \frac{n(n - 1)(n - 2) \cdots (n - r + 1)}{r!}. \tag{5-14}$$

Since, by definition, $0! = 1$, Formula (5-13) holds also for $r = 0$, in which case $C_{n,0} = n!/n! = 1$, so that Formula (5-13) is valid for $r = 0, 1, 2, \ldots, n$. Formula (5-14), on the other hand, holds only for $r = 1, 2, \ldots, n$.

The following notations are also used for $C_{n,r}$: C_r^n, C_r^n, ^nC_r, $_nC_r$, $C(n, r)$, $\binom{n}{r}$.

Example 5-18. Solve part (a) of Example 5-14 by use of Formula (5-13).

Solution. Here $n = 4$, $r = 3$, so by use of Formula (5-13)

$$C_{4,3} = \frac{4!}{3! \, 1!} = \frac{4 \cdot 3 \cdot 2 \cdot 1}{3 \cdot 2 \cdot 1 \cdot 1} = 4,$$

which agrees with the result found directly from the definition in the solution to Example 5-14 (a).

The main difficulty encountered in problems involving permutations and combinations is deciding whether the problem asks for the number of permutations or the number of combinations. In solving such problems it is advisable to determine first whether a change of order in the objects would make a difference in the desired result; if it does, the number of permutations is desired; if it does not, the number of combinations is wanted.

Example 5-19. A student body president is asked to appoint a committee consisting of 5 boys and 3 girls. He is given a list of 10 boys and 7 girls from which to make the appointments. From how many possible committees must he make his selection?

Solution. In a particular committee, if the order of the appointees is changed, the result would still be the same committee, so this is a problem asking for combinations. From 10 boys the student body president can select

$$C_{10,5} = \frac{10!}{5! \, 5!} = 252$$

groups of 5 boys. From 7 girls he can select

$$C_{7,3} = \frac{7!}{3! \, 4!} = 35$$

groups of 3 girls. Therefore, for each of the 252 selections of boys there are 35 possible choices of girls; thus the total number of possible committees is equal to

$$252 \cdot 35 = 8820.$$

Note that in Example 5-19 we assumed no distinction between the various positions in the committee. However, if the committee is to consist of a president, vice-president, secretary, treasurer, etc., that is, if each of the eight positions of the committee has a different designation and we wish to differentiate also between the various possible assignments of the members of the committee to these positions, we must, of course, calculate the number of permutations.

5.6 REPEATED TRIALS

If, during repeated trials, the probability of success in a single trial remains fixed, the calculation of the probability of obtaining a given number, say r, of successes in these n trials can be greatly simplified. We shall demonstrate this for the case where we toss a coin several times (or toss several coins once) by listing the probabilities as in Table 5-2, where H represents a head, T a tail, and $p = \frac{1}{2}$ the probability of obtaining a head, $q = \frac{1}{2}$ the probability of obtaining a tail.

Table 5-2 clearly suggests that, in n tosses of a coin, the probabilities of the various possible events (i.e. obtaining 0, 1, 2, ..., n heads) are given by the successive terms of the binomial expansion of $(q + p)^n$, which is

$$(q + p)^n = q^n + C_{n,1}q^{n-1}p + C_{n,2}q^{n-2}p^2 + \cdots + C_{n,r}q^{n-r}p^r + \cdots + p^n.$$

The coefficient $C_{n,r}$ represents the number of ways in which n tosses of a coin can produce exactly r heads and $n - r$ tails, which, of course, is precisely the same as the number of combinations of n things taken r at a time. The term

$$C_{n,r} \cdot q^{n-r}p^r = C_{n,r} \cdot \left(\frac{1}{2}\right)^{n-r}\left(\frac{1}{2}\right)^r$$

then represents the probability that exactly r heads and $n - r$ tails occur in the n tosses.

Thus, for example, the probability of obtaining exactly 2 heads in 3 tosses of a coin is given by

$$C_{3,2} \cdot q^{3-2}p^2 = 3\left(\frac{1}{2}\right)\left(\frac{1}{2}\right)^2 = \frac{3}{8},$$

where the coefficient $C_{3,2} = 3$ corresponds to the 3 cases THH, HTH, HHT (see Table 5-2) in which 2 heads can be obtained in 3 tosses of a coin.

Theorem 5-5 is a generalization of this discussion applying to any case of a sample or series of n trials, in each of which the probability of success is fixed and remains fixed during the entire procedure.

Table 5-2
Results of repeated trials in coin tossing

Number of tosses	Enumeration of possible results	Probabilities corresponding to these results	Expansion whose terms correspond to these probabilities
1	T, H	q, p	$(q + p)^1 = q + p$
2	TT, TH, HT, HH	q^2, qp, qp, p^2	$(q + p)^2 = q^2 + 2qp + p^2$
3	TTT, TTH, THT, HTT, THH, HTH, HHT, HHH	$q^3, q^2p, q^2p, q^2p,$ qp^2, qp^2, qp^2, p^3	$(q + p)^3 = q^3 + 3q^2p + 3qp^2 + p^3$
. . .	. . .	. . .	
n			$(q + p)^n = q^n + nq^{n-1}p + \cdots + p^n$

Theorem 5-5. If p is the probability of the success of an event in a single trial and q is the probability of its failure, then the probability P that there will be exactly r successes in n trials is the term in the binomial expansion of $(q + p)^n$ for which the exponent of p is r, that is, $C_{n,r} \cdot q^{n-r} p^r$.

Before proving Theorem 5-5, let us explain how to use it. Recall that

$$(q + p)^n = q^n + nq^{n-1}p + \frac{n(n-1)}{2} q^{n-2}p^2 + \frac{n(n-1)(n-2)}{2 \cdot 3} q^{n-3}p^3 + \cdots$$

$$+ \frac{n(n-1)(n-2) \cdots (n-r+1)}{2 \cdot 3 \cdots r} q^{n-r}p^r + \cdots + p^n$$

$$= q^n + C_{n,1}q^{n-1}p + C_{n,2}q^{n-2}p^2 + \cdots + C_{n,r}q^{n-r}p^r + \cdots + C_{n,n}p^n.$$

A convenient way to obtain the coefficients of the binomial expansion is to note that each coefficient of the expansion can be obtained from the previous term as follows: Multiply the coefficient of the previous term by the exponent of q and divide the result by one more than the exponent of p.

Thus, for example, in the expansion

$$(q + p)^8 = q^8 + 8q^7p + 28q^6p^2 + 56q^5p^3 + 70q^4p^4$$
$$+ 56q^3p^5 + 28q^2p^6 + 8qp^7 + p^8,$$

we obtain the coefficient of q^6p^2 from the term $8q^7p$ as $(8 \cdot 7)/2 = 28$, or the coefficient of q^5p^3 as $(28 \cdot 6)/3 = 56$, etc.

Example 5-20. Find the probability of getting 5 exactly twice in 7 throws of a die.

Solution. The probability of success (getting a 5) in a single trial is $p = \frac{1}{6}$, from which $q = \frac{5}{6}$. Since the number of trials n is 7, and the required number of successes r is 2, the desired probability is the term containing p^2 in the expansion of $(q + p)^7$; that is, $C_{7,2} \cdot q^5 p^2$.
Since

$$(q + p)^7 = q^7 + 7q^6p + 21q^5p^2 + \cdots + p^7,$$

we find

$$P = 21q^5p^2 = 21 \left(\frac{5}{6}\right)^5 \left(\frac{1}{6}\right)^2 = \frac{21,875}{93,312}.$$

Example 5-21. Find the probability of throwing at most 2 sixes in 6 throws with a single die.

Solution.   To throw at most 2 sixes means that we must throw 0, 1, or 2 sixes. Therefore, the desired probability is the sum of the terms in the binomial expansion of $(q + p)^6$ for which the exponents of p are 0, 1, and 2. Since

$$(q + p)^6 = q^6 + 6q^5 p + 15q^4 p^2 + \cdots + p^6,$$

we find

$$P = \left(\frac{5}{6}\right)^6 + 6\left(\frac{5}{6}\right)^5 \left(\frac{1}{6}\right) + 15\left(\frac{5}{6}\right)^4 \left(\frac{1}{6}\right)^2 = \frac{21{,}875}{23{,}328}.$$

Example 5-22.   It is known that 10% of certain articles manufactured are defective. What is the probability that in a random sample of 12 such articles at least 9 are defective?

Solution.   In this case $p = \frac{1}{10}$, $q = \frac{9}{10}$, $n = 12$. Since we are interested in at least 9 defective articles, we need the terms for which the exponents of p are 9, 10, 11, and 12. In the expansion of $(q + p)^{12}$, these terms will occur at the end. In order to obtain these terms at the beginning of the expansion we note that $(q + p)^{12} = (p + q)^{12}$ and find

$$(p + q)^{12} = p^{12} + 12p^{11}q + 66p^{10}q^2 + 220p^9 q^3 + \cdots + q^{12};$$

thus

$$P = \left(\frac{1}{10}\right)^{12} + 12\left(\frac{1}{10}\right)^{11}\left(\frac{9}{10}\right) + 66\left(\frac{1}{10}\right)^{10}\left(\frac{9}{10}\right)^2 + 220\left(\frac{1}{10}\right)^9 \left(\frac{9}{10}\right)^3$$

$$= \frac{165{,}835}{10^{12}} = \frac{33{,}167}{2 \cdot 10^{11}}.$$

Proof of Theorem 5-5.   First determine the probability of obtaining r consecutive successes, followed by $n - r$ consecutive failures. These n events are independent; therefore, the desired probability is

$$\overbrace{p \cdot p \cdots p}^{r} \cdot \overbrace{q \cdot q \cdots q}^{n - r} = p^r q^{n-r}.$$

Since the r successes and $n - r$ failures may occur in any order, it is necessary to count the number of such orders, which is the same as finding the number of ways of writing the product of n factors, r of which are p, and $n - r$ of which are q. This is precisely the number of combinations of n things taken r at a time, and it equals $C_{n,r}$. The desired probability is therefore

$$C_{n,r} \cdot p^r q^{n-r},$$

which is the term in the binomial expansion of $(q + p)^n$ for which the exponent of p is r.

Note that Theorem 5-5 is not applicable if, during the repeated trials, the probability of success does not remain constant. This is the case if the probability of success in a trial is dependent upon the result of one of the previous trials, which makes the events dependent.

EXERCISES

Express all answers representing probabilities as fractions in lowest terms.

1. In a single throw of two dice, what is the probability of getting
   (a) a total of 8;
   (b) a total of at most 8;
   (c) a total of at least 8?

2. In a single throw of three dice, what is the probability of getting
   (a) a total of 5;
   (b) a total of at most 5;
   (c) a total of at least 5?

3. A card is drawn at random from a deck of 52 playing cards. What is the probability that it is
   (a) an ace or a 10;
   (b) a face card;
   (c) an ace or a spade?

4. If a die is rolled, what is the probability that the roll yields a 3 or 4?

5. If a die is rolled twice, what is the probability that the first roll yields a 4, a 5 or a 6, and the second anything but 3?

6. In a single throw of two dice, what is the probability that a doublet (two of the same number) or a 6 will appear?

7. In a single throw of two dice, what is the probability that neither a doublet nor a 7 will appear?

8. In a single throw of two dice, what is the probability that neither a doublet nor an 8 will appear?

9. Solve Example 5-3 for the case in which 4 men go to the theater.

10. Ten balls, numbered 1 to 10, are placed in a bag. One ball is drawn and not replaced, and then a second ball is drawn. What is the probability that
    (a) the balls numbered 3 and 7 are drawn;
    (b) neither of these two balls is drawn?

11. An urn contains 40 balls numbered 1 through 40. Suppose that numbers 1 through 10 are considered " lucky." Two balls are drawn from the urn without replacement. Find the probability that
(a) both balls drawn are "lucky";
(b) neither ball drawn is "lucky";
(c) at least one of the balls drawn is "lucky";
(d) exactly one of the balls drawn is "lucky."

12. In a bag there are 5 white and 4 black balls. If they are drawn out one by one without replacement, what is the chance that the first will be white, the second black, and so on alternately?

13. A bag contains 4 red, 3 green, and 5 black balls. We draw 5 balls in succession, replacing each ball before the next is drawn. Find the probability that we draw 2 red and 3 green balls.

14. On a slot machine there are three reels with the 10 digits 0, 1, 2, 3, 4, 5, 6, 7, 8, 9, plus a star on each reel. When a coin is inserted and the handle is pulled, the three reels revolve independently several times and come to rest. What is the probability of getting
(a) 3 stars;
(b) 1 on the first reel, 2 on the second reel, and 3 on the third reel;
(c) 1, 2, and 3 in any order on the reels;
(d) the same number on each reel?

15. One man draws three cards from a well-shuffled deck without replacement, and at the same time another man tosses two dice. What is the probability of obtaining
(a) three cards of the same suit and a total of 6 on the dice;
(b) three aces and a doublet?

16. Four men in turn each draw a card from a deck of 52 cards without replacing the card drawn. What is the probability that the first man draws the ace of spades, the second the king of diamonds, the third a king, and the fourth an ace?

17. A man owns a house in town and a cabin in the mountains. In any one year the probability of the house being burglarized is 0.01, and the probability of the cabin being burglarized is 0.05. For any one year what is the probability that
(a) both will be burglarized;
(b) one or the other (but not both) will be burglarized;
(c) neither will be burglarized?

18. From a box containing 6 white balls and 4 black balls, 3 balls are drawn at random. Find the probability that 2 are white and 1 is black
(a) if each ball is returned before the next is drawn;
(b) if the 3 balls are drawn successively without replacement.

19. A person draws 3 balls in succession without replacement from a box containing 8 black, 8 white, and 8 red balls. If he is to receive $5 if he does not draw a black ball, what is his expectation?

20. Three people work independently at deciphering a message in code. The respective probabilities that they will decipher it are $\frac{1}{5}$, $\frac{1}{4}$, and $\frac{1}{3}$. What is the probability that the message will be deciphered?

21. After a battle, three soldiers agree to meet and celebrate the event 20 years later if at least two of them are still alive. What is the probability that they can carry out their agreement if the probabilities of their living 20 years are $\frac{12}{13}$, $\frac{9}{10}$, and $\frac{10}{11}$?

22. Coins A and B are weighted so that A comes up heads $\frac{2}{3}$ of the time and B comes up heads $\frac{3}{4}$ of the time. Coin A is tossed; if it comes up heads, B is tossed twice; if it comes up tails, B is tossed once. What is the probability that a total of exactly 2 tails comes up?

23. The probability that a certain door is locked is $\frac{1}{2}$. The key to the door is one of 12 keys in a cabinet. If a person selects two keys at random from the cabinet and takes them to the door with him, what is the probability that he can open the door without returning for another key?

24. A father and son practice shooting at a target. If, on the average, the father hits the target 3 times out of 4, and the son hits it 4 times out of 7, what is the probability, when both shoot simultaneously, that the target is hit at least once?

25. A pair of dice is tossed twice. What is the probability that either a 7 or a 12 appears on the first throw, and that neither a doublet nor an 8 appears on the second throw?

26. Two phenotypically similar but genotypically different flower seeds are planted, one in each of two pots numbered 1 and 2. The different genotypes lead us to expect 3 red-flowered plants to 1 white-flowered plant from one of the seeds, and 1 red-flowered plant to 1 white-flowered plant from the other. Find the probability that
    (a) pot number 1 contains a white flower;
    (b) both pots contain red flowers;
    (c) pot number 1 contains a red flower and pot number 2 a white flower.

27. If 3 dice are thrown, find the probability that
    (a) all 3 will show fours;
    (b) all 3 will be alike;
    (c) 2 will show fours and the third something else;
    (d) only 2 will be alike;
    (e) all 3 will be different.

28. What is the probability of throwing at least 4 sevens in 5 throws of a pair of dice?

29. A bag contains 50 envelopes of which 10 contain $5 each, 10 contain $1 each, and the others are empty. What is our expectation in drawing a single envelope?

30. A, B, and C in the order named are allowed one throw each with three dice. The first one, if any, who throws an 11 gets $583.20. If each pays an amount equal to his expectation to participate in the game, how much should each pay?

31. A class in mathematics is composed of 15 juniors, 30 seniors, and 5 graduate students; 3 of the juniors, 5 of the seniors, and 2 of the graduate students received an *A* grade in the course. If a student is selected at random from the class and is found to have received an *A* grade, what is the probability that he is a junior?

32. Of 100 boxes of fuses containing 5 fuses per box, 20 boxes contain fuses from factory *A*, 30 boxes from factory *B*, and 50 boxes from factory *C*. Fuses from factory *A* are, on the average, 5% defective; from factory *B*, 4%; and from factory *C*, 2%. The fuses and boxes all look alike and are piled without regard to place of manufacture. A box is selected at random, and a fuse in it is tested and found to be defective. What is the probability that it was produced at factory *B*?

33. Of the freshmen in a certain college, it is known that 18% attended private secondary schools and 82% attended public schools. The Dean's office reports that 30% of all students who attended private secondary schools and 15% of those who attended public schools attained grades high enough to be entered on the honors list. If one student is chosen at random from the freshman class and is found to be on the honors list, what is the probability that the student attended a public shool?

34. Of those people voting in an election, 45% are registered Republicans, 40% Democrats, and 15% Independents. There are three candidates for office, a Republican *R*, a Democrat *D*, and an Independent *I*, and the party votes were distributed as follows:

	Republican	Democrat	Independent
For *R*	80%	10%	15%
For *D*	5%	85%	10%
For *I*	15%	5%	75%

   (a) If a voter is selected at random and is found to have voted for *I*, what is the probability that he is a Republican?
   (b) Which candidate received the largest percentage of the vote?

35. Tests employed in the detection of a particular disease are 90% effective; they fail to detect it in 10% of the cases. In persons free of the disease, the tests indicate 1% to be affected and 99% not to be affected. From a large population, in which only 0.2% have the disease, one person is selected at random, is given the tests, and presence of the disease is indicated. What is the probability that the person is affected?

36. An urn contains 3 red marbles and 7 white marbles. A marble is drawn from the urn and a marble of the other color is then put into the urn. A second marble is drawn from the urn. If the two marbles drawn in this process are of the same color, what is the probability that they are both white?

37. Urn A contains 3 red and 5 white marbles; urn B contains 2 red and 1 white marble; urn C contains 2 red and 3 white marbles. An urn is selected at random, and a marble is drawn from the urn. If the marble is red, what is the probability that it came from urn A?

38. Suppose that the probability of a student passing an examination is $\frac{4}{5}$, that of the student sitting to his left $\frac{3}{5}$, and that of the student sitting to his right $\frac{1}{5}$. Assuming that none of these students looks at his neighbor's paper, find the probability that, of these three students,
    (a) exactly two pass the examination;
    (b) the student sitting in the middle passes the examination, given that the student on his right passes it.

39. Suppose that 5% of a student body ride bicycles and have accidents. If 75% of that student body ride bicycles, what is the probability of a bicycle rider from that student body having an accident?

40. How many four-digit numbers can be formed from the digits 1, 3, 5, 7, 8, and 9 if none of these appears more than once in each number?

41. In how many ways can 3 identical jobs be filled by selections from 12 different people?

42. A girl has invited 5 friends to a dinner party. After locating herself at the table how many different seating arrangements are possible?

43. From a bag containing 7 black and 5 white balls, how many sets of 5 balls of which 3 are black and 2 are white can be drawn?

44. From 7 men and 4 women, how many committees can be selected consisting of
    (a) 3 men and 2 women;
    (b) 5 people of which at least 3 are men?

45. From a bag containing 5 black and 7 white balls, how many sets of balls can be selected which include
    (a) exactly 2 black balls;
    (b) at most 2 black balls?

46. A company has 7 men qualified to operate a machine which requires 3 operators for each shift.
    (a) How many shifts are possible?
    (b) In how many of these shifts will any one man appear?

47. How many committees consisting of 3 representatives and 5 senators can be selected from a group of 5 representatives and 8 senators?

48. Three election judges are to be chosen from a group of 4 Republicans and 6 Democrats.
    (a) In how many ways can this be done?
    (b) In how many ways if both parties must be represented among the judges?

49. In how many seating arrangements can 8 men be placed around a table if there are 3 who insist on sitting together? Consider two seating arrangements different unless each man sits in the same chair in the two arrangements.

50. A certain college has only 3-unit courses in 13 of which a freshman may enroll. Find the number of 15-unit programs a freshman can consider if
    (a) there are no specific requirements;
    (b) English 1 and History 1 are required courses.

51. A delegation of 4 students is selected each year from a college to attend the National Student Association annual meeting.
    (a) In how many ways can the delegation be chosen if there are 12 eligible students?
    (b) In how many ways if there are 2 particular students among these 12 who refuse to attend the meeting together?
    (c) In how many ways if 2 of the eligible students are married and will attend the meeting only if they can both go together?

52. There are 10 chairs in a row.
    (a) In how many ways can two people be seated?
    (b) In how many of these ways will the two people be sitting alongside one another?
    (c) In how many ways will they have at least one chair between them?

53. A committee of 10 is to be selected from 6 lawyers, 8 engineers, and 5 doctors. If the committee is to consist of 4 lawyers, 3 engineers, and 3 doctors, how many such committees are possible?

54. In how many ways can a set consisting of 5 different books on mathematics, 3 on physics, and 2 on chemistry be arranged on a straight shelf that has space for 10 books if the books on each subject are to be kept together?

55. (a) Find the number of ways in which 4 boys and 4 girls can be seated in a row of 8 seats if the boys and girls are to have alternate seats.
    (b) Find the number of ways if they sit alternately and if there is a boy named Carl and a girl named Jane among this group who cannot be put into adjacent seats.

56. A stamp collector has 8 different Canadian stamps and 9 different United States stamps, all of the same size. Find the number of ways in which he can select 4 Canadian stamps and 3 United States stamps and arrange them in 7 numbered spaces in his stamp album.

57. A railway coach has 12 seats facing backward and 12 facing foward. In how many ways can 10 passengers be seated if 2 of these people are known to refuse to ride facing forward and 4 refuse to ride facing backward?

58. A coin is tossed 6 times. Find the probability of getting
    (a) exactly 4 heads;
    (b) at most 4 heads.

59. Seven dice are rolled. Calling a 5 or 6 a success, find the probability of getting
    (a) exactly 4 successes;
    (b) at most 4 successes.

60. In the manufacture of a certain article it is known that 1 out of 10 of the articles is defective. What is the probability that a random sample of 4 of the articles will contain
    (a) no defectives;
    (b) exactly 1 defective;
    (c) exactly 2 defectives;
    (d) not more than 2 defectives?

61. If a man having a batting average of 0.40 comes to bat 5 times in a game, what is the probability that he will get
    (a) exactly 2 hits;
    (b) less than 2 hits?

62. In tossing 8 coins, what is the probability of obtaining less than 6 heads?

63. According to records in the registrar's office, 5% of the students in a certain course fail. What is the probability that out of 6 randomly selected students who have taken the course less than 3 failed?

64. Records show that 3 out of 10 people recover from a certain disease. Of 5 people having the disease, what is the probability that
    (a) exactly 3 will recover;
    (b) the first 3 treated will recover and the next two will not;
    (c) the first 3 will recover;
    (d) more than 3 will recover?

65. A student can solve on the average half the problems given him. In order to pass he is required to solve 7 out of 10 problems on an examination. What is the probability that he will pass?

66. In target practice a man is able, on the average, to hit a target 9 times out of 10. Find the probability that he
    (a) hits exactly 4 times in 10 shots;
    (b) hits the first 4 shots and misses all the rest.
    (c) If 10,000 men, of skill equal to that of the man in parts (a) and (b), each have 4 shots, how many of them would you expect to make at least 3 hits?

67. Find the probability of obtaining the most probable number of heads, if a coin is tossed 6 times.

68. Find the probability of obtaining the most probable number of sixes, if a die is rolled 5 times.

69. One bag contains 6 oranges and 5 grapefruit, and a second bag contains 8 grapefruit and 6 oranges. If a bag is selected at random and then one fruit is chosen at random, find the probability that an orange will be selected.

70. A and B play a game in which the probability that A wins is $\frac{3}{8}$ and the probability that B wins is $\frac{5}{8}$. A and B play a series of games, and the winner is to be the one who first wins 4 games. If A and B have each won twice, find the probability that A wins the series.

71. In the World Series, teams A and B play until one team has won 4 games. If team A has probability $\frac{2}{3}$ of winning against team B in a single game, what is the probability that the Series will end only after 7 games are played?

72. A plays in a series of games against B such that the first to win 4 games wins the series. If A's chance of winning a game against B is $\frac{2}{3}$, and if B has already won 2 games, find the probability that A will win the series.

73. If the odds on every game between two players are two to one in favor of the winner of the preceding game, what is the probability that the person who wins the first game shall win at least two out of the next three?

74. One bag contains 6 white balls and 3 black balls, and a second bag contains 7 white balls and 4 black balls. Two balls are selected at random, one from each of the bags, and are placed in a third bag containing 3 white and 3 black balls. One ball is now selected from this third bag; what is the probability that it is white?

75. You are given the option of drawing a bill from one of two boxes. In one box are 9 one-dollar bills, and 1 ten-dollar bill, and in the other box are 5 one-dollar bills, 3 ten-dollar bills, and 7 five-dollar bills. What is the probability that you will draw a ten-dollar bill?

76. Fourteen white and 14 black balls are distributed on two trays. A blindfolded man selects a tray at random and picks a ball from the tray. Under which of the following two arrangements are his chances of obtaining a white ball better: on each tray there are as many black as white balls, or on one tray there are 2 white balls and all other balls are on the other tray?

77. Urn A contains 5 red balls and 5 black balls, urn B contains 4 red balls and 8 black balls, and urn C contains 3 red balls and 6 black balls. A ball is drawn from A, color unknown, and put into B. Then a ball is drawn from B, color unknown, and put into C. What is the probability that a ball now drawn from C will be red?

78. One urn contains 4 red balls and 1 black ball. A second urn contains 2 white balls. Three balls are drawn at random from the first urn and placed in the second, and then 4 balls are drawn at random from the second urn and placed in the first. Find the probability that the black ball is in the first urn after these transfers are completed.

79. A bag contains 2 white balls and 3 black balls. A second bag contains 3 white balls and 2 black balls. A ball is drawn at random from the first bag and placed in the second, then a ball is drawn at random from the second bag and placed in the first. This entire operation is repeated. What is the probability that after these two exchanges the first bag will contain only black balls?

80. Urn I contains 3 white and 5 red balls, and urn II contains 6 white and 2 red balls. An urn is selected at random and a ball withdrawn (color unknown) and put in urn III, which contains 3 white and 3 red balls. A ball is then withdrawn from urn III. What is the probability that it is red?

81. Urn A contains 5 red marbles and 3 white marbles. Urn B contains 1 red marble and two white marbles. A die is tossed. If a 3 or 6 appears, a marble is drawn from B and put into A and then a marble is drawn from A; otherwise, a marble is drawn from A and put into B and then a marble is drawn from B. What is the probability that the two marbles, the one moved from one urn to the other and the one finally drawn, are both red?

82. Six red poker chips and 2 white ones, all the same size, are thoroughly mixed and placed at random in a pile. What is the probability that
    (a) the 2 white ones are on top;
    (b) the white ones are third and fourth from the top;
    (c) the white ones are together;
    (d) the white ones are not together?

83. A man throws a die. He then tosses a coin as many times as the number of dots showing on the top of the die. What is the probability of getting exactly one head?

84. A bag contains 4 coins, one of which has been coined with two heads, and the others are normal coins with one head and one tail. A coin is chosen at random from the bag and tossed 5 times in succession. What is the probability of obtaining 5 heads?

85. A man chooses at random one of the integers 1, 2, 3, 4 and then throws as many dice as are indicated by the number chosen. What is the probability that he will throw a total of 5 points?

86. Eight persons in a room are wearing badges marked 1 through 8. Four persons are chosen at random and are asked to leave the room simultaneously. What is the probability that
    (a) the smallest badge number of those leaving the room is 4;
    (b) the largest badge number of those leaving is 4?

87. Find the probability that of the first 5 persons encountered on a given day at least 3 of them were born on a Saturday.

88. The probability that each of 5 fishermen will catch limits on any particular fishing trip is $\frac{1}{3}$. What is the probability that
    (a) at least 3 of these fishermen on any particular trip will return with limits;
    (b) the first 3 returning will each have a limit, and the others will not have limits?

89. On a certain multiple-choice examination, each question has 5 listed answers, of which one and only one is correct. There are 6 questions, all of equal weight, and a grade of $66\frac{2}{3}\%$ is required for passing. If a student selects the answer for each question at random, what is the probability that he will pass the examination?

90. A coin is tossed repeatedly. Find the probability that the fourth head appears on the eleventh toss.

91. Two dice are rolled repeatedly and a record is kept of the number of times a 9 is obtained. Find the probability that the third time a 9 is obtained occurs on the seventh roll of the two dice.

92. A die is rolled until a 1 occurs. What is the probability that
    (a) the first 1 occurs on the fourth toss;
    (b) more than two tosses are required.

93. A coin is tossed repeatedly. Find the probability that during this tossing it so happens that the fourth head appears on the eighth toss and the eighth head on the sixteenth toss.

94. A family has 6 children. Find the probability that there are fewer boys than girls. Assume that the probability of the birth of either sex is $\frac{1}{2}$.

95. Four statisticians arrange to meet at the Grand Hotel in Paris. It happens that there are 4 hotels with that name in the city. What is the probability that all the statisticians will choose different hotels?

96. If a die is rolled three times what is the probability of obtaining at least one 4 and at least one 5?

97. Six married couples are standing in a room. If 2 people are chosen at random, find the probability that
    (a) they are married to each other;
    (b) one is male and the other is female.

98. A committee of 5 is chosen at random from 6 married couples. What is the probability that the committee will not include 2 people who are married to each other?

99. A boy stands on a street corner and tosses a coin If it falls heads, he walks one block east; if it falls tails, he walks one block west. At his new position he repeats the procedure. What is the probability that after having tossed the coin 6 times he is
    (a) at his starting point;
    (b) 2 blocks from his starting point;
    (c) 4 blocks from his starting point?

100. Five red books and 3 green books are placed at random on a shelf. Find the probability that the green books will all be together.

101. In a roll of 5 dice, what is the probability of getting exactly 4 faces alike?

102. At golf, A defeats B on the average of 2 times out of 5; A defeats C 5 times out of 6; A defeats D 7 times out of 10; and C defeats D 3 times out of 8. To win a tournament A must defeat B and the winner of a match between C and D. If the prize for winning the tournament is $1000, what is A's expectation?

103. At the end of the season a nursery puts on sale 10 peach trees from which the tags have fallen. It is known that 4 of the trees are of one variety and the rest are of another variety. A customer purchases 3 of the trees. What is the probability that the 3 are the same variety?

104. It is known that 4 out of 5 zebra-finch eggs hatch, 6 out of 7 fire-finch eggs hatch, and 11 out of 12 mallard eggs hatch. If one egg each is incubated, what is the probability that something will hatch?

105. Two zebra-finch eggs are in one nest, and 2 fire-finch eggs are in another nest. An egg is taken at random from each nest simultaneously and placed in the other nest. This exchange is performed once more. What is the probability that both eggs in the first nest hatch? (Use the data of Exercise 104.)

106. A bag contains 5 white balls and 4 black balls. A and B take turns in drawing one ball at a time without replacement until all balls are drawn. If A draws first, what is the probability that A will be the first to draw a black ball?

107. A bag contains 3 white marbles and 4 black ones. A, B, and C in order withdraw a marble, without replacing it. The first to draw a white marble wins. What is A's probability of winning the game?

108. Suppose two bad light bulbs get mixed up with five good ones and that you start testing the bulbs, one by one, until you have found both defectives. What is the probability that you will find the second defective bulb on the seventh testing?

109. A box contains 6 good and 4 bad radio tubes. If the tubes are withdrawn one by one, what is the probability that the last of the bad tubes will be found on the 5th withdrawal?

110. Suppose we have four courts c_1, c_2, c_3, and c_4. An accused person must start in court c_1. If he is convicted in a lower court, he has the possibility of being heard in a higher court in the order c_1, c_2, c_3, c_4, but he cannot by-pass any court. The probabilities of winning a case are $\frac{1}{4}$, $\frac{1}{3}$, $\frac{1}{2}$, $\frac{1}{4}$ in the courts c_1, c_2, c_3, c_4, respectively. If a person loses his case in a court, the probabilities of a hearing in the next higher court are as follows: $\frac{1}{5}$ in c_2, $\frac{1}{4}$ in c_3, and $\frac{1}{8}$ in c_4. What is the probability that a person will be cleared of an accusation in these courts?

111. In Example 5-11, find the chances of each of B and C winning.

112. A, B, C, and D in order toss a coin. The first one to throw a head wins. What are their respective chances of winning?

113. A and B in order toss a pair of dice. The first to obtain a 10 wins. What are their respective chances of winning?

114. A, B, and C in order toss a pair of dice. The first to obtain a 7 wins. What are their respective chances of winning?

115. A professional tennis player, A, plays for a long period of time alternately against two amateurs, B and C. If he has a chance of $\frac{15}{16}$ of winning against B, and $\frac{9}{10}$ against C, what is the probability that C will be the first among the two amateurs to win against A, provided that the professional starts the series by playing against B?

116. If a die is thrown repeatedly, what is the probability that a 6 is obtained before a 1 turns up?

117. If two dice are thrown repeatedly, find the probability of obtaining a 9 before a 6 turns up.

118. The probability that a fertile egg hatches is $\frac{11}{12}$. Three eggs are taken for incubation from a box of 12 eggs, 4 of which are fertile and 8 infertile. What is the probability that something hatches from the 3 eggs?

119. The rules of the game of craps are the following: A player using two dice wins if he throws a 7 or 11 on the first throw. He loses if he throws 2, 3, or 12 on the first throw. If he throws, 4, 5, 6, 8, 9, 10 on the first throw, then he wins if he repeats the result of his first throw before throwing a 7. What is the probability of winning in the game of craps?

6

The Binomial Distribution (and Other Discrete Distributions)

6.1 THE BINOMIAL PROBABILITY DISTRIBUTION AND THEORETICAL FREQUENCY DISTRIBUTION

Theorem 5-5 states that, if p is the probability of success of an event in a single trial and $q = 1 - p$, then the probability P of obtaining exactly r successes in n trials is

$$P = C_{n,r} \cdot p^r q^{n-r} \qquad \text{for } r = 0, 1, 2, \ldots, n. \qquad (6\text{-}1)$$

Since, in this chapter, we shall keep n and p fixed but consider P for various values of r, we shall write P_r for P, so that (6-1) becomes

$$P_r = C_{n,r} \cdot p^r q^{n-r}. \qquad (6\text{-}2)$$

Now consider an experiment, call it E, consisting of performing n independent trials in which the object is to count the number of successes X. This number of successes will be one of the numbers $0, 1, 2, \ldots, n$. Let r be one of these numbers, then the probability that exactly r successes are obtained, that is, that $X = r$, is P_r, as given by (6-2).

Thus we have the following theorem, which is a restatement of Theorem 5-5.

Theorem 6-1. If the variable X denotes the number of successes in an experiment E, consisting of performing n independent trials, where p is the probability of success in a single trial and $q = 1 - p$, then the successive terms of the binomial expansion of $(q + p)^n$ give the respective probabilities of $X = 0, 1, 2, \ldots, n$ successes in these n trials. Table 6-1 shows for each value of r (listed in the first column), the probability (in the second column) that X assumes that value of r.

Table 6-1
Binomial distribution of a variable.

X (or r)	P_r
0	q^n
1	npq^{n-1}
2	$C_{n,2}\,p^2 q^{n-2}$
$\vdots$	$\vdots$
r	$C_{n,r}\,p^r q^{n-r}$
$\vdots$	$\vdots$
n	p^n

Definition 6-1. The set of X-values with their respective probabilities, shown in Table 6-1, is called a *binomial probability distribution* of the statistical (or random) variable X. We also say in this case that X is *binomially distributed* or has a *binomial distribution*.

Example 6-1. For an experiment in which a coin is tossed three times (or 3 coins are tossed once), construct the binomial distribution of X.

Solution. The probabilities for obtaining 0, 1, 2, or 3 heads when a coin is tossed three times is obtained by finding the successive terms of the binomial expansion of $(q + p)^3$ for $p = \frac{1}{2}$ and $q = \frac{1}{2}$. These probabilities are listed in Table 6-2.

Table 6-2
Binomial distribution for
$n = 3, p = \frac{1}{2}$.

X (or r)	P_r
0	$\frac{1}{8}$
1	$\frac{3}{8}$
2	$\frac{3}{8}$
3	$\frac{1}{8}$

When an experiment E, such as the one in Example 6-1, is performed repeatedly, say N times, we will frequently be interested in the expected number of times that each of the r successes will occur. For instance, in Example 6-1, we may be interested to know the expected number of times two heads will be obtained when the experiment is performed, say, 800 times. By Definition 5-3, this expected number will be $\frac{3}{8} \cdot 800 = 300$. Similarly, for any r, the expected number of successes in N trials of the experiment E will be $N \cdot P_r$, which we shall call the *expected frequency* or *theoretical frequency* of the value r. For each value of r, the expected frequency, therefore, is obtained by multiplying each of the entries of the second column of Table 6-1 by N. This results in Table 6-3.

Table 6-3
Expected (or theoretical) binomial
frequency distribution.

X (or r)	*Expected frequency*
0	Nq^n
1	$Nnpq^{n-1}$
2	$NC_{n,2}\,p^2q^{n-2}$
$\vdots$	$\vdots$
r	$NP_r = NC_{n,r}\,p^rq^{n-r}$
$\vdots$	$\vdots$
n	Np^n

Definition 6-2. The set of X-values with their respective frequencies, shown in Table 6-3, is called a *binomial frequency distribution* with total frequency N.

Example 6-2. For the experiment of Example 6-1 when repeated 32 times, that is, when three coins are tossed 32 times, construct the binomial frequency distribution of X.

Solution. The expected frequencies are calculated by multiplying each of the entries in the second column of Table 6-2 by 32, resulting in the binomial frequency distribution given in Table 6-4.

Table 6-4
Binomial frequency distribution for
$n = 3, p = \frac{1}{2}, N = 32$.

X (or r)	Expected frequencies
0	4
1	12
2	12
3	4

Example 6-3. For an experiment in which 9 coins are tossed N times (N a given number), construct the theoretical binomial frequency distribution.

Solution. The expected frequencies for 0, 1, 2, ..., 9 heads in N tosses are listed in the second column of Table 6-5.

In the third column, the expected frequencies for $N = 512$ are listed ($N = 512$ was arbitrarily chosen in order to obtain simple numbers, i.e. integers) and in the fourth column are shown the actual frequencies obtained when 9 coins were tossed 512 times. It can be seen that there is good agreement between the actual and theoretical frequencies.

If we were interested to determine from Table 6-5 the probability of obtaining between 3 and 6 successes (3 and 6 both included), we would find it, in accordance with Theorem 5-1, by adding the probabilities for 3, 4, 5, and 6 successes to obtain

$$\frac{84}{512} + \frac{126}{512} + \frac{126}{512} + \frac{84}{512} = \frac{420}{512} = \frac{105}{128}.$$

Table 6-5
Expected (binomial) and actual frequencies obtained in tossing
9 coins N times.

X (or r)	Expected f	Expected f for $N = 512$	Actual f for $N = 512$
0	$\dfrac{1}{512} N$	1	2
1	$\dfrac{9}{512} N$	9	11
2	$\dfrac{36}{512} N$	36	37
3	$\dfrac{84}{512} N$	84	88
4	$\dfrac{126}{512} N$	126	131
5	$\dfrac{126}{512} N$	126	119
6	$\dfrac{84}{512} N$	84	82
7	$\dfrac{36}{512} N$	36	32
8	$\dfrac{9}{512} N$	9	9
9	$\dfrac{1}{512} N$	1	1

In general, if we want the probability of between a and b successes (a and b both included), we calculate it as $P_a + P_{a+1} + P_{a+2} + \cdots + P_b$.

In this connection, note that since the number of successes, X, in a binomial distribution, must be some number between 0 and n (0 and n included), the sum of the P_r for $r = 0, 1, 2, \ldots, n$ must always equal 1. That this is so is evident, since

$$\sum_{r=0}^{n} P_r = \sum_{r=0}^{n} C_{n,r} \cdot p^r q^{n-r} = (p + q)^n = 1^n = 1.$$

6.2 PROBABILITY DISTRIBUTIONS AND THEORETICAL FREQUENCY DISTRIBUTIONS IN GENERAL

A binomial distribution of a variable X, as discussed in Section 6.1, is an example of a more general distribution to be discussed in this section. The result of an experiment E may not be one of the values $X = 0, 1, 2, \ldots, n$ as is true of the binomial distribution, but could be some other prescribed set of values of X. For example, if the experiment consisted in tossing two dice and if we let X be the sum of the two numbers showing on the dice, then X can take on the values 2, 3, $\ldots$, 12. The probabilities corresponding to each of these values of X can easily be calculated by the methods of Chapter 5 and are shown in the second column of Table 6-6.

Table 6-6

Probability distribution for the sum of the numbers showing on two dice.

X	Probability
2	$\frac{1}{36}$
3	$\frac{2}{36}$
4	$\frac{3}{36}$
5	$\frac{4}{36}$
6	$\frac{5}{36}$
7	$\frac{6}{36}$
8	$\frac{5}{36}$
9	$\frac{4}{36}$
10	$\frac{3}{36}$
11	$\frac{2}{36}$
12	$\frac{1}{36}$

This leads to the following generalization of Definition 6-1.

Definition 6-3. If in an experiment E the possible values of a discrete variable are X_1, X_2, ..., X_h (not necessarily integers) with respective probabilities P_1, P_2, ..., P_h (it is assumed here that in a single experiment E one and only one of the values X_1, X_2, ..., X_h can occur), then the set of X-values with their respective probabilities shown in Table 6-7 is called the *probability distribution* of the statistical (or random) variable X.

Table 6-7
Probability distribution of a
variable X.

X	Probability
X_1	P_1
X_2	P_2
⋮	⋮
X_i	P_i
⋮	⋮
X_h	P_h

If, as in Section 6.1, the Experiment E is performed N times, the expected number of times a particular X_i occurs is $N \cdot P_i$, so that the expected frequencies are obtained by multiplying each of the entries in Table 6-7 by N. This results in Table 6-8.

Table 6-8
Expected (or theoretical) frequency
distribution.

X	Expected frequency
X_1	NP_1
X_2	NP_2
⋮	⋮
X_i	NP_i
⋮	⋮
X_h	NP_h

Definition 6-4. The set of X-values with their respective frequencies shown in Table 6-8 is called an *expected* (*or theoretical*) *frequency distribution* for N repetitions of an experiment E.

As in Section 6-1, if we want the probability that the number of successes is between X_a and X_b (X_a and X_b both included), that is, that $X_a \leq X \leq X_b$, we calculate it as $P_a + P_{a+1} + \cdots + P_b = \sum_{i=a}^{b} P_i$.

Recalling Formula (4-2), which gives the mean of h distinct values X_i, each occurring with frequency f_i, namely

$$m = \frac{\sum_{i=1}^{h} X_i f_i}{N},$$

we find the mean of the theoretical frequency distribution of Table 6-8 to be

$$m = \frac{\sum_{i=1}^{h} X_i NP_i}{N} = \frac{N \sum_{i=1}^{h} X_i P_i}{N} = \sum_{i=1}^{h} X_i P_i.$$

Note that the result is independent of N, which means that, in order to calculate the mean of a theoretical frequency distribution, we do not need to know N.

Recalling the definition of the variance as given by (4-15), we find the variance of the theoretical frequency distribution of Table 6-8 to be

$$\sigma^2 = \frac{\sum_{i=1}^{h} (X_i - m)^2 NP_i}{N} = \frac{N \sum_{i=1}^{h} (X_i - m)^2 P_i}{N} = \sum_{i=1}^{h} (X_i - m)^2 P_i.$$

Again, we note that the result is independent of N. For this reason, we shall also call the quantities just obtained for m and σ the mean and variance of the probability distribution of X.

Definition 6-5. If X is a discrete variable with probability distribution given by Table 6-7, then the *mean of the probability distribution* is defined by

$$m = \sum_{i=1}^{h} X_i P_i \tag{6-3}$$

and the *variance of the probability distribution* by

$$\sigma^2 = \sum_{i=1}^{h} (X_i - m)^2 P_i. \tag{6-4}$$

The *standard deviation*, σ, of the distribution is defined as the square root of the variance.

It can easily be shown that the computing form of the variance for the theoretical frequency (and also the probability) distribution is given by

$$\sigma^2 = \sum_{i=1}^{h} X_i^2 P_i - m^2. \tag{6-5}$$

6.3 THE MEAN AND STANDARD DEVIATION OF A BINOMIAL DISTRIBUTION

Applying Formulae (6-3), (6-4), and (6-5) to obtain the mean and standard deviation of a binomial distribution, we find

$$m = \sum_{r=0}^{n} r P_r \tag{6-6}$$

and

$$\sigma^2 = \sum_{r=0}^{n} (r - m)^2 P_r = \sum_{r=0}^{n} r^2 P_r - m^2. \tag{6-7}$$

Example 6-4. Calculate the mean and standard deviation for the binomial distribution of Example 6-3.

Solution. The values of rP_r, r^2P_r, and their sums, needed in this calculation, are shown in Table 6-9.

Table 6-9
Calculation of mean and standard deviation of binomial distribution of Example 6-3.

r	P_r	rP_r	r^2P_r
0	$\dfrac{1}{512}$	0	0
1	$\dfrac{9}{512}$	$\dfrac{9}{512}$	$\dfrac{9}{512}$
2	$\dfrac{36}{512}$	$\dfrac{72}{512}$	$\dfrac{144}{512}$
3	$\dfrac{84}{512}$	$\dfrac{252}{512}$	$\dfrac{756}{512}$
4	$\dfrac{126}{512}$	$\dfrac{504}{512}$	$\dfrac{2016}{512}$

Table 6-9 (*continued*)

r	P_r	rP_r	r^2P_r
5	$\dfrac{126}{512}$	$\dfrac{630}{512}$	$\dfrac{3150}{512}$
6	$\dfrac{84}{512}$	$\dfrac{504}{512}$	$\dfrac{3024}{512}$
7	$\dfrac{36}{512}$	$\dfrac{252}{512}$	$\dfrac{1764}{512}$
8	$\dfrac{9}{512}$	$\dfrac{72}{512}$	$\dfrac{576}{512}$
9	$\dfrac{1}{512}$	$\dfrac{9}{512}$	$\dfrac{81}{512}$
	1	$\dfrac{2304}{512}$	$\dfrac{11,520}{512}$

Accordingly,

$$m = \sum_{r=0}^{n} rP_r = \frac{2304}{512} = 4.50$$

and

$$\sigma^2 = \frac{11,520}{512} - (4.50)^2 = 22.50 - 20.25 = 2.25,$$

so that

$$\sigma = 1.50.$$

We shall now state a theorem by means of which the mean and standard deviation of a binomial distribution may be calculated directly from n, p, and q without having to calculate the sums shown in Formulae (6-6) and (6-7).

Theorem 6-2. A variable X, which is binomially distributed, with the probability of success in a single trial being p, and the experiment consisting of n independent trials, has mean

$$m = np \tag{6-8}$$

and standard deviation

$$\sigma = \sqrt{npq}. \tag{6-9}$$

These quantities represent the mean and standard deviation of both the probability distribution and the theoretical frequency distribution for any total frequency N. A proof of the theorem will follow Example 6-5.

Example 6-5. Calculate the mean and standard deviation for the binomial distribution of Example 6-3 by use of Theorem 6-2.

Solution.

$$m = 9 \cdot \frac{1}{2} = 4.50, \qquad \sigma = \sqrt{9 \cdot \frac{1}{2} \cdot \frac{1}{2}} = \sqrt{\frac{9}{4}} = \frac{3}{2} = 1.50.$$

These results agree with those obtained in Example 6-4.

Proof of Theorem 6-2. To prove Theorem 6-2, we need to make the calculations that were made in Example 6-4 for certain given values of n, p, and q, so that they are applicable for any n, p, and q. Table 6-10 shows the values of r, P_r, rP_r, and $r^2 P_r$ needed in the calculation of the mean and standard deviation. The totals of the columns for rP_r and $r^2 P_r$ are listed in the last row of the table. Thus the total of the rP_r column is found as

$$\sum_{r=0}^{n} rP_r = np\left[q^{n-1} + (n-1)q^{n-2}p + \frac{(n-1)(n-2)}{2} q^{n-3}p^2 \right.$$
$$\left. + \cdots + (n-1)qp^{n-2} + p^{n-1} \right] = np(q+p)^{n-1} = np,$$

since $q + p = 1$.

To calculate the total of the $r^2 P_r$ column, the algebra is greatly simplified by writing each value of $r^2 P_r$ as the sum of two terms, the first one of which is rP_r. Consequently, in the $r^2 P_r$ column of Table 6-10, each entry is written as the sum of two terms whose separate sums are those shown at the bottom of the column.

Using the sums of the rP_r and $r^2 P_r$ columns, respectively, as shown in Table 6-10, and using (6-6), we find for the mean of the binomial distribution,

$$m = \sum_{r=0}^{n} rP_r = np$$

and using (6-7), we find for the standard deviation

$$\sigma^2 = \sum_{r=0}^{n} r^2 P_r - m^2$$
$$= np + n(n-1)p^2 - (np)^2$$
$$= np + n^2 p^2 - np^2 - n^2 p^2$$
$$= np - np^2$$
$$= np(1 - p)$$
$$= npq.$$

Table 6-10
Frequency table for calculation of mean and standard deviation of binomial distribution.

r	P_r	rP_r	r^2P_r
0	q^n	0	$0 = 0P_0$
1	$nq^{n-1}p$	$nq^{n-1}p$	$nq^{n-1}p = 1P_1$
2	$\dfrac{n(n-1)}{2}q^{n-2}p^2$	$n(n-1)q^{n-2}p^2$	$2n(n-1)q^{n-2}p^2 = 2P_2 + n(n-1)q^{n-2}p^2$
3	$\dfrac{n(n-1)(n-2)}{2\cdot 3}q^{n-3}p^3$	$\dfrac{n(n-1)(n-2)}{2}q^{n-3}p^3$	$\dfrac{3n(n-1)(n-2)}{2}q^{n-3}p^3 = 3P_3 + n(n-1)(n-2)q^{n-3}p^3$
$\vdots$	$\vdots$	$\vdots$	$\vdots$
$n-1$	nqp^{n-1}	$n(n-1)qp^{n-1}$	$n(n-1)^2qp^{n-1} = (n-1)P_{n-1} + n(n-1)(n-2)qp^{n-1}$
n	p^n	np^n	$n^2p^n = nP_n + n(n-1)p^n$
		$np(q+p)^{n-1} = np$	$\displaystyle\sum_{r=0}^{n} rP_r + n(n-1)p^2(q+p)^{n-2} = np + n(n-1)p^2$

Hence

$$\sigma = \sqrt{npq}.$$

Sometimes, in an experiment E, the probability p of success in a single trial is unknown. Then, by performing the experiment E a large number of times, we can use (6-8) to estimate p. This is illustrated in Example 6-6.

Example 6-6. In a marigold seed viability test, 450 seeds are placed on a filter paper in 90 rows of 5 seeds each. The number of seeds germinating in each row, denoted by X, is then determined and the frequency with which each of the values of $X = 0, 1, \ldots, 5$ is obtained is recorded as in Table 6-11.

Table 6-11
Observed frequencies of numbers
of germinating seeds

X	Observed frequencies
0	0
1	1
2	11
3	30
4	38
5	10
	90

Assuming that the frequencies can be expected to follow a binomial distribution, estimate the probability p that a single marigold seed germinates.

Solution. From Table 6-11, we calculate the mean number of germinating seeds per row as

$$m = \frac{\sum_{i=1}^{h} X_i f_i}{N} = \frac{315}{90} = 3.50.$$

Assuming this to be an approximation of the mean of the binomial distribution, we have, by (6-8),

$$3.50 = 5p$$

and, therefore,

$$p = \frac{3.50}{5} = 0.70.$$

6.4   HISTOGRAM OF A BINOMIAL DISTRIBUTION

For a theoretical frequency table of a binomial distribution, one can construct a corresponding histogram.

Example 6-7. Construct the histogram corresponding to the theoretical frequency distribution of Example 6-3.

Solution. The histogram is shown in Figure 6-1.

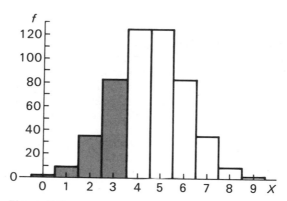

Figure 6-1

Histogram for expected (binomial) frequencies in 512 tosses of 9 coins.

It should be noted that the histogram for a binomial distribution is symmetric if and only if $p = \frac{1}{2}$.

Since the width of each rectangle in the histogram of a binomial distribution is equal to 1, the area of each rectangle is equal in magnitude to the altitude and, consequently, is also a measure of the expected frequency for the class. From this it follows that the frequency with which X can be expected to be between two given integer values a and b (a and b both included) is represented by the sum of the areas of the rectangles corresponding to $X = a, a + 1, \ldots, b$. For example, the shaded area in Figure 6-1 represents the expected frequency of obtaining at most 3 heads in 512 tosses of 9 coins.

If, in Figure 6-1, we had marked on the vertical axis the probability, instead of the expected frequency, we would have constructed the *probability* histogram corresponding to the binomial probability distribution. Since the width of each rectangle remains as 1, the area of each rectangle in this histogram is equal in magnitude to the *probability* corresponding to the class.

The advantages of using area to measure frequencies and probabilities will be explained in Chapter 7.

EXERCISES

1. (a) Find the binomial distribution obtained by rolling 6 dice, considering a 1 a success and everything else a failure.
   (b) Find the theoretical frequency distribution, if these 6 dice are rolled 46,656 times.
   (c) Construct the histogram for part (b).

2. Five coins are tossed 64 times.
   (a) Construct the theoretical frequency table for 0, 1, ..., 5 heads.
   (b) Draw the corresponding histogram.
   (c) Find the expected number of times of getting at least 3 heads.
   (d) In the histogram, shade the area that is a measure of the result in part (c).
   (e) Toss 5 coins 64 times and record the actual frequencies of 0, 1, ..., 5 heads.
   (f) Calculate the mean and standard deviation of the frequency distribution obtained in part (a) by use of Formulae (6-6) and (6-7), as was done in Table 6-9.
   (g) Calculate the mean and standard deviation of part (f) by use of Theorem 6-2.
   (h) Calculate the mean and standard deviation of the observed frequency distribution obtained in part (e).
   (i) Draw the histogram of the observed frequency distribution on the same page the histogram of the theoretical frequency distribution in part (b) was drawn on.

3. If 5 coins are tossed 3000 times, answer the same questions asked in Exercise 2(a), (b), (c), (d), (f), and (g).

4. Four dice are rolled 81 times. If a 1 or a 6 is considered a success and everything else a failure
   (a) construct the theoretical frequency table for 0, 1, 2, 3, 4 successes;
   (b) draw the corresponding histogram;
   (c) find the expected number of times of obtaining at least 3 successes among the 4 dice.
   (d) In the histogram, shade the area that is a measure of the result in part (c).
   (e) Roll 4 dice 81 times and record the actual frequencies for 0, 1, 2, 3, 4 successes.
   (f) Calculate the mean and standard deviation of the frequency distribution obtained in part (a) by use of Formulae (6-6) and (6-7), as was done in Table 6-9.
   (g) Calculate the mean and standard deviation of part (f) by use of Theorem 6-2.

5. (a) Find the theoretical binomial frequency distribution with $n = 5$, $p = 0.7$, and $N = 90$.
   (b) Find its mean and standard deviation.
   (c) Compare the theoretical frequencies obtained in part (a) with the actual ones given in Table 6-11.

7

The Normal Distribution
(and Poisson Distribution)

7.1 THE NORMAL DISTRIBUTION AS AN APPROXIMATION OF THE BINOMIAL DISTRIBUTION

In Chapter 6 we saw that the expected frequencies of the occurrence of r heads in tossing n coins follow a binomial distribution. The histograms representing expected frequencies in 64 tosses of 1–6 coins are shown in Figure 7-1, on the following page.

If the number of trials n is large, the calculation of frequencies and probabilities by means of the binomial theorem becomes tedious. Since many practical problems involve large numbers of repeated trials, it is important to find a more rapid method of calculating probabilities. Such a method is furnished by the normal distribution, which is the most frequently occurring continuous probability distribution and one on which much statistical theory is based. From Figure 7-1 it is apparent that as n increases, the tops of the rectangles of the histograms approach a bell-shaped curve. This limiting frequency curve obtained as n becomes larger and larger is called the *normal frequency curve*.

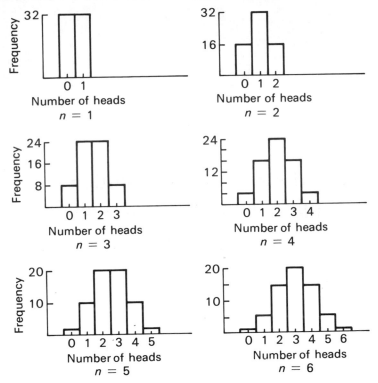

Figure 7-1

Histograms for expected frequencies of heads in tossing from 1 to 6 coins 64 times.

Theorem 7-1. In the binomial distribution for N samples of n trials each, where the probability of success in a single trial is p, if the value of n is increased, the histogram approaches a curve, called the *normal curve*, whose equation is

$$Y = \frac{N}{\sigma\sqrt{2\pi}} e^{-(X-m)^2/2\sigma^2}, \tag{7-1}$$

where

  m = mean of the binomial distribution = np,

  σ = standard deviation of the binomial distribution = $\sqrt{npq}$,

  e = base of the natural logarithms = $2.71828 \cdots$,

  π = $3.14159 \cdots$,

  Y = frequency with which any X occurs.

The proof of this theorem is omitted, since it requires a knowledge of mathematical techniques beyond the scope of this text.

In Figure 7-2 the histogram of Figure 6-1 has been approximated by the corresponding normal curve. A surprising agreement can be observed in spite of the fact that $n = 9$ is not a large number.

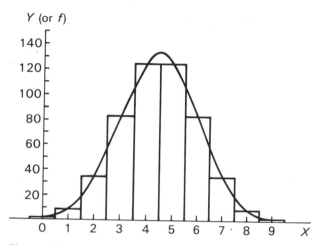

Figure 7-2

Histogram and normal curve for expected frequencies of heads in tossing 9 coins 512 times.

If the relative frequency (or probability) with which a variate X occurs is denoted by P, then $P = Y/N$, and Equation (7-1) can be written as

$$P = \frac{1}{\sigma\sqrt{2\pi}}\, e^{-(X-m)^2/2\sigma^2}, \tag{7-2}$$

the graph of which is called the *normal probability curve*.

If we multiply both sides of Equation (7-2) by σ and let $y = P\sigma$, thereby making σ the unit of measurement on the vertical axis, we can rewrite (7-2) as

$$y = \frac{1}{\sqrt{2\pi}}\, e^{-(X-m)^2/2\sigma^2} \tag{7-3}$$

In order to measure also the deviations $X - m$ in terms of σ, we introduce a new variable

$$z = \frac{X - m}{\sigma}, \tag{7-4}$$

which represents a measure of the deviation in terms of the standard deviation or in so-called *standard units*. The expression given by (7-4) is also frequently referred to as the *normal deviate*.

By use of (7-4) we can now rewrite (7-3) as

$$y = \frac{1}{\sqrt{2\pi}}\, e^{-z^2/2}, \qquad (7\text{-}5)$$

the graph of which is called the *standard normal probability curve*.

From Equation (7-2) and its equivalent (7-5) it is evident that y and P reach their maximum values when $z = 0$, that is, when $X = m$. It is also clear from (7-5) that since z is squared the substitution of z or $-z$ into the equation results in the same value of y. This shows that the curve represented by Equation (7-5) is symmetrical about the line drawn at $z = 0$ perpendicular to the horizontal axis. Furthermore, it is observed that as X increases more and more, y becomes smaller and smaller without, however, ever becoming equal to zero. This means that the curve approaches the horizontal axis but never touches it. The values of y corresponding to given values of z are listed in the Appendix in Table Ia. In Figure 7-3 is shown the normal probability curve for $\sigma = 1.5$

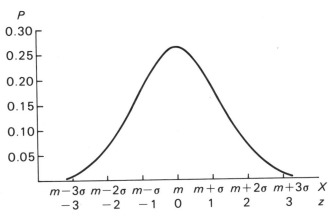

Figure 7-3
Normal probability curve for expected frequencies of heads in tossing 9 coins 512 times.

corresponding to the normal curve of Figure 7-2. The P-values in this case can easily be obtained by dividing the values of y obtained from Table Ia by $\sigma = 1.5$. Thus, the value of P corresponding to $z = 1$ is found as

$$P = \frac{0.2420}{1.5} = 0.1613.$$

Since the normal frequency curve is always symmetric, whereas the binomial histogram is symmetric only if $p = q = \frac{1}{2}$, it is clear that the normal curve is a better approximation of the binomial histogram if both p and q are equal to or nearly equal to $\frac{1}{2}$. The more p and q differ from $\frac{1}{2}$, the worse is the approximation. In fact, if p (or q) is very small even though n is large, the normal approximation should not be used. It is generally suggested that if p is so small that np is less than 5 (or if q is so small that nq is less than 5), the normal approximation to the binomial distribution should ordinarily not be used. In this case, fortunately, there exists another distribution—the Poisson distribution—which serves as an excellent approximation of the binomial distribution.

Let us suppose that the normal curve can be used to approximate a binomial distribution and that we wish to use the normal probability curve to find the probability that a variate X will fall between two given integral values a and b (a and b both included and b greater than a). We recall from Chapter 6 that the sum of the areas of the rectangles whose centers are at $X = a$, $a + 1, \ldots, b$ represents a measure of the expected frequency with which X will fall between a and b. We need then to find the area under the binomial histogram from $X = a - \frac{1}{2}$ to $X = b + \frac{1}{2}$ or, since the normal frequency curve is assumed to approximate the binomial distribution, the area between the same two values of X under the normal frequency curve. Since, however, we are interested here in the probability (rather than the frequency), we need to divide the result by N, or, what is the same thing, we need to find the area under the curve where frequencies have already been divided by N, i.e. the area under the normal probability curve as given by (7-5). The areas under this curve in terms of z are given in Appendix Table I where the first column gives the z-value, the second the area between the line of symmetry of the normal curve (i.e. where $z = 0$) and the given z-value. Since these areas are tabulated, the probabilities can be determined directly without making use of the formulae for the normal frequency (or probability) curves. The procedure is shown in Example 7-1. It should be noted of course that the total area under the normal probability curve, representing the probability that X will fall anywhere, is equal to 1.

Example 7-1. Using the normal curve approximation, find the probability that between 3 and 6 heads are obtained in a toss of 9 coins, i.e. that X lies between 3 and 6 (both 3 and 6 included).

Solution. In Figure 7-2 the sum of the areas of the rectangles with centers at $X = 3, 4, 5,$ and 6 represents the expected number of times we would get 3 to 6 heads; this is the area from $X = 2.5$ to $X = 6.5$ under the histogram of Figure 7-2. To obtain the desired probability, we divide this area by $N = 512$. Since the normal frequency curve approximates the binomial frequency curve, we find the desired probability with a good degree of accuracy by finding the area under the corresponding normal probability curve (Figure 7-3) from $X = 2.5$ to $X = 6.5$. Since the areas in Appendix Table I are given in terms of z, we need to find the z-values corresponding to the two X-values. Since

$$m = np = 9 \cdot \frac{1}{2} = 4.5,$$

and

$$\sigma = \sqrt{9 \cdot \frac{1}{2} \cdot \frac{1}{2}} = 1.5,$$

we have for

$$X = 2.5: \qquad z = \frac{2.5 - 4.5}{1.5} = \frac{-2}{1.5} = -1.33,$$

and for

$$X = 6.5: \qquad z = \frac{6.5 - 4.5}{1.5} = \frac{2}{1.5} = 1.33.$$

We therefore need to find the area from $z = -1.33$ to $z = 1.33$. This we find by looking up the area from $z = 0$ to $z = +1.33$ (which is the same as the area from $z = 0$ to $z = -1.33$) and multiplying it by 2. The desired probability is therefore equal to

$$P = 0.4082 \cdot 2 = 0.8164.$$

From Table 6-1 we find that if we had used the binomial distribution for this example we would have obtained

$$\frac{84 + 126 + 126 + 84}{512} = 0.8203.$$

The discrepancy is not surprising, in view of the small value of n.

In order to be able to make use of the tables in the appendix without having to resort to interpolation, we shall calculate all standard deviations and z-values to two decimals only, which may occasionally cause a slight loss of accuracy in the result.

In Table I, areas are listed only for z-values up to $z = 3.89$, since for values of z larger than 3.89, the area, when rounded to four decimal places, equals 0.5000. This does not mean that the area equals exactly $\frac{1}{2}$, but, as is evident in Figure 7-3, the area approaches $\frac{1}{2}$ closer and closer, the larger the z-value.

Example 7-2.   If a die is thrown 36 times, what is the probability, using the normal curve approximation, that a 5 is obtained at least 7 times?

Solution.   This probability is clearly the same as that of obtaining at least 7 times a 5 in a throw of 36 dice. Here, $p = \frac{1}{6}$, $q = \frac{5}{6}$, $n = 36$; thus

$$m = 36 \cdot \frac{1}{6} = 6,$$

$$\sigma = \sqrt{36 \cdot \frac{1}{6} \cdot \frac{5}{6}} = \sqrt{5} = 2.24.$$

To determine which area is needed, it is important to sketch the binomial histogram involved and the normal curve approximating it. This sketch need not be very accurate nor complete. It only needs to indicate where the peak of the binomial histogram occurs, which is always at the mean (in this case, $X = 6$), and the values of X involved in the problem (in this case $X \geq 7$). Such a sketch is shown as Figure 7-4. Note that although the binomial histogram is not symmetric, the normal curve approximating it is a symmetric curve.

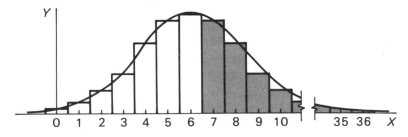

Figure 7-4

Histogram and normal curve for expected frequencies of data of Example 7-2.

Since we wish to obtain at least 7 successes, we shade the desired area—that is, the rectangles with centers at $X = 7, 8, \ldots, 36$—and find the area under the corresponding part of the normal curve from $X = 6.5$ to $X = 36.5$; i.e. from

$$z = \frac{6.5 - 6}{2.24} = 0.22 \quad \text{to} \quad z = \frac{36.5 - 6}{2.24} = \frac{30.5}{2.24} = 13.62.$$

But since the area under the curve to the right of $z = 13.62$ is so small, we can just as well include it without introducing any significant error. We find the area to the right of $z = 0.22$ by subtracting the area from $z = 0$ to $z = 0.22$ from 0.5, the total area to the right of the line $z = 0$; therefore,

$$P = 0.5 - 0.0871 = 0.4129.$$

From the solution of Example 7-2 it is evident why the following procedure is appropriate. If a normal curve approximates a binomial distribution, and the probability that X is greater than or equal to a (an integer) is desired, we simply take the total area under the normal probability curve to the right of $X = a - \frac{1}{2}$; on the other hand, if the probability that X is less than or equal to a (an integer) is desired, we take the total area under the normal probability curve to the left of $X = a + \frac{1}{2}$. The adoption of this procedure is also motivated by the desire to have the probabilities that X is greater than or equal to a and the probability that X is less than a add up to 1, which would not be true had we not accepted the preceding convention.

7.2 THE POISSON DISTRIBUTION AS AN APPROXIMATION OF THE BINOMIAL DISTRIBUTION

In applying the normal curve approximation to the calculation of binomial probabilities, the approximation is best when n is large and p close to $\frac{1}{2}$. If for an event, p is very small and $np < 5$, it is termed *rare*; in that case, the normal approximation is unsatisfactory, since the binomial distribution is markedly skew. In many applications, however, rare events are encountered, and it is desirable to have a good approximation available for these cases. Fortunately, there is such an approximation, which is given by the so-called Poisson distribution. This is stated in the following theorem.

Theorem 7-2. In the binomial distribution, if the number of trials, n, becomes larger and larger while p becomes smaller and smaller, in such a manner that $m = np$ remains constant, then the binomial probability for r successes in n trials, $P = C_{n,r} \cdot q^{n-r} p^r$, approaches the Poisson probability given by

$$P = \frac{e^{-m} m^r}{r!}. \tag{7-6}$$

For small values of m the Poisson distribution is skew but tends to become more symmetrical as m increases, becoming nearly symmetrical for $m = 5$, as indicated in Figure 7-5.

Proof of Theorem 7-2. If in the binomial distribution, for a given value of successes r, we let n approach infinity and p approach zero in such a way that $m = np$ always remains constant, we can put $p = m/n$ and $q = 1 - (m/n)$ in

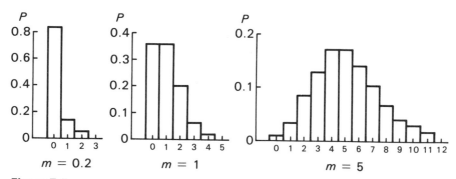

Figure 7-5

Frequency polygons of the Poisson distribution for various values of m.

the binomial probability $P = C_{n,r} \cdot q^{n-r} p^r$ and obtain

$$P = \frac{n(n-1) \cdots (n-r+1)}{r!} \left(1 - \frac{m}{n}\right)^{n-r} \left(\frac{m}{n}\right)^r.$$

The right hand side of the equation can be rewritten as

$$\left(1 - \frac{m}{n}\right)^n \left(1 - \frac{m}{n}\right)^{-r} \left(\frac{m^r}{r!}\right) \frac{n(n-1) \cdots (n-r+1)}{n^r} =$$

$$\left(1 - \frac{m}{n}\right)^n \left(\frac{m^r}{r!}\right) \left[\left(1 - \frac{m}{n}\right)^{-r} \cdot 1 \cdot \left(1 - \frac{1}{n}\right)\left(1 - \frac{2}{n}\right) \cdots \left(1 - \frac{r-1}{n}\right)\right].$$

As n approaches infinity, each term inside the brackets approaches 1, and the limit of $[1 - (m/n)]^n$ can be shown to equal e^{-m}, where e, as before, is equal to $2.71828 \cdots$, the base of the natural logarithms. Hence as n becomes larger and larger, P gets closer and closer to

$$\frac{e^{-m} m^r}{r!}.$$

Since we are usually interested in calculating the probability of obtaining at least X successes (that is, that $r \geq X$), it would be convenient to have a table giving the sum of the probabilities for the Poisson distribution for various specified values of X. Such a table is given as Table II in the Appendix. The entries in that table represent the area under the histogram to the right of and including the rectangle whose center is X. As in the binomial distribution, the total area under the histogram is equal to 1.

Example 7-3. The mortality rate for a certain disease is 5 deaths per 1000 cases. In a group of 360 cases of this disease, what is the probability of
(a) 3 or more deaths;
(b) exactly 3 deaths?

Solution. In this case, $p = 0.005$, $n = 360$, so that $m = np = 360(0.005) = 1.8$ and $X = 3$.
(a) From Table II, we find for $m = 1.8$ and $X = 3$, $P = 0.269$.
(b) The probability of exactly 3 deaths occurring is determined from Table II as the difference between the probabilities for $X = 3$ and $X = 4$:

$$P = 0.269 - 0.109 = 0.160.$$

In the binomial distribution as p tends toward zero, q tends toward 1, and the variance $\sigma^2 = npq$ approaches np. Therefore, in the Poisson distribution the variance equals the mean.

The Poisson distribution is not limited to serving as an approximation to the binomial distribution, but also serves as a model for dealing with events distributed randomly over time or space. In this context it will not be discussed here.

This distribution is applicable in many situations where rare events occur, such as the incidence of the tire failures per week for a fleet of delivery trucks, the number of atoms disintegrating per second in a quantity of radioactive material, the number of hits on a target, the number of reported cases of a rare disease, the number of automobile accidents in a specified time at a particular street crossing, and the number of parasites on an individual host. The Poisson distribution is especially useful in the inspection and quality control of manufactured goods where the proportion of defective articles in a large lot can be expected to be small.

7.3   THE NORMAL FREQUENCY DISTRIBUTION AS A LIMIT OF A FREQUENCY DISTRIBUTION OF A CONTINUOUS VARIABLE

The normal frequency curve serves as a good approximation not only for a histogram of a binomial frequency distribution but for many other histograms as well. Frequently a histogram of a frequency distribution with mean m and standard deviation σ is well approximated for a large number of data N by the curve whose equation is

$$Y = \frac{Nc}{\sigma\sqrt{2\pi}} \, e^{-(X-m)^2/2\sigma^2}, \tag{7-7}$$

where c is the difference between two consecutive class midpoints. The smaller the value of c, the better the approximation usually will be.

The normal curve is indeed the most important frequency curve. Many measurements in science and industry follow distributions that can be approximated well by normal distributions.

In order to find from the normal curve the number of variates falling between $X = a$ and $X = b$, consider Figure 7-6, which shows the histogram and corresponding normal frequency curve for the total seasonal rainfall at Sacramento, California (see data of Table 2-4).

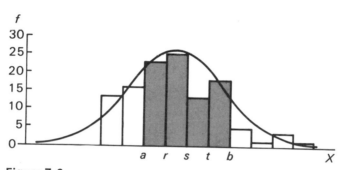

Figure 7-6

Histogram and normal curve for 120 years of rainfall at Sacramento, California.

From Figure 7-6 the number of seasonal rainfall measurements falling between $X = a$ and $X = b$ is simply the shaded area divided by c, since the shaded area is $f_a c + f_r c + f_s c + f_t c$, where f_a is the frequency for the interval for which $X = a$ is the lower class boundary, i.e. the height of the rectangle whose base is $ar = c$, etc. The normal curve in Figure 7-6 is the graph of (7-7) for $m = 17.92$, $\sigma = 6.09$, and $Nc = 360$, or it can be considered as the result of increasing N and decreasing c in the histogram of Figure 7-6. To use the normal curve to determine the number of variates (rainfall measurements) falling between $X = a$ and $X = b$, we find the corresponding area under the normal curve of Figure 7-6 and divide by c, which is equivalent to finding the area under the curve for which each ordinate has been divided by c, that is,

$$Y = \frac{N}{\sigma\sqrt{2\pi}} e^{-(X-m)^2/2\sigma^2}$$

from $X = a$ to $X = b$.

Example 7-4. Given a normal frequency distribution of a continuous variable with $N = 2000$, $m = 20$, $\sigma = 5$. Find
(a) the number of variates between $X = 12$ and $X = 22$;
(b) the number of variates larger than $X = 30$.

Solution. (a) We need first the area under the normal probability curve from $X = 12$ to $X = 22$, i.e. from

$$z = \frac{12 - 20}{5} = -1.6 \quad \text{to} \quad z = \frac{22 - 20}{5} = 0.4,$$

which is found from Appendix Table I to be

$$0.4452 + 0.1554 = 0.6006.$$

Therefore the desired number of variates is

$$0.6006 \cdot 2000 = 1201.$$

(b) We first find the total area to the right of $X = 30$, i.e. to the right of

$$z = \frac{30 - 20}{5} = 2,$$

which is 0.0227, and the desired number of variates is therefore

$$0.0227 \cdot 2000 = 45.$$

In a given problem the reader may have difficulty in deciding whether the normal frequency curve arises as an approximation to a binomial frequency distribution or as a frequency curve approximating a histogram of a continuous variable. But since the first case arises only for variables taking on only non-negative integral values, there should be no difficulty in making a decision.

7.4 PROBABILITY DENSITY FUNCTIONS OF A CONTINUOUS VARIABLE IN GENERAL

Just as the binomial distribution is an example of a distribution of a discrete variable, so the normal distribution is an example of a distribution of a continuous variable. The probability distribution of a discrete variable was defined in Definition 6-3. We shall now define the distribution of a continuous variable or, more precisely, its density function.

If we denote the right-hand side of (7-3), which is a certain function of X, by $P(X)$, then (7-3) becomes

$$y = P(X), \tag{7-8}$$

where $P(X)$ stands for

$$P(X) = \frac{1}{\sqrt{2\pi}} e^{-(X-m)^2/2\sigma^2}. \tag{7-9}$$

$P(X)$ is called the normal probability density function, the term "density" being used since the properties of this function are identical to those of the density as a physical quantity.

A probability density function of a continuous variable in general then is a function of the type of (7-8), where $P(X)$ must have certain properties similar to those satisfied by the general normal probability density function (7-3). These are stated precisely in the following definition.

Definition 7-1. If X is a continuous variable and $y = P(X)$ is a function such that $P(X) \geq 0$ for all X, then $y = P(X)$ is called the *probability density function* of the continuous statistical (or random) variable X if for any numbers a and b, with $b > a$, the area under the graph of $y = P(X)$ from $X = a$ to $X = b$ equals the probability that X lies between a and b.

It immediately follows from Definition 7-1, by taking $a = -\infty$ and $b = +\infty$, that the total area under $y = P(X)$ for any probability density function must equal 1.

The general normal probability density function, as given by (7-3), is clearly an example of a probability density function of a continuous variable, where $P(X)$ is given by (7-9); another is the standard normal probability density function given by (7-5), where the random variable is z (instead of X) and $P(z)$ is given by the right-hand side of (7-5). Other important probability density functions of a continuous variable are the so-called Student's t-distribution, the F-distribution and others, which will be considered in later chapters. Strictly speaking, we should refer to these as Student's t-density function, the F-density function, etc., but the tradition of referring to them as distributions rather than as density functions has been so widely adopted in statistics that we shall also follow this convention henceforth and use "distribution," even though "probability density function" would be more precise.

An obvious question arises: For a probability density function of a continuous variable, what is the probability that $X = a$? For example, in Example 7-4, we might ask: What is the probability that $X = 12.0$? There are two possible interpretations of this question:

(a) What is the probability that $X = 12.0$, accurate to one decimal place? This is equivalent to asking: What is the probability that X is between 11.95 and 12.05? The answer to this question is obtained, exactly as in Example 7-4, by finding the area under the normal probability curve from $X = 11.95$ to $X = 12.05$.

(b) What is the probability that $X = 12.0$ exactly, that is, to infinitely many decimal places? The answer to this question is obtained by observing that the probability that X lies between 12.0 and b, where b is any number greater than 12, is equal to the area under the normal probability curve from $X = 12.0$ to $X = b$. If we now let b approach 12, the resulting area becomes smaller and smaller and equals 0 as b coincides with 12, so that we conclude that the probability that $X = 12.0$ exactly is 0.

Accordingly, for a probability density function of a continuous variable, the probability that $X = a$ *exactly* is always 0.

7.5 SPECIAL AREAS UNDER THE NORMAL CURVE

Because certain areas under the normal curve arise in so many applications we shall make special mention of them. The area under the normal probability curve from $X = m - \sigma$ to $X = m + \sigma$ (i.e. from $z = -1$ to $z = 1$) is found from Appendix Table I to be $0.3413 \cdot 2 = 0.6826$, or, from more extensive tables, 0.68268, which means that in a normal distribution 68.268% (or about two-thirds) of the data fall in this interval. This explains the property of the standard deviation referred to in Chapter 4, where we remarked that in many frequency distributions (in particular, the normal distribution) about two-thirds of the data fall within one standard deviation on either side of the mean.

The area under the normal probability curve from $X = m - 2\sigma$ to $X = m + 2\sigma$ is 0.95450; thus about 95% of the data in a normal distribution fall within two standard deviations on either side of the mean. Finally, the area from $X = m - 3\sigma$ to $X = m + 3\sigma$ is equal to 0.99730. These special areas are shown in Figure 7-7.

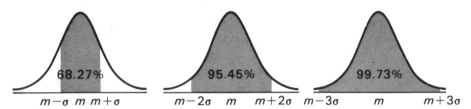

$$m-\sigma \;\; m \;\; m+\sigma \qquad\qquad m-2\sigma \;\; m \;\; m+2\sigma \qquad m-3\sigma \qquad m \qquad m+3\sigma$$

Figure 7-7
Special areas under the normal curve.

In many applications, notably in physics, it is of interest to determine the interval around the mean which contains 50% of the data of the distribution. Clearly, the z-value at the upper limit of this interval must be such that the corresponding area is equal to $\frac{1}{2}(0.50) = 0.25$. Determining then from Appendix Table I the value of z corresponding to an area of 0.25, we find to two decimal places $z = 0.67$, or, from more extensive tables, $z = 0.6745$. Consequently, in a normal distribution, half of the data fall within 0.6745σ on either side of the mean.

Definition 7-2.   The *probable error* (*P.E.*) of a distribution is defined such that the probability that X falls between $m - P.E.$ and $m + P.E.$ is $\frac{1}{2}$.

From the discussion above it follows that for a normal distribution the probable error is given as

$$P.E. = 0.6745\sigma. \tag{7-10}$$

7.6   GRADING ON THE CURVE

As most readers undoubtedly are aware, many instructors, especially those teaching large classes, use a system of grading referred to as "grading on the curve," which of course refers to the normal curve. While there are many such systems that depend upon the percentage of A, B, C, D, and F grades the instructor desires to give, the basic idea is the same in all of them. If the grades are assumed to be normally distributed and if intervals of the distribution (given in terms of the mean and standard deviation) are assigned to each of the various grades, the percentage for each grade is specified. Example 7-5 illustrates the procedure for a system that is, perhaps, most frequently used.

Example 7-5. An instructor assigns grades in an examination according to the following procedure:

   A if score exceeds $m + 1.5\sigma$;
   B if score is between $m + 0.5\sigma$ and $m + 1.5\sigma$;
   C if score is between $m - 0.5\sigma$ and $m + 0.5\sigma$;
   D if score is between $m - 1.5\sigma$ and $m - 0.5\sigma$;
   F if score is less than $m - 1.5\sigma$.

What percentage of each grade does this instructor give, assuming that the scores are normally distributed?

Solution. To determine the percentage of A's we need to determine the area under the normal probability curve to the right of $z = 1.5$, which from Appendix Table I is found to be

$$P_A = 0.5000 - 0.4332 = 0.0668.$$

Consequently, the instructor is assigning 6.68% A's. To determine the percentage of B's, we determine the area under the normal probability curve from $z = 0.5$ to $z = 1.5$ from Appendix Table I and find

$$P_B = 0.4332 - 0.1915 = 0.2417;$$

thus, he gives 24.17% B's. Similarly, we find that he assigns 38.30% C's 24.17% D's, and 6.68% F's.

Note that in Example 7-5 it was assumed that the intervals are arranged symmetrically with respect to the mean. This, however, need not always be the case. Sometimes instructors might wish to use a system in which the lower limit of the C's is farther removed from the mean than the upper limit.

EXERCISES

Calculate standard deviations and z-values to two decimal places.

1. A coin is tossed 100 times. Use the normal curve approximation to find the probability of obtaining

(a) exactly 50 heads, and compare the result with the value 0.0796 obtained from the binomial distribution;

(b) 60 or more heads. The actual value from the binomial distribution is 0.028.

2. Use the normal curve approximation to find the probability of obtaining exactly 16 sixes in 96 tosses of a die. Compare the result with the value 0.110 obtained from the binomial distribution.

3. Assume that one-half the people in a certain community are regular viewers of television. Of 100 investigators, each interviewing 10 individuals regarding their viewing habits, how many would you expect to report that 3 people or fewer were regular television viewers?
(a) Calculate by an exact method (using binomial distribution).
(b) Calculate by an approximate method (using normal distribution).

4. For a binomial frequency distribution for which $p = \frac{1}{4}$, find the probability of obtaining 25 or more successes in 80 trials.

5. A rare blood type occurs in 6 percent of the U.S. population. What is the probability that a team checking the blood type of 100 individuals in a certain city will find more than 10 individuals with this blood type?

6. A manufacturer of light bulbs finds that on the average 2% are defective. What is the probability that out of 1000 such bulbs selected at random 15 or more are defective?

7. A pair of unbiased dice is rolled 45 times. What is the probability that a 6, 7, or 8 will appear 25 or more times?

8. According to American Experience Mortality Tables the probability of a man of age 41 dying within a year is 0.01. If an insurance company has 50,000 policies in force on men of this age, estimate by use of a normal curve the probability that the company will have to pay more than 525 death claims on this group of men within a year.

9. A bag contains a very large number of red, white, and green beads, all of uniform size and weight. There are the same numbers of red and green beads, and the number of white beads equals twice the sum of those of the other two colors. If a bead is drawn at random and replaced, what is the probability that in 18 such drawings at least 14 are white? Use the normal curve approximation.

10. A pump has a failure, on the average, once in every 5000 hours of operation. Use the Poisson distribution to find
(a) the probability that more than one failure will occur in a 1000-hour period;
(b) the probability that no failure will occur in 10,000 hours of service.

11. A book contains 100 typographical errors distributed randomly throughout its 500 pages. Assuming a Poisson distribution find the probability that
(a) a page selected at random contains at least two errors;
(b) 10 pages selected at random contain 5 or more errors.

12. A purchaser will accept a large shipment of articles if in a sample of 1000 from that shipment there are at most 10 defective articles. If the entire shipment is 0.5% defective, find the probability that the shipment will not be accepted
   (a) by use of the Poisson distribution;
   (b) by use of the normal curve approximation to the binomial.

13. A telephone switchboard handles on the average 4 calls per minute. If the calls follow a Poisson distribution, what is the probability of
   (a) exactly 8 calls per minute;
   (b) more than 10 calls per minute?

14. The number of white blood corpuscles on a slide follows a Poisson distribution If the mean number of white corpuscles per slide is 4, on 100 such slides how many will have 10 or more white corpuscles?

15. If a normal distribution of a continuous variable has $m = 21.2$ and $\sigma = 3.10$, find the probability that a variate selected at random will be larger than 30 or less than 15.

16. In a certain building trade the average wage is $3.60 per hour and the standard deviation is 45 cents. If the wages are assumed to follow a normal distribution, what percent of the workers receive wages between $3.00 and $3.50 per hour?

17. The weight of grapefruit from a large shipment averages 15.00 ounces with a standard deviation of 1.62 ounces. If these weights are normally distributed, what percent of all these grapefuit would be expected to weigh between 15.00 and 18.00 ounces?

18. If the average life of a certain make of storage battery is 30 months with a standard deviation of 6 months, what percent of these batteries can be expected to last from 24 months to 36 months? Assume that the lifetimes follow a normal distribution.

19. Records indicate that the average life of television picture tubes is 3.0 years with a standard deviation of 1.5 years. Tubes lasting less than a year are replaced free. For every 100 sets (one tube to a set) sold, how many tubes can be expected to have to be replaced free?

20. A manufacturer knows that on the average 2% of his products are defective. Using the normal curve approximation, what is the probability that a lot of 100 pieces will contain exactly 5 defectives?

21. How many times do two dice have to be rolled so that there is a 50% chance of getting a total of 5 at least 3 times?

22. By use of the normal curve find how many throws with one die will be required in order that the probability of getting at least 5 sixes will have the value $\frac{1}{2}$.

23. It has become known that 5% of all 1968 cars produced by a certain company have defects in their steering mechanism. How many of these cars must have been shipped to a town in order for there to be a 50% chance that it has received at least 40 cars from that company with a defective steering mechanism?

24. The average seasonal rainfall in a certain town in 18.75 inches with a standard deviation of 6.50 inches. Assume that the seasonal rainfall for this town satisfies a normal distribution. In how many years out of a period of 50 years would we expect between 15.00 and 25.00 inches of rain?

25. If the heights of 10,000 college men closely follow a normal distribution with a mean of 69.0 inches and a standard deviation of 2.5 inches,
    (a) how many of these men would you expect to be at least 6 feet in height;
    (b) how many of these men would you expect to be 70.0 inches in height to the nearest tenth of an inch;
    (c) how many of these men would you expect to be exactly 70.0 inches in height;
    (d) what range of heights would you expect to include the middle 75% of the men in this group?

26. Three students take different tests. A gets a score of 72, B one of 85, and C one of 17. All students taking the test with A had an average grade of 85, with B an average of 90, and with C an average of 25. The three standard deviations are 7, 3, and 7, respectively. Arrange the three students in order of excellence as you would judge them by these results, indicating your reasons.

27. In grading Satsuma plums whose weights are normally distributed, 20% are called small, 55% medium, 15% large, and 10% very large. If the mean weight of all Satsuma plums is 4.83 ounces with a standard deviation of 1.20 ounces, what are the lower and upper bounds for the weight of medium Satsuma plums?

28. In a normal distribution having a standard deviation of 2.00 the probability that a variate selected at random exceeds 28.00 is 0.03.
    (a) Find the mean of the distribution.
    (b) Find the variate above which lie 95% of all variates of this distribution.

29. The mean of a normal distribution is 32.00. If 69.5% of the variates are less than 34.04, what is the standard deviation of the distribution?

30. On an examination the average grade was 70.0, and the standard deviation was 10.0. The instructor gave all students with grades from 61.0 to 79.0 the grade of C. There were 24 students who received a C grade. If the grades are assumed to satisfy a normal distribution, how many students took the examination?

31. Of a large group of men, 4% are under 63.3 inches in height and 52% are between 63.3 and 69.0 inches. Assuming a normal distribution of heights, what would be the mean and standard deviation of the distribution?

32. The standard deviation of a binomial distribution is 6, and the probability that any variate is equal to or exceeds 65 is 0.2266. Use the normal curve approximation to find p and n.

33. In a normal distribution with mean 15.00 and standard deviation 3.50 it is known that 647 variates exceed 16.25. What is the total number of variates in the distribution?

34. An instructor who assigns 10% A's, 20% B's, 40% C's, 20% D's, 10% F's gives an examination on which the average is 68.0. If the borderline between the C's and B's is 78.0 on that examination and if the grades are normally distributed, what is the standard deviation of the class?

35. A physical education instructor announces that he will give an A grade to 20% of his high-jump class based entirely on ability to jump. If experience in his classes has shown that the height of the jump is normally distributed with mean 4 feet 8 inches and standard deviation 4 inches, at least how high must a student jump if he is to get an A?

36. The heights of a number of students are normally distributed with a mean of 68.50 inches. If 12% are at least 71.20 inches tall,
    (a) what is the standard deviation of this distribution;
    (b) what is the probable error?

37. A set of 10,000 variates of a continuous variable is found to be normally distributed with mean 450. If 1700 of them lie between 450 and 460,
    (a) what is the standard deviation;
    (b) what is the probable error?

38. The diameters of ball bearings are normally distributed with standard deviation 2 mm. If it is known that 4% of the bearings have diameters larger than 23.5 mm., what is the mean of the distribution?

39. In a normal distribution of a continuous variable with mean 100 and standard deviation 53 there are 135 variates greater than 200. How many variates should we expect between 150 and 200?

40. In a restaurant on a particular morning the amounts spent for breakfast by all patrons follow a normal distribution with mean 87.2 cents and standard deviation 12.0 cents. On that morning if 420 people spend 85 cents or more for breakfast, what is the total number of people served?

41. In a normal distribution with mean 72.0 and standard deviation 12.0 there are 220 variates between 40.0 and 90.0. How many variates are there in the whole distribution?

42. In a normal distribution with mean 120.0 and standard deviation 30.0 there are 300 variates between 130 and 150. How many variates are there between 130 and 145?

43. It is known that a variate originates either from a normal population with mean 10 and standard deviation 2 or from a normal population with mean 11 and standard deviation 3. If it is considered twice as likely that the variate comes from the second population than from the first, find the probability that the variate exceeds 12.5.

44. In an examination 13% of the class receives a grade of A; 20% B; 48% C; 10% D; and 9% F. The C grade ranges from 55 to 79. Assuming a normal distribution, what are the mean and the standard deviation of the grades?

45. Genetic theory indicates that in a certain cross of two varieties of peas, half the seeds obtained will be smooth and half will be wrinkled. How many seeds have to be examined in order that at least 100 smooth seeds are obtained with a probability of 90%?

46. According to Mendelian theory of inheritance, certain crosses of peas give yellow and green peas in the ratio of 3 : 1. How many seeds from such a cross have to be planted in order that there is a 95% probability that at least 50 green peas are obtained?

8

Random Sampling.
Large Sample Theory

8.1 INTRODUCTION

Sampling is one of the most important concepts in the study of statistics. It is basic to statistical theory and to applications of statistical theory in all fields of the physical, biological, and social sciences, in economics, in medicine, in agriculture, and in business and industry. Before proceeding, we must clarify certain fundamental ideas about populations and samples.

Definition 8-1. A *population* (or universe) is the totality of the measurements or counts obtainable from all objects possessing some common specified characteristic.

In the statistical sense, therefore, a population is never a set of objects, but always a set of measurements or counts. A population may consist of finitely or infinitely many variates. For example, in a study of the size of a particular variety of fruit at some specified stage of development we may be interested only in the fruits on a certain limb of a tree; the sizes of these fruits, then, constitute the population. But if we are interested in all fruits on the tree, their sizes form the population. Or we may wish to study all the fruits on all the trees of a particular orchard; in this case the population consists of the sizes of all fruits in the orchard. Frequently, we wish to extend the applicability of our conclusions beyond such finite populations to encompass, say, all fruits at a specified stage of development on all such trees that now exist or will exist in the future. In this population the number of measurements of size may be inexhaustible, and this constitutes an infinite population.

Another example of a finite population is the set of "maximum blowing currents" of a lot of 1000 fuses, all manufactured under similar conditions. Since for each fuse there is one, and only one, maximum current at which it will blow, these currents constitute a finite population of 1000 items.

Since we can rarely investigate a whole population we are obliged to formulate conclusions regarding a population from samples selected from it.

Definition 8-2. A *sample* is a set of measurements that constitute part or all of a population.

In Chapter 5, in evaluating probabilities, some exercises involved a series of drawings in which the object drawn was returned before the next drawing was made, while in other exercises the withdrawn objects were not returned. Sampling also can be carried out either with or without replacement. In studying the size of fruit on a particular limb of a tree, we might select 10 fruits for measurement. If the fruits are picked one by one and measured, the sampling is effected without replacement; if, however, the fruits are measured on the limb, without picking, the procedure is equivalent to sampling with replacement, since after each measurement the population remains the same as before and a fruit which has already been measured might be selected again. Much sampling is done with replacement; however, some populations must be sampled without replacement. The latter is necessary when the items of the sample are detroyed in the process of measurement. For example, since a fuse is destroyed in measuring the maximum blowing current, it cannot be used for further measurement.

The main object in taking a sample is to draw some conclusion about the population from which it is obtained. The relation of sample to population is one of the most important problems in statistical theory and practice, since good estimates concerning a population necessitate good samples. Securing good samples is not always easy, but generally it can be done. A precise definition of a good sample, a so-called *random sample*, will not be given here, but the following will suffice.

Definition 8-3. A *random sample* is a sample in which any one individual measurement in the population is as likely to be included as any other.

Definition 8-4. A *biased* sample is a sample in which certain individual measurements have a greater chance to be included than others.

Since from now on frequent reference will be made to samples and populations, it is necessary to distinguish between the mean of a sample and the mean of a population, also between the standard deviation of a sample and that of a population. Table 8-1 indicates the notation which will be used to distinguish between samples and populations.

Table 8-1
Notation for samples and populations.

Characteristic	Sample	Population
Number of variates	n	N
Mean	$\bar{X}$	m
Standard deviation	$\hat{s}$	σ

According to Table 8-1, the mean of a sample of n variates $X_1, X_2, X_3, \ldots, X_n$ is given by

$$\bar{X} = \frac{\sum\limits_{i=1}^{n} X_i}{n}. \tag{8-1}$$

For this sample, we would calculate the standard deviation, as we have done before for any set of data, by taking the square root of the sum of the squares of the deviations of the variates from their mean, $\bar{X}$, divided by the number of variates, n; that is,

$$\hat{s} = \sqrt{\frac{\sum\limits_{i=1}^{n} (X_i - \bar{X})^2}{n}}. \tag{8-2}$$

In a number of textbooks on statistics the standard deviation of a sample is defined as

$$s = \sqrt{\frac{\sum\limits_{i=1}^{n} (X_i - \bar{X})^2}{n-1}}. \tag{8-3}$$

This definition is made so that the standard deviation from a sample represents an appropriate estimate of the standard deviation of the population. When only a single sample from a population is available, the best estimate of the mean m of the population is given by $\overline{X}$, the mean of the sample. The most appropriate estimate of the standard deviation σ of the population, however, is not $\hat{s}$, the standard deviation of the sample, but rather s as defined by (8-3), as will be explained in Chapter 10. For consistency, however, we shall define the standard deviation of a sample by (8-2) with n in the denominator, since, if the mean of a set of data is the same whether the set is considered as a sample or as a population, the same should be true for the standard deviation; that is the standard deviation of a set of data should not depend on whether the data are considered as a sample or as a population. The use of $\hat{s}$ is consistent with this reasoning. In Chapter 10 we shall use s as defined in (8-3) to represent an appropriate estimate of the standard deviation of a population determined from a sample.

In preceding chapters m and σ were used, sometimes for samples, sometimes for populations. For simplicity we made no distinction. Usually, Roman letters are used for measurements obtained from samples, and the corresponding Greek letters are used for population measurements. We shall follow this convention in later chapters. If we were to follow it strictly, we should of course have denoted by Greek letters the number of variates and the mean of a population, as is done in some texts (μ being used for the mean). However, since by tradition N and m are so generally used for number of variates and mean of a population, we shall adhere to this widely accepted practice.

Definition 8-5.   A numerical characteristic of a population, such as its mean or standard deviation, is called a *population parameter* or simply a *parameter*.

Definition 8-6.   A quantity calculated from a sample, such as its mean or standard deviation, is called a *sample statistic* or simply a *statistic*.

Since parameters of a given population are based upon all its variates, they are fixed for that population. On the other hand, since statistics are based upon only a part of a population, they usually vary from sample to sample. In order to develop valid statistical procedures we must first know something about the distribution of certain statistics.

Throughout the rest of this book we shall assume that populations consist of finitely many variates in order to simplify many of the discussions. Actually, all of the results that are derived will, with occasional slight changes in wording, be true also for infinite populations. The assumption that the size N of a population must be finite should not cause any concern, since there is no restriction on how large N can be. Indeed, in most of the discussions N is assumed to be a very large (although finite) number.

8.2   DISTRIBUTION OF THE SAMPLE MEAN

Consider a population of N variates with mean m and standard deviation σ, and draw all possible samples of n variates from this population, denoting their means by $\bar{X}_1, \bar{X}_2, \ldots, \bar{X}_M$. (See Figure 8-1.) Assume here that the

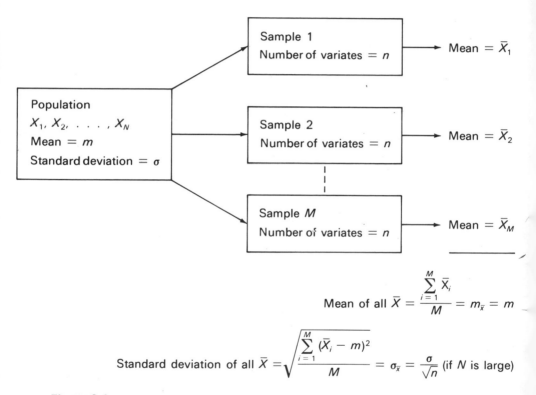

$$\text{Mean of all } \bar{X} = \frac{\sum\limits_{i=1}^{M} \bar{X}_i}{M} = m_{\bar{x}} = m$$

$$\text{Standard deviation of all } \bar{X} = \sqrt{\frac{\sum\limits_{i=1}^{M} (\bar{X}_i - m)^2}{M}} = \sigma_{\bar{x}} = \frac{\sigma}{\sqrt{n}} \text{ (if } N \text{ is large)}$$

Figure 8-1
Relation between sample means and population parameters.

samples are replaced, so that the total number of samples (M) which can be drawn is equal to the number of combinations of N things taken n at a time, that is, $M = C_{N,n}$. The mean of all these sample means is denoted by $m_{\bar{x}}$, and the standard deviation of these sample means by $\sigma_{\bar{x}}$. Strictly speaking, capital letters $\bar{X}$ should follow m and σ as subscripts, but from here on, where no confusion is possible, small letters will be used as subscripts. Then we can prove that

$$m_{\bar{x}} = m \qquad (8\text{-}4)$$

and

$$\sigma_{\bar{x}} = \sigma \sqrt{\frac{N-n}{n(N-1)}}. \qquad (8\text{-}5)$$

Since in general N is very large when compared with n, a simpler relationship can be obtained for a very large value of N. The expression under the square root of (8-5) may be written

$$\frac{N-n}{n(N-1)} = \frac{N}{n(N-1)} - \frac{1}{N-1} = \frac{1}{n\left(1 - \dfrac{1}{N}\right)} - \frac{1}{N-1},$$

which for very large N is close to $1/n$. Therefore, for large N Formula (8-5) becomes

$$\sigma_{\bar{x}} = \frac{\sigma}{\sqrt{n}}. \qquad (8\text{-}6)$$

Formula (8-4) does not seem surprising, since we would expect the mean of all the sample means of a population to be about the same as the mean of all the variates of the population, although it may seem surprising that this mean should always be precisely equal to the population mean.

Formula (8-6) is reasonable, since we would expect the sample means to vary less than the variates of the population, and to vary less and less as n becomes larger and larger (i.e. have a much smaller standard deviation than the variates of the population).

The proofs of (8-4) and (8-5) will not be included here. It can be shown in a simple case, where $N = 3$ and $n = 2$, how easy it is to verify these formulae.

If $N = 3$, there are 3 variates in the population—call them X_1, X_2, X_3. Since $n = 2$, all possible samples of 2 variates from this population are: X_1, X_2, and X_1, X_3, and X_2, X_3. Let the mean of the first sample be denoted by $\bar{X}_1$, that of the second sample by $\bar{X}_2$, and so on. Then

$$\bar{X}_1 = \frac{X_1 + X_2}{2}, \qquad \bar{X}_2 = \frac{X_1 + X_3}{2}, \qquad \bar{X}_3 = \frac{X_2 + X_3}{2}.$$

To verify (8-4) we must show that $m_{\bar{x}}$, the mean of $\bar{X}_1$, $\bar{X}_2$, $\bar{X}_3$—that is, $(\bar{X}_1 + \bar{X}_2 + \bar{X}_3)/3$—is equal to m, the mean of X_1, X_2, X_3—that is $(X_1 + X_2 + X_3)/3$. Now

$$m_{\bar{x}} = \frac{1}{3}(\bar{X}_1 + \bar{X}_2 + \bar{X}_3) = \frac{1}{3}\left(\frac{X_1 + X_2}{2} + \frac{X_1 + X_3}{2} + \frac{X_2 + X_3}{2}\right)$$

$$= \frac{1}{3}\left(\frac{2X_1 + 2X_2 + 2X_3}{2}\right) = \frac{1}{3}(X_1 + X_2 + X_3) = m.$$

To verify (8-5) for $N = 3$ and $n = 2$, it is necessary to show that either $\sigma_{\bar{x}} = \sigma\sqrt{1/4}$ or $\sigma_{\bar{x}}^2 = (1/4)\sigma^2$. The variance σ^2 is obtained as

$$\sigma^2 = \frac{\sum\limits_{i=1}^{3}(X_i - m)^2}{3} = \frac{2}{9}\left[\sum_{i=1}^{3}X_i^2 - (X_1X_2 + X_1X_3 + X_2X_3)\right],$$

where the latter step is easily verified as required in Exercise 5, and

$$\sigma_{\bar{x}}^2 = \frac{\sum\limits_{i=1}^{3}(\bar{X}_i - m)^2}{3} = \frac{1}{18}\left[\sum_{i=1}^{3}X_i^2 - (X_1X_2 + X_1X_3 + X_2X_3)\right],$$

which can also be easily checked as required in Exercise 6. Therefore,

$$\sigma_{\bar{x}}^2 = \frac{1}{18}\left(\frac{9}{2}\sigma^2\right)$$

$$= \frac{1}{4}\sigma^2.$$

Theorem 8-1. If X is normally distributed with mean m and standard deviation σ and if all possible samples of n variates are drawn from the population (as shown in Figure 8-1), then the sample mean $\bar{X}$ will be normally distributed with mean m and standard deviation $\sigma/\sqrt{n}$.

In other words, if the distribution of X is given by (7-3), whose graph is the general normal probability curve with mean m and standard deviation σ, then the distribution of $\bar{X}$ has as its graph the general normal probability curve with mean $m_{\bar{x}} = m$ and standard deviation $\sigma_{\bar{x}} = \sigma/\sqrt{n}$. This means that, if we let

$$z = \frac{\bar{X} - m_{\bar{x}}}{\sigma_{\bar{x}}}, \qquad\qquad (8\text{-}7)$$

then z will satisfy the standard normal probability curve (7-5), and the probability that $\overline{X}$ will lie between two given numbers a and b, with $b > a$, is determined, as in Chapter 7, by finding the area under the normal probability curve between the two values of z corresponding to a and b, as calculated from (8-7). As before, we use Appendix Table I to look up these areas.

The distributions of sample means for samples of 4 and 16 variates together with the population distribution as shown in Figure 8-2 clearly indicate the

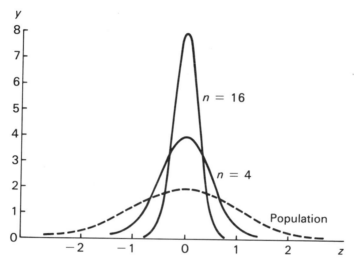

Figure 8-2

Distribution of the means of samples of 4 and 16 variates from a normally distributed population.

way in which the type of resulting normal distribution varies with a change in size of sample. As we might expect, with increasing values of the sample size (n) the normal probability curve becomes narrower and taller, since the means of large samples will obviously vary less than those of small samples, and the area under all probability curves will always be equal to unity.

Although Theorem 8-1 is applicable only if the population has a normal distribution, Theorem 8-2 is applicable for the much more general case where the distribution of the population is unknown. It is known as the *Central Limit Theorem*.

Theorem 8-2.   If X has a distribution with mean m and standard deviation σ and if all possible samples of n variates are drawn, then the sample mean

$\overline{X}$ will have a distribution which for larger and larger n approaches the normal distribution with mean m and standard deviation $\sigma/\sqrt{n}$.

Theorem 8-2 states therefore that for "large n" (which is assumed to mean that n is greater than 30) $\overline{X}$ will be approximately normally distributed, even if the population does not satisfy a normal distribution.

The proofs of Theorems 8-1 and 8-2 are dependent on more advanced mathematical procedures than are assumed here, and will not be given.

Example 8-1. The mean height of a group of 1000 college students is 68.20 inches, and the standard deviation is 2.50 inches. Find the probability that in a sample of 100 students the mean will be greater than 68.90 inches.

Solution. Since by Theorem 8-2 the mean of the samples is approximately normally distributed with

$$m_{\bar{x}} = m = 68.20$$

and

$$\sigma_{\bar{x}} = \frac{\sigma}{\sqrt{n}} = \frac{2.50}{10} = 0.25,$$

we must find the area under the normal probability curve to the right of $\overline{X} = 68.90$, i.e. to the right of

$$z = \frac{\overline{X} - m}{\sigma_{\bar{x}}} = \frac{68.90 - 68.20}{0.25} = 2.80.$$

From Appendix Table I this area is found to be 0.0026; therefore in a sample of 100 students the probability that the mean will be greater than 68.90 inches is 0.0026.

8.3   DISTRIBUTION OF THE SAMPLE STANDARD DEVIATION

Consider a population of N variates with mean m and standard deviation σ, and draw all possible samples of n variates from this population. Suppose there are M such samples having standard deviations $\hat{s}_1, \hat{s}_2, \ldots, \hat{s}_M$. We shall denote the mean of all these standard deviations by $m_{\hat{s}}$, and the standard deviation of all these standard deviations by $\sigma_{\hat{s}}$. Then for large n it can be proved that

$$m_{\hat{s}} = \sigma \qquad\qquad (8\text{-}8)$$

and

$$\sigma_{\hat{s}} = \frac{\sigma}{\sqrt{2n}}.$$
(8-9)

Since these formulae as well as theorems describing the distribution of the standard deviation have little practical application, we consider it sufficient to state these formulae without further discussion. They are given merely for reasons of completeness.

8.4   DISTRIBUTION OF THE DIFFERENCE BETWEEN TWO SAMPLE MEANS

Consider two populations with means m_x and m_y, and standard deviations σ_x and σ_y, respectively. Let N denote the number of all possible pairs of samples which can be taken from these populations such that the first of each pair is from the first population and has n_1 variates and the other is from the second population and has n_2 variates. Let the means of the samples from the first population be denoted by $\overline{X}_1, \overline{X}_2, \ldots, \overline{X}_N$, and those from the second by $\overline{Y}_1, \overline{Y}_2, \ldots, \overline{Y}_N$. Then consider for each pair the difference between the mean of the first sample and that of the second; thereby N such differences are obtained, as shown in Figure 8-3.

In many practical problems it is of great importance to know what the mean and standard deviation of these N differences are in terms of the means and the standard deviations of the two populations. Denoting by $m_{\overline{x}-\overline{y}}$ the mean and by $\sigma_{\overline{x}-\overline{y}}$ the standard deviation of these N differences, it can be proved that

$$m_{\overline{x}-\overline{y}} = m_x - m_y,$$
(8-10)

and that if the populations are independent (i.e. there is no relationship whatsoever between specific variates of the two populations),

$$\sigma_{\overline{x}-\overline{y}} = \sqrt{\sigma_{\overline{x}}^2 + \sigma_{\overline{y}}^2},$$
(8-11)

where

$$\sigma_{\overline{x}} = \frac{\sigma_x}{\sqrt{n_1}}, \qquad \sigma_{\overline{y}} = \frac{\sigma_y}{\sqrt{n_2}}.$$

The proof of (8-10) should be carried out as an exercise. The proof of (8-11) will be given in Chapter 12 [see Formula (12-39)].

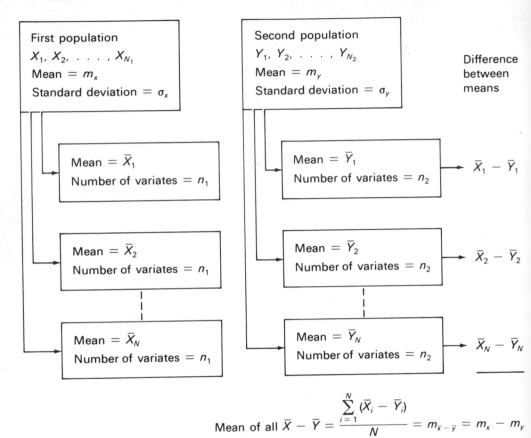

Figure 8-3
Relation between differences of sample means and population parameters.

Theorem 8-3. If X and Y are normally and independently distributed with means m_x and m_y and standard deviations σ_x and σ_y, respectively, and if all possible pairs of samples are drawn from these populations (as shown in Figure 8-3), then the difference between sample means $\bar{X} - \bar{Y}$ is normally distributed with mean $m_x - m_y$ and standard deviation

$$\sigma_{\bar{x}-\bar{y}} = \sqrt{\sigma_{\bar{x}}^2 + \sigma_{\bar{y}}^2}.$$

In other words, the variable $\bar{X} - \bar{Y}$ has as its distribution the general normal distribution with mean $m_{\bar{x}-\bar{y}} = m_x - m_y$ and standard deviation $\sigma_{\bar{x}-\bar{y}} = \sqrt{\sigma_{\bar{x}}^2 + \sigma_{\bar{y}}^2}$. This means that, if we let

$$z = \frac{(\bar{X} - \bar{Y}) - m_{\bar{x}-\bar{y}}}{\sigma_{\bar{x}-\bar{y}}}, \qquad (8\text{-}12)$$

then z will satisfy the standard normal probability curve (7-5).

Again, since the proof of Theorem 8-3 depends on more advanced mathematics, it will be omitted.

Example 8-2. From two normal and independent populations, where the mean of the second population is 0.50 less than that of the first, pairs of samples are drawn such that each pair contains one sample from each population. If the samples from the first population contain 90 variates, and those from the second 120 variates, and if the standard deviation of the first population is 10, and that of the second is 8, find the probability that in a pair of samples the difference between the first and second sample means exceeds 1.00 in absolute value.

Solution. Denoting by X the variable of the first population, by Y that of the second, we have

$$m_x - m_y = 0.50.$$

Now, since

$$\sigma_x = 10, \qquad \sigma_y = 8, \qquad n_1 = 90, \qquad n_2 = 120,$$

we have

$$\sigma_{\bar{x}} = \frac{10}{\sqrt{90}}, \qquad \sigma_{\bar{y}} = \frac{8}{\sqrt{120}}.$$

The calculations for $\sigma_{\bar{x}}$ and $\sigma_{\bar{y}}$ should not be carried any further since the computations from here on become simpler if these quantities are left in this form; in particular, since in the next step the squares of $\sigma_{\bar{x}}$ and $\sigma_{\bar{y}}$ are required, the square-roots appearing in the denominators of $\sigma_{\bar{x}}$ and $\sigma_{\bar{y}}$ should not be computed. Using the values of $\sigma_{\bar{x}}$ and $\sigma_{\bar{y}}$ in the above form, we then obtain

$$\sigma_{\bar{x}-\bar{y}} = \sqrt{\frac{10^2}{90} + \frac{8^2}{120}} = \sqrt{\frac{100}{90} + \frac{64}{120}} = \sqrt{1.1111 + 0.5333}$$

$$= \sqrt{1.6444} = 1.28.$$

From Theorem 8-3 we know that the difference between means for all pairs of samples of sizes 90 and 120 is normally distributed with mean $m_{\bar{x}-\bar{y}} = m_x - m_y = 0.50$ and standard deviation $\sigma_{\bar{x}-\bar{y}} = 1.28$. We wish to find the probability that this difference exceeds 1.00 in absolute value, i.e. that the difference is either greater than 1.00 or less than -1.00. Now, we have for

$$\bar{X} - \bar{Y} = 1.00: \qquad z_1 = \frac{1.00 - 0.50}{1.28} = \frac{0.50}{1.28} = 0.39,$$

and for

$$\bar{X} - \bar{Y} = -1.00: \qquad z_2 = \frac{-1.00 - 0.50}{1.28} = \frac{-1.50}{1.28} = -1.17,$$

so we need the area under the normal probability curve to the right of 0.39 and to the left of -1.17, which is found from Appendix Table I to be

$$P = (0.5000 - 0.1517) + (0.5000 - 0.3790) = 0.3483 + 0.1210 = 0.4693.$$

Important applications of Theorem 8-3 will be given in Chapter 9.

8.5   STANDARD ERROR

Definition 8-7.   The *standard error* of a statistic is the standard deviation of the sampling distribution of that statistic.

For example, the standard error of the mean is another designation for the standard deviation of the mean. From the previous discussion in this chapter we can therefore say: The standard error of the mean is $\sigma/\sqrt{n}$; the standard error of the standard deviation is $\sigma/\sqrt{2n}$; and the standard error of the difference between the means of samples from two independent populations is

$$\sqrt{\sigma_{\bar{x}}^2 + \sigma_{\bar{y}}^2}.$$

EXERCISES

1. Verify Formula (8-4) for $N = 4$ and $n = 2$.

2. Verify Formula (8-4) for $N = 4$ and $n = 3$.

3. Verify Formula (8-4) for $N = 5$ and $n = 4$.

4. Verify Formula (8-4) for $N = 5$ and $n = 3$.

5. For the case where $N = 3$ and $n = 2$, show that

$$\sum_{i=1}^{3} (X_i - m)^2 = \frac{2}{3}\left[\sum_{i=1}^{3} X_i^2 - (X_1 X_2 + X_1 X_3 + X_2 X_3)\right].$$

6. For the case where $N = 3$ and $n = 2$, show that

$$\sum_{i=1}^{3} (\bar{X}_i - m)^2 = \frac{1}{6} \left[\sum_{i=1}^{3} X_i^2 - (X_1 X_2 + X_1 X_3 + X_2 X_3) \right].$$

7. Verify Formula (8-5) for $N = 4$ and $n = 2$.

8. Verify Formula (8-5) for $N = 4$ and $n = 3$.

9. In a normal population with mean 72.10 and standard deviation 3.10, find the probability that in a sample of 90 variates the mean will be less than 71.70.

10. In a savings bank the average account is $159.32 with a standard deviation of $18.00. What is the probability that a group of 400 accounts taken at random shows an average deposit of $160.00 or more?

11. In a particular area the daily wages of coal miners are normally distributed with a mean of $16.50 and a standard deviation of $1.50. What is the probability that a representative sample of 25 miners will have an average daily wage below $15.75?

12. The heights of a certain group of adults have a mean of 67.42 inches and a standard deviation of 2.58 inches. If these heights are normally distributed, and if 25 people are taken at random from the group, what is the probability that their mean will be 68.00 inches or more?

13. A random sample 100 variates is drawn from a normal population with mean 50 and standard deviation 8, and a random sample of 400 variates is drawn from a normal population with mean 40 and standard deviation 12. Find the probability that
   (a) the mean of the first sample exceeds the mean of the second sample by 8 or more;
   (b) the two means differ in absolute value by 12 or more.

14. From each of two normal and independent populations with identical means and with standard deviations of 6.40 and 7.20, a sample of 64 variates is drawn. Find the probability that the difference between the means of the samples exceeds 0.60 in absolute value.

15. If a random sample of 36 variates is taken from a population, how many variates should be taken in another sample from the same population to make the standard error of the mean of the second sample $\frac{2}{3}$ of the standard error of the mean of the first sample?

16. If two random samples are taken from a population, and if the standard error of the mean of one of them is k times the standard error of the mean of the other, how are the sizes of the samples related?

17. In one restaurant the average amount spent for breakfast is 98.3 cents with a standard deviation of 15.0 cents, and in a second restaurant the corresponding figures are 92.4 and 12.0 cents. If a random sample of 80 breakfast charges is taken from the first restaurant and a random sample of 60 from the second,

what is the probability that the difference between the average amounts of these two samples is less than 1 cent in absolute value?

18. It is known that in a certain large city persons eating in a restaurant spend, on the average, 98.3 cents for breakfast with a standard deviation of 15.0 cents. If each of 50 restaurants is asked to select at random the breakfast bills of 100 persons and to report the average amount spent by these 100 persons, how many restaurants may be expected to report the average bill as $1.00 or more?

19. The average gasoline mileage for cars of make A is 20 miles per gallon with a standard deviation of 6 miles per gallon. Comparable figures for cars of make B are 25 and 5.5 miles per gallon. Assume that the gasoline mileage for each of the two makes is normally distributed. What is the probability in a gasoline mileage contest that the average gasoline mileage for 10 cars of make A is greater than that for 9 cars of make B?

20. On the average, students from university A get up 50 minutes after sunrise with a standard deviation of 15 minutes, students from university B get up 60 minutes after sunrise with a standard deviation of 18 minutes. A group of 25 students from university A takes a trip with a group of 20 students from university B. Find the probability that the average time of rising for the group from university B is earlier than that for the group from university A.

21. Past experience shows that, on the average, men and women students perform equally well in tests on statistics. The standard deviation of the men students however, is found to be 15; that of the women students 12. What is the probability that, in an examination in statistics taken by 120 students of which 69 are men, the average of the women students will exceed that of the men by more than 3.0 points?

22. Two normal populations with the same variance $\sigma^2 = 4$ have means of 66.0 and 65.5. What is the probability that the mean of a sample of 50 variates from the first population exceeds the mean of a sample of 50 variates from the second population?

23. If all possible samples of size 25 are drawn from a normally distributed population with mean equal to 20 and standard deviation equal to 4, within what range will the middle 90% of the sample means fall?

24. The daily wages in a particular industry are normally distributed with a mean of $13.20. If 9% of the mean daily wages of samples of 25 workers fall below $12.53, what is the standard deviation of daily wages in this industry?

25. From a normal distribution all possible samples of size 10 are selected. If 2% of these samples have means which differ from the mean of the population by more than 4.00 in absolute value, find the standard deviation of the population.

26. In a normal distribution of a continuous variable with mean 9.00 the probability that the mean of a random sample of 25 variates exceeds 10.00 is 0.33. What is the probability that a single variate selected at random from this distribution is greater than 10.00?

27. In a normal population with mean 8.00, the probability that the mean of a random sample of 25 variates exceeds 9.76 is 0.33. What is the probability that a single variate selected at random from the population is greater than 3.00?

28. If it is assumed that the heights of men are normally distributed with a standard deviation of 2.5 inches, how large a sample should be taken in order to be fairly sure (probability 0.95) that the sample mean does not differ from the true mean (population mean) by more than 0.50 in absolute value?

29. Referring to the data of Exercise 28, how large a sample should be taken if we want to be 99% sure that the sample mean does not differ from the true mean by more than 0.50 in absolute value?

30. A normal population has a standard deviation of 2.0. How large a sample should be taken in order to be 99% sure that the sample mean differs from the population mean by less than 0.8?

31. In measuring reaction time, a psychologist estimates that the standard deviation of the population is 0.05 seconds. How large a sample of measurements must he take in order to be 99% sure that he will estimate the mean of the population with an error not exceeding 0.01 seconds?

32. IQ scores are usually accepted to be normally distributed with a standard deviation of 15. At least how large a sample should be taken in order to be 95% sure that the sample mean does not differ from the population mean by more than 2 IQ points?

33. On a nation-wide examination the scores followed a normal distribution with a mean of 72 and a standard deviation of 10. How large a sample of candidates from University Y must be taken in order that there be a 90 percent chance that its mean score is more than 70?

9

Testing Hypotheses, Significance Levels, Confidence Limits. Large Sample Methods

9.1 TESTING HYPOTHESES

We shall now consider one of the most important problems, that of testing statistical hypotheses.

Definition 9-1.   A *statistical hypothesis* is an assumption made about some parameter, i.e. about a statistical measure of a population.

From this definition it follows that not every hypothesis is a statistical hypothesis. The following hypotheses are not statistical hypotheses:

(a) Mars is inhabited by human beings.
(b) Mr. E. is the candidate who would be the best president for the United States.

Neither of these statements is an assumption about a parameter. On the other hand, only a slight change in wording of statement (b) would result in the following statistical hypothesis:

(c) Mr. E. is the candidate most favored for the presidency of the United States.

The last statement is a statistical hypothesis, since it states an assumption about a parameter, the percentage of all people favoring Mr. E.

In order to establish the truth or falsity of a statistical hypothesis with complete certainty it may be necessary to examine the entire population. Since this is frequently impossible or impractical, we are forced to take a sample from the population and use the sample in deciding whether the hypothesis is true or false. A statistical procedure or decision rule that leads to establishing the truth or falsity of a hypothesis is called a *statistical test*. The hypothesis being tested is often denoted by H_0. (Read "H sub zero"). Example 9-1 will serve as an illustration.

Example 9-1.   Suppose we wish to test the "honesty" of a coin. This is a statistical hypothesis, since the assumption of "honesty" is equivalent to the assumption that for the population of a very large number of tosses of this coin, the probability p of obtaining heads is $\frac{1}{2}$. Suppose now in tossing this coin 100 times we obtain 60 heads, and from this information we are forced to reach a decision regarding the honesty of the coin.

Solution.   If we had obtained 50 or close to 50 heads we would not have doubted the honesty of the coin. The obtained number, 60, however, is sufficiently far from 50 to cause some doubt. Consequently, it seems reasonable to calculate—on the assumption that the coin is honest—the probability of obtaining 60 or more heads in 100 throws.

Using the normal curve approximation to the binomial distribution, we find:

Hypothesis: $H_0: p = \dfrac{1}{2}.$

Then,

$$q = \frac{1}{2};$$

$$n = 100;$$

$$m = \frac{1}{2} \cdot 100 = 50;$$

$$\sigma = \sqrt{100 \cdot \frac{1}{2} \cdot \frac{1}{2}} = \sqrt{25} = 5;$$

$$z = \frac{59.5 - 50}{5} = \frac{9.5}{5} = 1.90;$$

$$P = 0.5000 - 0.4713 = 0.0287.$$

Therefore, if an honest coin is tossed 100 times, the probability of obtaining 60 or more heads is 0.0287, certainly not a high probability. From this result we are justified in drawing one of two conclusions:

1. The hypothesis is correct, but a rare event has occurred.
2. The hypothesis is not correct.

In statistics it is customary to choose the second of these two alternatives whenever the probability is less than a given value α, called the significance level.

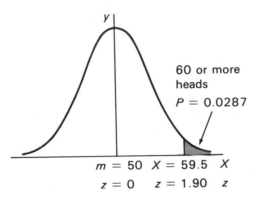

Figure 9-1
Graphical representation of the probability involved in Example 9-1.

Definition 9-2. In testing a statistical hypothesis, the hypothesis is declared to be true if the calculated probability exceeds a given value α, called the *significance level*; and it is considered false if the calculated probability is less than α.

Definition 9-3. If the calculated probability is less than α, indicating that the hypothesis is false, the result is termed *significant*.

It is a commonly accepted convention to consider a result significant if the calculated probability is less than $\alpha = 0.05$, and to term it *highly significant* if the calculated probability is less than $\alpha = 0.01$.

In Example 9-1 we conclude that, since 0.0287 is less than 0.05, the result is significant (although not highly significant), and consequently on the basis of the 5% level of significance ($\alpha = 0.05$) we reject the hypothesis that the coin is honest. It should be noted that, if the hypothesis is assumed to be correct, by chance alone once in 20 trials the calculated probability will be less than $\alpha = 0.05$, and once in 100 trials it will be less than $\alpha = 0.01$. Therefore, in using the 5% level of significance we will be in error on the average once in 20 trials, and at the 1% level once in 100 trials. If we are not willing to draw the wrong conclusion that frequently, we must select a smaller level of significance, but no matter how small the level selected, we still run the risk of being in error in our conclusion.

Why, in Example 9-1, did we calculate the probability of obtaining 60 or more heads and not the probability of obtaining a deviation of 10 or more from the expected number of heads (i.e. 60 or more or 40 or fewer heads)? The answer as to which of these two probabilities to calculate depends on what we wish to consider as the alternative to the hypothesis $H_0: p = \frac{1}{2}$. If we consider the *alternative hypothesis*, denoted by H_1, to be $p > \frac{1}{2}$, then the appropriate probability to calculate is that of obtaining 60 or more heads, since, by considering $p > \frac{1}{2}$ as the alternative hypothesis, we have indicated that we do not wish to reject the hypothesis if fewer than 50 heads are obtained. If, on the other hand, we consider the alternative hypothesis to be $p \neq \frac{1}{2}$, then the appropriate probability to calculate is that of obtaining 60 or more or 40 or fewer heads. The latter probability is twice the former; i.e. $2(0.0287) = 0.0574$, so in this case, using $\alpha = 0.05$ as the significance level, we do not consider this result significant.

Definition 9-4. If the area in only one tail of a curve is used in testing a statistical hypothesis, we speak of a *one-tailed test*; if the areas of both tails are used, we refer to the test as *two-tailed*.

The decision about whether a one-tailed or a two-tailed test is to be used depends entirely on what is considered to be the alternative hypothesis, and what the alternative hypothesis is depends on the question to be answered in the problem: If the interest is restricted to the fact that a very low result was obtained, the alternative hypothesis clearly is that the parameter is smaller than the one stated in the hypothesis, and a one-tailed test is called for; if on the other hand, the interest in the problem is that the result obtained was different (that is, either lower or higher) from the expected one, the alternative hypothesis is that the parameter is different from the one stated in the hypothesis, and a two-tailed test is called for. In a given problem, the statement of the question usually indicates what should be considered as the alternative hypothesis and accordingly whether a one- or a two-tailed test should be applied. In case of doubt, a two-tailed test is recommended.

When $\alpha = 0.05$ is used, a result in a two-tailed test is significant if z falls in the shaded part of Figure 9-2, i.e. if $z > 1.96$ or $z < -1.96$, where 1.96 is

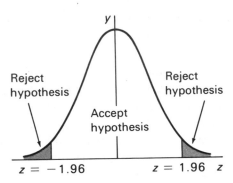

Figure 9-2
Two-tailed test for $\alpha = 0.05$.

the value of z obtained from Appendix Table I for which each of the shaded areas is 0.0250; similarly, a result in a one-tailed test is significant if z falls in the shaded part of Figure 9-3, i.e. if $|z| > 1.64$, where 1.64 is the z-value for which the shaded area is 0.0500. These values of z are called critical values.

Why, in Example 9-1, in testing the honesty of the coin, did we not calculate the probability of obtaining exactly 60 heads in 100 throws? This probability

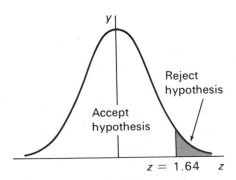

Figure 9-3
One-tailed test for $\alpha = 0.05$.

obviously is very small. Indeed, even the probability of obtaining the most likely event (50 heads) for an honest coin is quite small (0.0796). Therefore, in testing hypotheses it is never proper to use the probability of obtaining precisely the event which occurred, but rather the probability of the occurrence of that event or one which deviates even more from the hypothesis.

Example 9-2. From Mendelian theory of inheritance, we know that certain crosses of peas should give yellow and green peas in the ratio 3 : 1. In a particular experiment, 179 yellow and 49 green peas are obtained. Is this a significant deviation from the theory on the basis of the 5% level of significance?

Solution. Here we are testing the statistical hypothesis that the probability p of obtaining yellow peas is $p = \frac{3}{4}$. Since the sample contains 228 peas, we would expect, according to the hypothesis, $\frac{3}{4} \cdot 228 = 171$ yellow peas. Since the question concerns a deviation from the theory, and a deviation can occur on either side of the expected result, the alternative hypothesis H_1 should be taken here as $p \neq \frac{3}{4}$, which means that a two-tailed test is applicable. (If the question had been "Is this a significantly large number of yellow peas?" we would have considered the alternative hypothesis to be $p > \frac{3}{4}$ and would have used a one-tailed test.) Consequently, we calculate here the probability of obtaining in a sample of 228 peas a deviation of 8 or more from the expected number of peas (i.e. 179 or more or 163 or fewer yellow peas).

The calculations for the normal curve approximation of the binomial distribution involved are shown below.

Hypothesis: $\quad H_0: p = \dfrac{3}{4}.$

Then,

$$q = \frac{1}{4};$$

$$n = 228;$$

$$m = \frac{3}{4} \cdot 228 = 171;$$

$$\sigma = \sqrt{228 \cdot \frac{3}{4} \cdot \frac{1}{4}} = \sqrt{42.75} = 6.54;$$

$$z = \frac{178.5 - 171}{6.54} = 1.15;$$

$$P = (0.5000 - 0.3749)2 = (0.1251)2 = 0.2502.$$

Since $P > 0.05$, the result does not represent a significant deviation from the theory. The same conclusion also follows from the fact that $1.15 < 1.96$, the critical value of z.

It should be emphasized that the testing of statistical hypotheses does not constitute a mathematical proof of the truth or falsity of the hypothesis. Unfortunately, there is no absolute certainty that the conclusion reached will be correct. In fact, in testing statistical hypotheses, two types of incorrect conclusions are possible.

Definition 9-5.   If it happens that the hypothesis being tested is actually true, and if from the sample we reach the conclusion that it is false, we say that a *type 1 error* has been committed.

In Example 9-1, where 60 heads were obtained in 100 throws of a coin, a one-tailed test resulted in the conclusion that the hypothesis should be rejected, which meant that the coin was not honest. If now it happens that this coin actually is honest, we have committed a type 1 error. The probability of committing a type 1 error is the relative frequency with which we reject a correct hypothesis, and this is precisely equal to the significance level α. The smaller the value of α, the smaller the chance of committing a type 1 error.

Definition 9-6.   If it happens that the hypothesis being tested is actually false, and if from the sample we reach the conclusion that it is true, we say that a *type 2 error* has been committed.

In Example 9-2 we concluded that the hypothesis of a $3:1$ ratio was correct. If now it happens that this hypothesis actually is not true, and some other segregation ratio is applicable, we have committed a type 2 error.

The probability of committing a type 2 error is usually denoted by β. For a given number of observations it can be shown that if α is given β can be determined, and that if α is decreased β is increased. This means that, for a fixed sample size, if the probability of committing a type 1 error is decreased, the probability of committing a type 2 error is increased, and vice versa. If we wish to decrease the chances of both types of error at the same time, it can be accomplished only by increasing the size of the sample.

Depending upon whether the stated hypothesis is true or false, and depending upon the conclusion reached from the sample, there are four different possibilities as summarized in Table 9-1.

Whenever we want to disprove a hypothesis, we naturally hope to obtain the smallest possible value for the probability that the hypothesis is true. The smaller that probability, the more convinced we are that the stated hypothesis is false and should be rejected. The procedure of testing hypotheses, therefore, is analogous to that applied in trying to convict in a court trial

Table 9-1
Hypotheses and conclusions reached from sample.

		Hypothesis stated is	
Conclusion		True	False
Sample indicates	Reject hypothesis	Type 1 error	Correct conclusion
	Accept hypothesis	Correct conclusion	Type 2 error

a person accused of a crime, but claiming innocence. In such trials the hypothesis is always that the accused is innocent. The prosecuting attorney tries to demonstrate that the probability of this being the case is extremely small. Although, in most cases, he will not actually calculate the probability involved, there are trials in which this probability is actually calculated. We illustrate this in the following case.

Some years ago, in Berkeley, California, a 14-year old girl disappeared on her way from school. A search for her proved fruitless until her body was found buried in a grave in the mountains. Shortly thereafter an arrest was made of a person living not far from Berkeley, in Oakland. He denied any involvement in the case. In the trial which followed, the following facts were presented: It was known what kind of sweater the girl wore on the day of her disappearance. The car of the accused was carefully vacuumed, which resulted in the recovery of some yarn, which appeared to be from the sweater of the girl. By determining the number of sweaters in existence made of exactly the same fabric, and of the same color as that of the sweater worn by the girl, a calculation was made of the probability that this yarn did not come from the girl's sweater. This probability, which we shall denote by p_1, was found to be very small, but not small enough to prove the guilt of the accused.

Accordingly, further calculations were made. A pair of shoes belonging to the accused and found in his basement were covered with soil similar to that near the gravesite of the girl, which, incidentally, happened to be close to the summer home of the accused. A calculation was then made of the probability that the soil on the soles of the shoes was not the same as that found at the gravesite of the girl. That probability, which we shall denote by p_2, was also found to be very small. The probability that neither the yarn found in the car was from the girl's sweater nor the soil on the shoes was from the gravesite of the girl is $p_1 \cdot p_2$, which is an extremely small number, since both p_1 and p_2 are small. This means that the hypothesis that the accused is innocent can now be rejected with a great deal of confidence, but not with certainty.

Since the life of the accused was at stake, further probabilities were calculated until the probability of his innocence became so small that there could no longer be any reasonable doubt as to his guilt of the crime. Indeed, on this basis, he was convicted.

Testing a hypothesis, then, is a method of determining the truth or falsehood of a hypothesis in the face of uncertainty, in the same sense that a court trial is a method of determining the innocence or guilt of an accused in the face of his denial of involvement in the crime.

The reader is urged to keep this analogy in mind in the following examples.

Example 9-3. The mean breaking strength of a certain type of cord has been established from considerable experience at 18.3 ounces with a standard deviation of 1.2 ounces. A new machine is purchased to manufacture this type of cord. A sample of 100 pieces obtained from the new machine shows a mean breaking strength of 17.0 ounces. Would you say that this sample is inferior on the basis of the 1 % level of significance?

Solution. Here we wish to determine whether the new machine's apparent poor performance can be ascribed to the fact that there is a certain natural variability in the means of samples of 100 pieces taken from any given population or to the fact that the new machine's output cannot be assumed to come from the same population as that established from prior experience. In other words, we want to determine whether the new machine is "innocent" or "guilty" of the charge of a poorer performance than that of the machines previously used. Accordingly, we test here the hypothesis $H_0: m = 18.3$ and, since we are only interested in determining whether the new machine has produced an inferior sample, we consider as the alternative hypothesis $H_1: m < 18.3$.

Accordingly, we calculate the probability of obtaining from a population with a mean of 18.3 and a standard deviation of 1.2 a sample of 100 pieces with a mean of 17.0 or less. If the probability turns out to be less than 1 %, we reject the hypothesis; otherwise, we accept it.

Recalling Theorem 8-2 concerning the distribution of the means of samples, the necessary calculations can be made:

Hypothesis: $H_0: m = 18.3$.

Then,

$$\sigma = 1.2;$$

$$\bar{X} = 17.0;$$

$$n = 100;$$

$$\sigma_{\bar{x}} = \frac{1.2}{\sqrt{100}} = \frac{1.2}{10} = 0.12;$$

$$z = \frac{17.0 - 18.3}{0.12} = \frac{-1.3}{0.12} = -10.83.$$

Since $|z| = 10.83$ is larger than any value of z listed in Appendix Table I, the corresponding tail area is much less than 0.01 and, consequently, the result is highly significant.

Since the probability found here is so small, we are virtually certain that our conclusion is correct; that is, that the machine is "guilty" of the charge of a poorer performance. This means that the chance of having committed a type 1 error is almost negligible.

A frequently tested hypothesis in statistics is the assumption that the means of two populations are identical. Suppose we are given two samples of sizes n_1 and n_2, and that we have calculated the means of these samples ($\bar{X}$ and $\bar{Y}$, respectively). We then wish to test the hypothesis that the two populations from which these samples were selected have identical means. This is determined by calculating the probability of selecting from two populations with identical means two samples for which the means differ by $|\bar{X} - \bar{Y}|$ or more. A two-tailed test is normally used in testing the hypothesis that two means are equal (or that the difference between means is zero) although there are rare occasions when a one-tailed test may be preferable. Since in this case we are dealing with the distribution of differences between sample means, Theorem 8-3 is applicable.

Example 9-4. The gasoline mileage for 50 cars of make A has a mean of 17.0 miles per gallon with a standard deviation of 2.5 miles per gallon, while that for 70 cars of make B has a mean of 18.6 miles per gallon with a standard deviation of 3.0 miles per gallon. If this is the only information available on the gasoline consumption of these two makes of cars, can we conclude that cars of make B consume less gasoline than those of make A (5% level)?

Solution. Here we wish to test the statistical hypothesis that the means of the two populations (distances traveled per gallon of gasoline for makes A and B, denoted by X and Y, respectively) are identical, or $H_0 : m_x = m_y$. The alternative hypothesis H_1 is $m_x < m_y$, since we want to determine whether cars of make A travel a smaller distance per gallon than those of make B.

Accordingly, we calculate the probability of selecting from these two populations two samples of sizes 50 and 70, for which the difference in sample means is less than $17.0 - 18.6 = -1.60$.

As is almost always true in problems of this type, the standard deviations, σ_x and σ_y, of the two populations are unknown. Since, however, both samples are quite large, no serious error is committed by assuming that the standard deviation of each population is identical with the standard deviation of the sample taken from that population. Such an assumption is generally permissible if the numbers of observations in each of the samples exceed 30. For smaller samples, this assumption is not valid; the appropriate procedure for small samples will be discussed in Chapter 10.

The calculations for this example follow.

Hypothesis: $H_0: m_x = m_y$, or $m_x - m_y = 0$.

Alternative hypothesis: $H_1: m_x < m_y$.

Then,

$$\bar{X} = 17, \qquad \bar{Y} = 18.6;$$

$$\hat{s}_x = \sigma_x = 2.5, \qquad \hat{s}_y = \sigma_y = 3.0;$$

$$n_1 = 50, \qquad n_2 = 70;$$

$$\sigma_{\bar{x}} = \frac{\sigma_x}{\sqrt{n_1}} = \frac{2.5}{\sqrt{50}}, \qquad \sigma_{\bar{y}} = \frac{\sigma_y}{\sqrt{n_2}} = \frac{3.0}{\sqrt{70}};$$

$$\sigma_{\bar{x}-\bar{y}} = \sqrt{\sigma_{\bar{x}}^2 + \sigma_{\bar{y}}^2} = \sqrt{\frac{(2.5)^2}{50} + \frac{(3.0)^2}{70}} = \sqrt{\frac{6.25}{50} + \frac{9.00}{70}}$$

$$= \sqrt{0.1250 + 0.1286} = \sqrt{0.2536} = 0.50;$$

$$z = \frac{(\bar{X} - \bar{Y}) - (m_x - m_y)}{\sigma_{\bar{x}-\bar{y}}} = \frac{(17.0 - 18.6) - 0}{0.50}$$

$$= \frac{-1.60}{0.50} = -3.20;$$

$$P = 0.5000 - 0.4993 = 0.0007.$$

Since the probability is less than 0.05, the result is significant. This means that if the populations actually have identical means, the probability of obtaining a difference of -1.60 or less between the two sample means is very small ($P = 0.0007$), and this is an excellent indication that cars of make B consume less gasoline than those of make A or that the latter are "guilty" of a higher gasoline consumption.

In Example 9-4, we tested the hypothesis that $m_x - m_y = 0$ or that there was no difference, or a difference of null, between the two means. Such a hypothesis is frequently referred to as a *null-hypothesis*. Indeed the term "null-hypothesis" has come to be used for any hypothesis that is tested, regardless of whether the hypothesis involves a test for a parameter to be zero.

Definition 9-7.   The hypothesis being tested is usually called the *null-hypothesis*.

As previously indicated, the null-hypothesis is denoted by H_0, and the alternative hypothesis by H_1.

9.2   CONFIDENCE LIMITS

Suppose we are given a sample of n variates having mean $\bar{X}$ and standard deviation $\hat{s}$. We shall now consider the problem of estimating the mean of the population from which the sample was selected.

To illustrate, suppose a random sample of 100 variates is taken from a normal population and is found to have a mean of 40 and a standard deviation of 11. It will not be possible to determine precisely the mean of the population, since this sample of 100 variates with mean 40 could have been taken from any population with mean at or near 40. Consequently, the best we can hope for is to establish limits within which the mean of the population will fall with a specified probability or confidence (usually taken as 95%).

Definition 9-8.   The limits that will contain a parameter with a probability of 95% (or some other given percent) are called the 95% (or that other percent) *confidence limits* for the parameter.

Definition 9-9.   The interval between the confidence limits is called the *confidence interval*.

The larger the specified confidence, the longer the confidence interval.

Let us now calculate the 95% confidence limits for the mean of the population from which the sample of 100 variates was selected.

From Theorem 8-1 we know that $\bar{X}$ is normally distributed with mean m and standard deviation $\sigma_{\bar{x}}$. To calculate $\sigma_{\bar{x}}$ we shall assume that, since the sample is large, the standard deviation of the population can be well approximated by the standard deviation of the sample, that is, $\sigma = \hat{s}$. Then we find

$$\sigma_{\bar{x}} = \frac{\sigma}{\sqrt{n}} = \frac{11}{10} = 1.10.$$

Therefore, 95% of the sample means lie between the values of $\bar{X}$ obtained by solving for $\bar{X}$ the two equations

$$\frac{\bar{X} - m}{1.10} = \pm 1.96.$$

The value of 1.96 is obtained from Appendix Table I as the value of z corresponding to an area of $\frac{1}{2}(0.95) = 0.475$. The solutions are

$$\overline{X}_1 = m - 2.16, \quad \text{and} \quad \overline{X}_2 = m + 2.16.$$

However, since in the illustration we are given $\overline{X}$ and wish to determine m, we must solve the above equations for m, which gives

$$m_1 = \overline{X} - 2.16 = 40 - 2.16 = 37.84,$$

and

$$m_2 = \overline{X} + 2.16 = 40 + 2.16 = 42.16.$$

These values, m_1 and m_2, constitute the lower and upper 95% confidence limits for the mean of the population. Frequently they are written more briefly as

$$m = 40 \pm 2.16.$$

These limits, then, define an interval that would contain the mean of the population with a probability of 95%. Figure 9-4 shows the distributions of the sample means $\overline{X}$ for samples of 100 variates from the populations with these two extreme values of the mean.

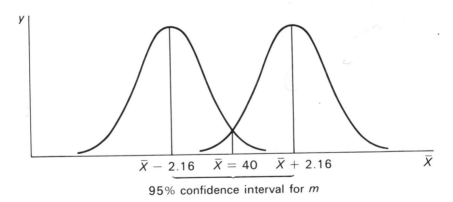

Figure 9-4
The distributions corresponding to the two extreme values of the mean.

Note that this method of obtaining confidence limits is applicable only to large samples, where it is permissible to assume that the standard deviation of a sample can be taken as an approximation of the standard deviation of the population. The procedure to be followed for small samples (30 or fewer variates) will be discussed in Chapter 10.

Example 9-5.   A sample of 70 variates has a mean of 65 with a standard deviation of 4.2. Find the 98 % confidence limits for the mean of the population.

Solution.   Here we are given $\bar{X} = 65$, $\hat{s} = 4.2$, $n = 70$.
   Assuming $\sigma = \hat{s}$, we find

$$\sigma_{\bar{x}} = \frac{\sigma}{\sqrt{n}} = \frac{4.2}{\sqrt{70}} = \frac{4.2}{8.37} = 0.50.$$

From Appendix Table I, the value of z corresponding to an area of $\frac{1}{2}(0.98) = 0.49$ is 2.33; thus

$$\frac{\bar{X} - m}{0.50} = \pm 2.33.$$

Therefore

$$\bar{X} - m = \pm 1.16,$$

and

$$m = \bar{X} \pm 1.16 = 65 \pm 1.16,$$

which are the 98 % confidence limits for the mean of the population.

9.3   DETERMINING THE SIZE OF THE SAMPLE IN SURVEYS

Let us assume that we are interested in making a poll to determine what percentage of the population favors one of two candidates in an upcoming election. Our first problem, clearly, is to decide on the number of people to be interviewed in order to find this percentage with a certain prescribed accuracy. In many such problems—including, for example, the interviews conducted by the Gallup Poll—the accuracy is prescribed by requiring that the percentage of the population favoring the candidate be within 3 % of the percentage obtained from the sample at the 95 % confidence limits. This means that if, in the sample, 43 % favor the candidate, then the actual percentage of the total population favoring the candidate would be somewhere between 40 % and 46 %. To find the sample size as precisely as possible it is necessary to have a rough idea of what the percentage in the population might be. Such an estimate need not be accurate at all, however, as we shall see later in this section.

We shall now show how in actual practice the size of the sample is determined.

Example 9-6.   A poll is to be conducted to determine the percentage of the population favoring candidate A in an election in which there are two candidates. Determine the size of the sample that should be polled in order to obtain the percentage in the population within 3% at the 95% confidence limits. A quick survey conducted on a small sample leads to the belief that the percentage of voters favoring candidate A is 43%.

Solution.   From results of the survey sample we will assume that the percentage of voters in the population favoring candidate A is 43%. We are interested in determining the size of the sample so that with a probability of 0.95 the percentage in the sample does not differ by more than 3% from that of the population.
We then have

$$p = 0.43, q = 0.57;$$

$$m = 0.43n;$$

$$\sigma = \sqrt{n(0.43)(0.57)} = \sqrt{0.2451n} = 0.495\sqrt{n}.$$

Since we wish the number of people favoring candidate A to differ from that in the population by at most $0.03n$ with a probability of 0.95, we have, by using the normal curve approximation to the binomial distribution,

$$\frac{0.03n}{0.495\sqrt{n}} = 1.96$$

or

$$\sqrt{n} = \frac{1.96(0.495)}{0.03} = 32.3$$

or

$$n = (32.3)^2 = 1043.29,$$

and, rounding the resulting value of n to the next larger integer to assure a probability of at least 95%, we find $n = 1044$.

The reader may feel uncomfortable about the fact that we had to have some idea about the percentage favoring candidate A before we could determine the sample size required to obtain the actual percentage favoring candidate A. It turns out that the initial estimate of this percentage does not have to be accurate at all for the determination of the sample size; that is, if in Example 9-6 we had assumed that the percentage of voters favoring the candidate had not been 43%, but rather 36%, the value of n which would have been obtained is $n = 986$, which differs only slightly from the value of $n = 1044$, obtained in Example 9-6.

If no preliminary survey data are available to indicate what percentage of the population favors a candidate, it is clearly desirable to estimate n as conservatively as possible, that is, to determine it for that value of p which results in the largest value of n. Let us, for example, suppose that in Example 9-6 we had no idea what percentage of the voters favored candidate A. Let us call that percentage p. Then the equation to be solved for n would become

$$\frac{0.03n}{\sqrt{np(1-p)}} = 1.96$$

or

$$\sqrt{n} = \frac{1.96\sqrt{p(1-p)}}{0.03}.$$

To find the value of p for which n (and consequently $\sqrt{n}$) assumes its largest value, we note that the expression under the square root sign can be written as

$$p(1-p) = p - p^2 = \frac{1}{4} - \left(p^2 - p + \frac{1}{4}\right) = \frac{1}{4} - \left(p - \frac{1}{2}\right)^2,$$

so that the largest value is obtained, when the second term on the right hand side is 0; that is, $p = \frac{1}{2}$.

The same reasoning shows that n will always assume its largest value when $p = \frac{1}{2}$, so that, when no preliminary survey data are available, the sample size should always be estimated for $p = \frac{1}{2}$.

Example 9-7.   Solve Example 9-6 if the last sentence is omitted.

Solution.   Solving for n with $p = \frac{1}{2}$, we have

$$\sqrt{n} = \frac{1.96\sqrt{\frac{1}{2} \cdot \frac{1}{2}}}{0.03} = \frac{0.98}{0.03} = 32.67,$$

so that $n = 1067.33$ and, rounding this value of n to the next larger integer, we find $n = 1068$.

In practice, to be on the safe side, such numbers are always increased substantially. Thus, the Gallup Poll in recent years has always taken polls of 1500 people, thereby assuring that their predictions would very rarely be off by as much as 3%. For this reason, in the national elections since 1948 Gallup has never been off by more than 2.7% and, since 1954, his election polls have been off by no more than 1.5%. Gallup selects his 1500 voters by picking 300 sections of the country and selecting five voters in each section.

Pollsters have found that they cannot significantly increase accuracy unless they go very far above 1500, and that is usually uneconomic. For example, to determine the percentage in the population within 1% of the percentage obtained from the sample with a 95% probability would require a substantially larger sample (see Exercise 32 of this chapter), and if in addition the probability is to be increased to 99%, an even larger sample would be required.

EXERCISES

Unless otherwise specified, use the 5% level of significance. Answer each question exactly in the form in which it is asked (i.e. do not merely state that a result is significant or not significant).

1. In 144 rolls of a die, a 6 is obtained 32 times. Is this number of 6's sufficiently large to cause doubt as to the honesty of the die?

2. A coin is tossed 200 times, and 118 heads are obtained. At the 1% level
   (a) is this considered a significantly large number of heads;
   (b) is this considered a significantly large deviation from the expected number of heads?

3. From experience it is known that 20% of a certain kind of seed will germinate. If in an experiment 60 out of 400 seeds germinate, is this considered a significantly poor germination on the basis of the 1% level of significance?

4. It is known from long experience that 5% of certain articles produced are defective and have to be discarded. A new man who has produced 600 of these articles has made 42 defective articles. Does this case doubt on the man's ability to perform the job?

5. A coin is tossed 15 times. For what values of X, the number of heads obtained, would you reject the hypothesis that the coin was honest (5% level)?

6. At a university, student records kept over a period of many years indicate that 64% pass the entrance examination in mathematics. During the fall of 1976, of 400 freshmen taking the test, 70% passed. Was this a significant improvement in the students' aptitude in mathematics?

7. If experience shows that 4% of all white males of exact age 65 die within a year, and it is found that 55 such males of a group of 1000 actually die within a year, should the group be regarded as being essentially different from the general mass, using the 1% level of significance?

8. A county is studying the question whether its juries are selected with proper representation of minority races. In this county, minority races constitute 15% of the population. During the first year of this study, it was found that the jury panel had 85 members of whom only 9 were of minority races. Is it valid to charge discrimination in the selection of juries (5% level)?

9. Under standard conditions, the percent defective in the manufacture of a given type of bearing is 3%. Would $4\frac{1}{2}$% defective in a sample of 400 be considered excessively high on the basis of the 5% level of significance?

10. A doctor asserts that he has a method that is 80 percent accurate in determining, several months before birth, the sex of an unborn child. He is challenged to prove this claim. As a test, he predicts the sex of 20 unborn children and is right in 13 of the predictions. Does this support the doctor's claim?

11. Of 64 offspring of a certain cross between guinea pigs, 8 are black and the others are not black. According to the genetic model, these numbers (black and not black) should be in the ratio 3 : 13. Are the observed data consistent with the model at the 5% level of significance?

12. A standard medication reduces reports of postoperative pains in 80% of the patients treated. A new medication for the same purpose is tested. In at least how many patients out of 100 tested must the medication be effective so that it can be considered a superior medication at the 1% level of significance?

13. The height of adults in a certain town has a mean of 65.42 inches with a standard deviation of 2.32 inches. A sample of 144 adults living in the slum district is found to have a mean height of 64.82 inches. Does this indicate that the residents of the slums are significantly retarded in growth on the basis of the 1% level of significance?

14. A manufacturer of string has established from long experience that the string he manufactures has a mean breaking strength of 15.9 pounds, with a standard deviation of 2.4 pounds. A change is made in the manufacturing process, after which a sample of 64 pieces is found to have a mean breaking strength of 15.3 pounds. Does this indicate that the new process produces string whose breaking strength is significantly different from that produced by the old process?

15. A car manufacturer claims that his cars use on the average 5.50 gallons of gasoline for each 100 miles. A salesman selling cars for this company tests 35 cars for gasoline mileage and finds that the average consumption of gasoline for these 35 cars is 5.65 gallons for each 100 miles with a standard deviation of 0.35 gallons. Do these results cast doubt on the claim of the company (1% level)?

16. An electric company has been selling a light bulb it advertises as having an average life of 1000 hours. Assume that the lifetime of these bulbs is normally distributed with a standard deviation of 200 hours (known from repeated past observation). The company's research staff claims to have improved the bulb so that its average life is now greater than 1000 hours. To test this claim, 20 new bulbs are allowed to burn and their lifetimes recorded. At least how long must

the average life of these 20 bulbs be so that the claim of the company's research staff can be considered valid at the 1% level of significance?

17. The daily wages in a particular industry are normally distributed with a mean of $13.20 and a standard deviation of $2.50. If a company in this industry employing 40 workers pays these workers on the average $12.20, can this company be accused of paying inferior wages at the 1% level of significance?

18. Two types of electric bulbs are observed for length of life, and the following data are obtained:

	Type I	Type II
Number in sample	46	64
Mean of sample	1070 hours	1041 hours
Sum of squares of deviations from mean	$\Sigma(X - \bar{X})^2 = 21{,}000$	$\Sigma(Y - \bar{Y})^2 = 23{,}200$

Are the mean lives of the two types of electric bulbs significantly different on the basis of the 1% level of significance?

19. The following data pertain to growth of wheat under two treatments, A and B:

	Treatment A	Treatment B
Numbers of stalks	44	36
Mean height	15.6	14.1
Sum of squares of deviations from mean	$\Sigma(X - \bar{X})^2 = 167.52$	$\Sigma(Y - \bar{Y})^2 = 159.89$

Is there a significant difference between these two means?

20. Two manufacturing processes are to be compared. A sample of size 100 of the first process is found to have a mean of 107 and a standard deviation of 17. For the second process a sample of size 90 has a mean of 103 and a standard deviation of 16. Is there a significant difference between the means of the two processes?

21. Two brands of frozen food are compared. A sample of 40 items of the first brand has mean 77 and variance 80, while a sample of 50 items of the second brand has mean 73 and variance 100. The means of the two samples are judged to be significantly different. What is the smallest level of significance (an integer) at which this judgment is correct?

22. It has been claimed that 50% of the people have exactly two colds per year. We decide to reject this claim if among 400 people 216 or more say that they have had two colds per year.

(a) What is the probability that we have committed a type 1 error?

(b) Is it possible in this investigation to commit a type 2 error? Explain.

23. A sample of 36 variates has a mean of 40 and a standard deviation of 2.1. Find the 95% confidence limits for the mean of the population.

24. In a sample of 60 variates the mean is found to be 35 and the standard deviation 4.2. Find limits within which the population mean is expected to be with a probability of 98%.

25. Suppose the standard deviation of height in males is 2.48 inches. One hundred male students chosen at random in a large university are measured, and their average height is found to be 68.52 inches. Determine the 99% confidence limits for the mean height of the men in this university.

26. From two samples the following data are calculated:

$$\bar{X} = 75, \quad n_1 = 36, \quad \Sigma(X - \bar{X})^2 = 1482;$$

$$\bar{Y} = 60, \quad n_2 = 64, \quad \Sigma(Y - \bar{Y})^2 = 1830.$$

Find the 98% confidence limits for the difference between the means of the populations from which these samples were selected.

27. Two different samples from the same population are used to find confidence limits for the mean of the population. The first sample gives confidence limits of 11 and 21, the second 13 and 25. If the same level of significance is used, what is the ratio of the size of the first sample to the size of the second sample?

28. The average monthly income of families in a certain city is to be estimated from a random sample selected from a directory. If the average income is to be estimated within $100.00 with a probability of 90% (that is, if we desire to be 90% confident that the sample mean and the mean income of all families do not differ by more than $100.00), find the least size of the sample required for such an estimate. Assume it is known that the standard deviation of the incomes of the families in the city is $200.00.

29. For how may days must a pollution study be conducted in a certain city in the United States if it is desired to determine the 99% confidence limits for the mean midday concentration of carbon monoxide in that city so that the confidence interval is less than 1 part per million. Note that it is assumed that one measurement is taken each day at midday. Also assume that it is known that the standard deviation of the midday concentration of carbon monoxide is 4.4 parts per million.

30. In a plant-breeding experiment in which two types of flowers (types A and B) can occur, the expected probability of occurrence of type A is $\frac{7}{16}$. A worker carrying out an experiment yielding these two types of flowers finds that exactly one-half of his flowers are of type A and, therefore, rejects, at the 1% level of significance, the hypothesis that $\frac{7}{16}$ of the flowers are of type A. What is the least number of flowers the experimenter could have investigated for the rejection of the hypothesis to be justified?

31. From a sample of size 10, taken from a normal distribution, the 99% confidence limits for the mean of the population are calculated to be $\bar{X} \pm 5.2$. Determine the standard deviation of the population.

32. Solve Example 9-6 if it is desired to obtain the percentage of the population within 1% at the 95% confidence limits.

33. Solve Example 9-6 if it is desired to obtain the percentage of the population within 1% at the 99% confidence limits.

34. In an election there are two candidates, A and B. If the actual percentage of the voters favoring candidate A is 29% and a poll interviews a sample of 1000 people, within what percent can the result of the sample be expected to be correct at the 95% confidence limits?

35. In an election in which there are two candidates, A and B, it is believed that 52% favor candidate A. In order to predict the winner in this election, an error of 3% can clearly not be tolerated. How large a sample of people needs to be interviewed to predict the winner with a probability of 98%?

36. In an election there are two candidates, A and B. Determine the size of the sample that should be polled in order to be able to predict within 2% the actual percentage of the voters who will vote for each of the two candidates with a 95% confidence.

37. Solve Exercise 36 if it is desired to make the prediction with a 98% confidence.

10

Student's *t*-Distribution.
Small Sample Methods

10.1 ESTIMATE OF THE STANDARD DEVIATION OF THE POPULATION

In the preceding chapter it was pointed out that if a sample is small (n equal to 30 or less), its standard deviation $\hat{s}$ can no longer be used as an approximation of the standard deviation of the population σ. Since in general a population will contain data that vary more greatly than do the data in a sample, it is to be expected that σ will usually be somewhat larger than $\hat{s}$. Of course, it is impossible to write a precise formula for σ in terms of $\hat{s}$, since different samples from the same population may have rather widely differing standard deviations. In most cases, however, the quantity

$$\sqrt{\frac{n}{n-1}}\,\hat{s}$$

will give an excellent approximation for σ, although it will rarely be precisely equal to σ. This quantity we shall denote by s; thus

$$s = \sqrt{\frac{n}{n-1}}\,\hat{s} \tag{10-1}$$

is the most appropriate estimate of the standard deviation of the population as obtained from a sample. Keep in mind that *s* may not always approximate σ very well. Indeed, it may happen that in a particular case $\hat{s}$ is closer than *s* to σ.

We note from (10-1) that, if *n* is large, the expression under the radical is close to 1; thus, as indicated before, $\hat{s}$ for large *n* affords a good approximation of σ.

Formula (10-1) can be justified as follows: Consider a population with a large number of variates; then the mean of all the sample variances $\hat{s}^2$ can be shown to equal

$$m_{\hat{s}^2} = \frac{n-1}{n}\,\sigma^2.$$

Now, since most of the $\hat{s}^2$ do not differ much from $m_{\hat{s}^2}$, let us replace $m_{\hat{s}^2}$ by $\hat{s}^2$ and σ by *s*, the estimate of the standard deviation; thus

$$\hat{s}^2 = \frac{n-1}{n}\,s^2,$$

from which Formula (10-1) follows.

Another justification for the use of (10-1) as an estimate of the standard deviation of the population is given in Section 10.3.

Suppose a sample of *n* variates $X_1, X_2, \ldots, X_n$ is given; then the estimate of the standard deviation of the population is given by

$$s = \sqrt{\frac{n}{n-1}}\,\hat{s} = \sqrt{\frac{n}{n-1}}\,\sqrt{\frac{\Sigma(X-\overline{X})^2}{n}} = \sqrt{\frac{\Sigma(X-\overline{X})^2}{n-1}},$$

and the formula

$$s = \sqrt{\frac{\Sigma(X-\overline{X})^2}{n-1}} \tag{10-2}$$

gives the estimate of the standard deviation of the population when the *n* variates of a sample are known. It should be noted that the formulae for *s* and $\hat{s}$ differ only in that the former has $n-1$ in the denominator whereas the latter has *n*.

As we indicated earlier, in Formula (8-3), many textbooks in statistics refer to the expression given by Formula (10-2) as the "standard deviation of the sample." This is merely a matter of terminology, which in no way affects any of the theorems which follow. The authors of this book prefer to denote the standard deviation of the sample by $\hat{s}$ for reasons explained in Chapter 8.

We shall now introduce a notation which is universally accepted; that is, whenever the standard deviation of a population is unknown and statistics have to be calculated from the estimate of the standard deviation, σ is replaced by s in the corresponding formulae. Thus, the estimate of the standard deviation of the mean of samples of n variates is denoted by $s_{\bar{x}}$ and is equal to

$$s_{\bar{x}} = \frac{s}{\sqrt{n}}. \tag{10-3}$$

10.2   STUDENT'S *t*-DISTRIBUTION

Consider the case where the standard deviation of the population is unknown and its estimate s, obtained from a sample, has to be used. We must first determine what effect the substitution of s for σ has on the theorems of Chapter 8. This problem was first studied by W. S. Gosset, a statistician at the Guinness Brewery in Dublin who wrote under the pseudonym "Student." He noted that the quantities resulting from this substitution no longer satisfied a normal distribution, but satisfy a different type of distribution which has since been called Student's *t*-distribution.

Definition 10-1.   If the distribution of a variable t is given by

$$y = c \left(1 + \frac{t^2}{v}\right)^{-(v+1)/2}, \tag{10-4}$$

the distribution is called a *Student's t-distribution*, where v is called the number of degrees of freedom, and c is a constant depending on v and is chosen in such a way that the total area under the probability curve (10-4) is equal to 1.

Number of degrees of freedom is a term frequently used in statistics, and it is defined below.

Definition 10-2.   For a given set of conditions, the *number of degrees of freedom* (frequently denoted by *d.f.* or *D.F.*) is the maximum number of variates that can freely be assigned (i.e. calculated or assumed) before the rest of the variates are completely determined; that is, it is the total number of variates minus the number of independent relationships existing among them.

To graph the curves given by (10-4) for various values of v, we note that for all values of v the curve reaches its maximum value at $t = 0$ and is symmetric about a line drawn at $t = 0$ perpendicular to the horizontal axis.

Figure 10-1 shows the graphs of (10-4) for $v = 3$ and $v = 7$, and also the standard normal probability curve.

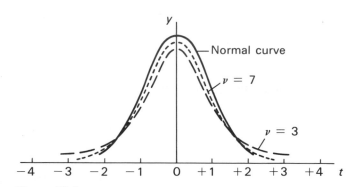

Figure 10-1
Typical Student's *t*-distributions.

The curves given by (10-4) approach the horizontal axis at a less rapid rate than does the normal probability curve, but it can be proved that as v becomes larger and larger the curves given by (10-4) approach the normal curve.

Since Student's *t*-distribution is merely another example of a probability density function (or distribution) of a continuous variable, as defined in general by Definition 7-1, we find the probability that t lies between two given values a and b, with $b > a$, by determining the appropriate area under the applicable curve (10-4). These areas can be found by use of Appendix Table III, which gives for various v-values (listed in the first column) the values of t corresponding to various two-tail areas. For example, for $v = 20$, we find the *t*-value corresponding to a value of 0.05 for the two tail areas to be $t = 2.086$. This means that the probability of obtaining in a Student's *t*-distribution a value of $t > 2.086$ or $t < -2.086$ for $v = 20$ is equal to 0.05.

10.3   THE DISTRIBUTION OF THE MEAN

We may now state the theorem that results from the substitution of s for σ in Theorem 8-1.

Theorem 10-1.   If a variable X satisfies a normal distribution and has mean m, and if all possible samples of n variates are taken from this population and their means denoted by $\overline{X}$, then the quantity

$$t = \frac{\overline{X} - m}{s_{\overline{x}}} \tag{10-5}$$

satisfies a Student's t-distribution with $n - 1$ degrees of freedom, where

$$s_{\overline{x}} = \frac{s}{\sqrt{n}}.$$

Note that the difference between this theorem and Theorem 8-1 is that $\sigma_{\overline{x}}$ is replaced by $s_{\overline{x}}$ and that the resultant quantity t satisfies a Student's t-distribution whereas the quantity z satisfies a normal distribution.

The value of s to be used in the formula above is, of course, that given by Formula (10-1) or its equivalent (10-2). This provides another justification for using (10-1) or (10-2) as an estimate of the standard deviation of the population: Only when this particular formula for s is used does the quantity given by (10-5) satisfy a Student's t-distribution. The proof of Theorem 10-1 is dependent upon more advanced mathematical procedures than are assumed here and will not be given.

Example 10-1.   A sample of 16 variates is found to have a mean of 28.0 and a standard deviation of 3.0. On the basis of the 5% level of significance, is there reason to reject the hypothesis that the mean of the population is 30.0?

Solution.   To test the hypothesis that $m = 30.0$ we calculate the probability of selecting from a population with $m = 30.0$ a sample of 16 variates whose mean is 28.0 or less or 32.0 or more (assuming a two-tailed test). Actually, since we are using the 5% level, it is sufficient to determine whether this probability is more or less than 0.05. The calculations follow.

Hypothesis: $H_0 : m = 30.0$.

Alternative hypothesis: $H_1 : m \neq 30.0$.

Then,

$$\hat{s} = 3.0;$$

$$s = \sqrt{\frac{16}{15}} \cdot 3.0 = \frac{12}{\sqrt{15}};$$

$$s_{\overline{x}} = \frac{s}{\sqrt{16}} = \frac{1}{\sqrt{16}} \cdot \frac{12}{\sqrt{15}} = \frac{3}{\sqrt{15}} = 0.78;$$

$$t = \frac{28.0 - 30.0}{0.78} = \frac{-2.0}{0.78} = -2.56.$$

For $\nu = 15$, $t_{.05} = 2.131$.

Since $|t| = 2.56 > 2.131$, the probability mentioned above is less than 0.05, so we conclude that the result is significant, which causes us to reject the hypothesis.

By linear interpolation, the probability of obtaining a mean of 28 or less or 32 or more is found to be 0.0226, which result, however, is not precise owing to the inaccuracy inherent in linear interpolation.

It might be interesting to determine the probability that would have been obtained if we had mistakenly used large sample methods, i.e. if we had approximated σ by $\hat{s}$ and used the normal curve. In that case we would have found

$$\sigma = \hat{s} = 3.0;$$

$$\sigma_{\bar{x}} = \frac{3}{\sqrt{16}} = 0.75;$$

$$z = \frac{28.0 - 30.0}{0.75} = \frac{-2.0}{0.75} = -2.67;$$

$$P = (0.5000 - 0.4962)2 = (0.0038)2 = 0.0076,$$

a value considerably smaller than 0.0226.

This demonstrates that the mistaken use of large sample methods, when, properly, Student's *t*-distribution should be applied, will cause us to underestimate the probability involved, and therefore will frequently cause us to call a result significant even though it may not actually be so. That the use of large sample methods (normal curve) will result in too small a probability can also be seen from Figure 10-1, which shows that for a given value of *t* (or *z*) the two tail areas are considerably smaller for the normal probability curve than for the Student's *t*-curves.

Example 10-2.   A company claims that the average life of batteries of a certain type is 21.5 hours. A laboratory tests 6 batteries manufactured by this company and obtains the following lives in hours: 19, 18, 22, 20, 16, 25. Do these results indicate that batteries of this type have a shorter life than that claimed by the company (5% level)?

Solution.   To test the hypothesis that $m = 21.5$, we calculate the probability of selecting from a population with $m = 21.5$ a sample of 6 variates whose mean is less than that obtained in the experiment. A one-tailed test is applicable here.

Hypothesis: $H_0: m = 21.5$.

Alternative hypothesis: $H_1: m < 21.5.$

Then,

X	$X - \bar{X}$	$(X - \bar{X})^2$
19	-1	1
18	-2	4
22	2	4
20	0	0
16	-4	16
25	5	25
120	0	50

And,

$$\bar{X} = \frac{120}{6} = 20.0;$$

$$s = \sqrt{\frac{50}{5}} = \sqrt{10};$$

$$s_{\bar{x}} = \frac{\sqrt{10}}{\sqrt{6}} = \sqrt{1.67} = 1.29;$$

$$t = \frac{20.0 - 21.5}{1.29} = \frac{-1.5}{1.29} = -1.16.$$

Since we are using a one-tailed test, we need to find the value of t such that the area to its left is equal to 0.05, which is the same as finding the value of t for which both tail areas total 0.10. Thus, we find for $\nu = 5$, $t_{.10} = 2.015$.

Now since $|t| = 1.16 < 2.015$, the probability mentioned above exceeds 5%, so the result is not significant. Consequently, the results do not indicate that batteries of this type have a shorter life than that claimed by the company.

It should be recalled from Chapter 4 that for purposes of machine calculation it is very convenient to use the fact that

$$\Sigma(X - \bar{X})^2 = \Sigma X^2 - \frac{(\Sigma X)^2}{n}, \tag{10-6}$$

whenever sums of squares of deviations have to be calculated. Formula (10-6) was not used in Example 10-2, since we were dealing with a very small sample and simple numbers. However, since such a situation ordinarily will not prevail, the use of (10-6) is encouraged.

Example 10-3. A sample of 25 variates is found to have a mean of 38.5 and a standard deviation of 2.3. Find the 95% confidence limits for the mean of the population.

Solution. In this case we find

$$s = \sqrt{\frac{25}{24}} \cdot 2.3 = \frac{11.5}{\sqrt{24}} \cdot$$

$$s_{\bar{x}} = \frac{1}{\sqrt{25}} \cdot \frac{11.5}{\sqrt{24}} = \frac{2.3}{\sqrt{24}} = \frac{2.3}{4.90} = 0.47.$$

To determine the 95% confidence limits for m, we find from Appendix Table III, for 24 degrees of freedom, the value of t for which the two tail areas total 0.05 to be $t_{.05} = 2.064$. Therefore

$$\frac{\bar{X} - m}{0.47} = \pm 2.064,$$

which we solve for m:

$$m = \bar{X} \pm 0.970 = 38.5 \pm 0.970.$$

A mistaken use of large sample methods would have resulted in our obtaining the confidence limits $m = 38.5 \pm 0.902$, which does not differ appreciably from the above result, owing to the relatively large size of the sample.

10.4 THE DISTRIBUTION OF THE DIFFERENCE BETWEEN MEANS

We shall now study the changes that result in Theorem 8-3 if the standard deviations of the two populations are unknown, as is frequently the case. Before introducing the appropriate estimates, we shall make the additional assumption that the two populations have identical standard deviations. The reader may be inclined to believe that this is a rather severe restriction. It will be shown, however, that in most applications this assumption is satisfied. If we are given two samples with variates $X_1, X_2, \ldots, X_{n_1}$ and $Y_1, Y_2, \ldots, Y_{n_2}$, from two different populations with identical standard deviations, it would seem natural to make use of all these variates to arrive at an estimate of this common standard deviation. This is done by replacing the numerator of Formula (10-2) by the total of the sums of the squares of the deviations of the variates of each sample from its mean, and the denominator by the total of the degrees of freedom of the two samples. The standard deviation evaluated in this way is not affected by any difference that may exist between the means of the populations. This substitution results in the formula

$$s = \sqrt{\frac{\Sigma(X - \overline{X})^2 + \Sigma(Y - \overline{Y})^2}{n_1 + n_2 - 2}}, \tag{10-7}$$

which is frequently referred to as a "pooled estimate of the standard deviation," since we have pooled the information available from two samples.

We can now state the theorem that results from the substitution of the estimate s as given by (10-7) for both σ_x and σ_y in Theorem 8-3.

Theorem 10-2. Let X and Y be the variables of two normally and independently distributed populations having means m_x and m_y, respectively, and identical standard deviations. Take all possible pairs of samples of sizes n_1 and n_2, respectively, from these populations, and denote their means by $\overline{X}$ and $\overline{Y}$, respectively, and consider for each pair the difference between sample means $\overline{X} - \overline{Y}$. Then the quantity

$$t = \frac{(\overline{X} - \overline{Y}) - m_{\bar{x}-\bar{y}}}{s_{\bar{x}-\bar{y}}} \tag{10-8}$$

satisfies a Student's t-distribution with $n_1 + n_2 - 2$ degrees of freedom, where

$$s_{\bar{x}-\bar{y}} = \sqrt{s_{\bar{x}}^2 + s_{\bar{y}}^2}, \qquad s_{\bar{x}} = \frac{s}{\sqrt{n_1}}, \qquad s_{\bar{y}} = \frac{s}{\sqrt{n_2}}, \tag{10-9}$$

and s is given by (10-7).

This theorem now makes it possible to solve the very practical problem of determining from two given samples whether or not the respective populations have the same or different means.

Example 10-4. Strength tests on two types of wool fabric give the following results (in pounds per square inch):

Type I: 138, 127, 134, 125

Type II: 134, 137, 135, 140, 130, 134

Is the difference in means of the two types sufficient to warrant the conclusion that one is better than the other (5% level)?

Solution. Denoting by X the variates of the sample of type I, by Y those of the sample of type II, and assuming that the populations from which these two samples were taken have identical standard deviations, estimated by s, we find

X	Y	$X - \bar{X}$	$(X - \bar{X})^2$	$Y - \bar{Y}$	$(Y - \bar{Y})^2$
138	134	7	49	−1	1
127	137	−4	16	2	4
134	135	3	9	0	0
125	140	−6	36	5	25
	130			−5	25
	134			−1	1
524	810	0	110	0	56

Hypothesis: $H_0 : m_x = m_y$,   or $m_x - m_y = 0$.

Alternative hypothesis: $H_1 : m_x \neq m_y$.

Then,

$$\bar{X} = \frac{524}{4} = 131;$$

$$\bar{Y} = \frac{810}{6} = 135;$$

$$s = \sqrt{\frac{110 + 56}{8}} = \sqrt{\frac{166}{8}} = \sqrt{20.75};$$

$$s_{\bar{x}} = \frac{\sqrt{20.75}}{\sqrt{4}} = \sqrt{5.19};$$

$$s_{\bar{y}} = \frac{\sqrt{20.75}}{\sqrt{6}} = \sqrt{3.46};$$

$$s_{\bar{x} - \bar{y}} = \sqrt{5.19 + 3.46} = \sqrt{8.65} = 2.94;$$

$$t = \frac{(131 - 135) - 0}{2.94} = \frac{-4}{2.94} = -1.36.$$

For $\nu = 8$, $t_{.05} = 2.306$.

Since $|t| = 1.36 < 2.306$, we conclude that the probability of selecting from two populations with identical means and identical standard deviations two samples whose means differ by more than 4 (in absolute value) is considerably more than 5%, which indicates that this result is not significant. Therefore, the difference between means of the two types is not sufficient to warrant the conclusion that one is better than the other.

The assumption of identical standard deviations of the two populations in problems of this type needs justification. Let us assume a population whose variates are $X_1, X_2, \ldots, X_N$; furthermore, let us assume that the effect of a treatment upon these variates is such that it raises each of them by a fixed amount d resulting in a new population $Y_1, Y_2, \ldots, Y_N$. The mean of the population of Y's is

$$m_y = \frac{\Sigma(X+d)}{N} = \frac{\Sigma X + Nd}{N} = \frac{\Sigma X}{N} + d = m_x + d;$$

thus the mean has been changed by the amount d. Let us now compare the standard deviations of these populations:

$$\sigma_y^2 = \frac{1}{N}[\Sigma(Y-m_y)^2] = \frac{1}{N}\{\Sigma[(X+d)-(m_x+d)]^2\}$$

$$= \frac{1}{N}[\Sigma(X-m_x)^2] = \sigma_x^2;$$

thus there has been no change in the standard deviations. This demonstrates that if treatments have purely additive effects they change the means but not the standard deviations.

On the other hand, there are treatments that have other than additive effects and consequently may influence the standard deviations. In the rare situations where this occurs, Theorem 10-2 is, of course, not applicable. The procedure applicable in such a case will not be discussed, although it requires only a relatively minor modification of Theorem 10-2.

The question of deciding from two given samples whether or not the assumption of identical standard deviations of the two populations is valid is answered by use of the so-called F-distribution, which topic we discuss in Chapter 16.

10.5   THE CASE OF PAIRED VARIATES

Many experiments are performed in such a way that each variate of one sample is paired with a particular variate of another sample. If, for example, the effectiveness of a fertilizer is to be determined, the experimenter might select 10 plots, treat half of each plot with the fertilizer, and leave the other half untreated to serve as a control. In this case, the yields from the treated and untreated halves of each plot constitute a set of paired variates. Such a procedure of carrying out the experiment is superior to selecting 20 plots, fertilizing half of them, and not fertilizing the other half. In the paired experiment, the chance that a possible significant difference between means is due to the treatments that were applied rather than some extraneous source such as soil or weather conditions is greatly increased. Consequently, wherever possible a paired experiment should be applied.

However, for paired experiments Theorem 10-2 is not applicable, since it requires that the populations be independent, which is not the case with pairing.

To analyze paired experiments, we consider for each pair the difference between the variates X and Y, and denote this difference by D,

$$D = X - Y.$$

We now consider the population of differences so obtained. Select all possible samples of size n from this population, denoting their means by $\bar{D}_1, \bar{D}_2, \ldots$, and consider the distribution of these sample means. This situation is completely analogous to that of Theorem 10-1, the difference D now playing the role of the variable X of Theorem 10-1. Consequently, a mere change of notation in Theorem 10-1 gives us the distribution of $\bar{D}$ and results in Theorem 10-3.

Theorem 10-3. Let X and Y be the variables of two normally distributed populations in which the variates are paired, and for each pair let the difference between the variates be denoted by D. Now if all possible samples of n variates are taken from this population of D values, and the sample mean is denoted by $\bar{D}$, then the quantity

$$t = \frac{\bar{D} - m_d}{s_{\bar{d}}} \tag{10-10}$$

satisfies a Student's t-distribution with $n - 1$ degrees of freedom, where

$$s_{\bar{d}} = \frac{s_d}{\sqrt{n}}, \qquad s_d = \sqrt{\frac{\Sigma(D - \bar{D})^2}{n - 1}}. \tag{10-11}$$

The subscript d in s_d indicates that we are dealing with a population of differences.

To determine whether two sample means from paired populations differ significantly, we set up the hypothesis that the two populations have identical means, in which case

$$m_d = \frac{\Sigma D}{N} = \frac{\Sigma(X - Y)}{N} = \frac{\Sigma X}{N} - \frac{\Sigma Y}{N} = 0,$$

and calculate the probability of obtaining from two such populations a pair of samples for which the value of $\bar{D}$ exceeds (in absolute value) that obtained in the experiment.

Example 10-5.   In a certain school, 16 girl beginners in the first grade were selected and paired on the basis of IQ, socioeconomic rating of family, general health, and family size. One member of each pair had attended 1 year of kindergarten and the other had not. On a certain first grade readiness test given to all 16 pupils the scores were as follows:

Pair	Kindergarten	No kindergarten
1	83	78
2	74	74
3	67	63
4	64	66
5	70	68
6	67	63
7	81	77
8	64	65

Using the 5% level of significance, is there evidence from this investigation that kindergarten is of benefit in preparing for the first grade?

Solution.   Denoting the variates of the first sample by X and those of the second sample by Y, we find

X	Y	$D = X - Y$	$D - \bar{D}$	$(D - \bar{D})^2$
83	78	5	3	9
74	74	0	-2	4
67	63	4	2	4
64	66	-2	-4	16
70	68	2	0	0
67	63	4	2	4
81	77	4	2	4
64	65	-1	-3	9
		16	0	50

Hypothesis: $H_0: m_d = 0$.

Alternative hypothesis: $H_1: m_d > 0$.

Then,

$$\bar{D} = \frac{16}{8} = 2.0;$$

$$s_d = \sqrt{\frac{50}{7}} = \sqrt{7.14};$$

$$s_{\bar{d}} = \frac{\sqrt{7.14}}{\sqrt{8}} = \sqrt{0.892} = 0.94;$$

$$t = \frac{2.0 - 0}{0.94} = 2.13.$$

For $\nu = 7$, $t_{.10} = 1.895$.

Since $t = 2.13 > 1.895$, the result is significant, and we conclude from these data that kindergarten is of benefit in preparing for the first grade.

It should be understood that Theorem 10-3 is applicable only when variates are paired. When variates are not paired, Theorem 10-2 applies. Consequently, in order to use the proper statistical analysis of a test for a significant difference between means, we must know whether or not pairing was employed in the experiment.

Example 10-6.   In a wheat-variety test conducted on 16 pairs of plots, the mean difference in yields between the two varieties is found to be 3.0 bushel per acre. The standard error of this mean difference is 1.1 bushels per acre.
(a) Determine the 98% confidence limits for the mean difference in yields between the two varieties.
(b) Use the answer to part (a) to decide whether there is a significant difference in yields between the two varieties (2% level).

Solution.   (a) $n = 16$; $\bar{D} = 3.0$; $s_{\bar{d}} = 1.1$.
We are asked to estimate the mean m_d of the population of differences in yields between the two varieties. Since this is a paired experiment, Theorem 10-3 is applicable. The t-value for 15 degrees of freedom for which the two tail areas add up to 0.02 is found from Appendix Table III to be $t_{.02} = 2.602$; thus

$$\frac{\bar{D} - m_d}{1.1} = \pm 2.602$$

and

$$m_d = \bar{D} \pm 2.862 = 3.0 \pm 2.862$$

or

$$0.138 \quad \text{and} \quad 5.862,$$

which are the 98% confidence limits.

(b) If there is no significant difference between the varieties at the 2% level, the 98% confidence limits for m_d must include 0, the mean difference between the two varieties. However, since, from part (a) m_d is between 0.138 and 5.862, 0 is not included, and a significant difference in yields is indicated.

EXERCISES

Unless otherwise specified, use the 5% level of significance. Answer each question exactly in the form in which it is asked. Also, in each case list the number of degrees of freedom and the appropriate tabular t-value.

1. In a sample of 23 variates the mean is 26.3 and the standard deviation 1.9. Find the 98% confidence limits for the mean of the population.

2. A sample of 14 variates has a mean of 34.86 and a standard deviation of 4.23. Find limits within which the population mean lies 99% of the time.

3. (a) Find the 90% confidence limits for the mean of a normally distributed population with standard deviation 3.83, if a sample taken from that population has the following variates: 18.5, 20.6, 12.9, 14.6, 19.8, 15.0.
   (b) What 90% confidence limits would have been found if in part (a) the standard deviation of the population had been unknown?

4. A sample of 10 observations has a mean of 45 and a standard deviation of 12.
   (a) Find the 98% confidence limits for the mean of the population.
   (b) Do the same for a sample of 15 observations, everything else being the same.
   (c) Do the same for a sample of 20 observations, everything else being the same.
   (d) What is the implication regarding the relation between the size of the sample and the length of the confidence interval, other things being the same?

5. The drained weights, in ounces, of a random sample of 8 cans of cherries are as follows: 12.1, 11.9, 12.4, 12.3, 11.9, 12.1, 12.4, 12.1.
   (a) Find the 99% confidence limits for the mean drained weight per can of cherries for the population from which the sample is taken.
   (b) Using the result of part (a), does this sample indicate (1% level) that the production standard of 12.35 ounces average drained weight per can of cherries is being maintained? Explain your answer.

6. A psychologist claims that the IQ of underprivileged children in a certain city is 62. A social worker believes the value is somewhat lower. To test this claim, 25 underprivileged children are selected by the social worker and their IQ's are

measured. The average IQ for those tested is 54 and the sum of the squares of the deviations for this sample is 5400. Should the psychologist's claim be rejected at the 5% level?

7. Experience shows that a fixed dose of a certain drug causes an average increase in pulse rate of 10 beats per minute with a standard deviation of 4. A group of 9 patients given the same dose showed the following increases: 13, 15, 14, 10, 8, 12, 16, 9, 20.
   (a) Is there evidence (5% level) that the mean increase of this group is significantly different from that of the population?
   (b) If, in this problem, the population standard deviation of 4 is not known, solve the problem.

8. A real estate broker who is anxious to sell a piece of property to a motel chain assures them that during the summer months, on the average, 4000 cars pass by the property each day. Being suspicious that this figure might be a bit high, the management of the motel chain conducts its own study and obtains a mean of 3838 cars and a standard deviation of 486 cars for the 10 days on which it made the survey. In the light of this survey, is there reason to be suspicious of the real estate broker's claim (5% level)?

9. A test of the breaking strengths of 10 safety belts manufactured by a company showed a mean breaking strength of 7910 lbs. and a standard deviation of 90 lbs., whereas the manufacturer claimed a mean breaking strength of 8000 lbs. In view of the lower mean breaking strength of the sample, can we still support the manufacturer's claim (1% level)?

10. Assume that, on the average, each student in a university owns 25 paperback books. A sample of 9 students is found to possess an average of 30 such books. This sample mean is found to be just barely significant at the 10% level. What is the standard deviation of the sample of 9 students?

11. From a sample of five grapefruit the 95% confidence limits for the mean weight of all grapefruit in a shipment are calculated to be 12.89 and 17.11 ounces. Find the mean and standard deviation of this sample of five grapefruit.

12. From a sample of 26 variates, the 95% confidence limits for the mean of the population whose standard deviation is unknown are determined to be 21.50 and 26.50. Find the 98% confidence limits for the mean of the population.

13. The results of quality control tests on two manufacturing processes are as follows:

$$\text{Process I:} \quad 1.5, 2.5, 3.5, 2.5$$
$$\text{Process II:} \quad 2.5, 3.0, 3.0, 4.0, 3.5, 2.0$$

Do these results allow the conclusion that the processes yield different qualities (10% level)?

14. A new method for making concrete has been proposed. To test whether the new method has increased the compressive strength, five sample blocks are made by

each method. The compressive strengths in pounds per square inch are as follows:

New method: 154, 143, 132, 147, 139
Standard method: 144, 131, 155, 126, 134

Would you say that the new method has increased the compressive strength?

15. The table below gives data on the gain in weight of two lots of rats:

Diet	High protein	Low protein
Number of rats	12	7
Mean gain (in grams)	120	101
Sum of squares of deviations from mean	5032	2552

Do these data indicate a significant difference between mean gains in weight for the two groups?

16. A cigarette manufacturer tests tobaccos of two different brands for nicotine content and obtains the following results (in milligrams):

Brand A: 24, 26, 25, 22, 23
Brand B: 27, 28, 25, 29, 26

Do these results indicate that there is a difference in the mean nicotine content of the two brands?

17. Below are figures for protein tests of the same variety of wheat grown in two districts. If these are the only figures available, test whether or not there is a significant difference between the average proteins for the two districts.

District 1: 12.6, 13.4, 11.9, 12.8, 13.0
District 2: 13.1, 13.4, 12.8, 13.5, 13.3, 12.7, 12.4

18. Solve Example 9-4 by the methods of this chapter, i.e. small sample methods.

19. Solve Exercise 18, Chapter 9, by the methods of this chapter.

20. Solve Exercise 19, Chapter 9, by the methods of this chapter.

21. Two samples of 15 specimens each of two types of wool fabric were subjected to strength tests. The mean of the first sample was 131 pounds per square inch, that of the second 136 pounds per square inch. The respective standard deviations of the two samples were 6.25 pounds per square inch and 4.65 pounds per square inch. Does this indicate that the second type of wool fabric is superior?

22. A certain test is given to a 45-student statistics class, in order to determine the difference in accomplishment of men and women. The 20 women students have an average score of 60 with a standard deviation of 19, and the 25 men students

have an average score of 66 with a standard deviation of 16. What are the 98% confidence limits for the difference between the mean scores of men and women students?

23. Two analysts, *A* and *B*, study the melting point of a metal. Five readings by *A* show a variance of 12 and 4 readings by *B* a variance of 20. What is the least absolute value of the difference between the means of the readings that would be considered significant at the 5% level?

24. In an experiment on 10 plots of wheat, half of each plot is sown with seed of a rust-resistant variety, and the other half is sown with ordinary seed. Do the results listed below (weight of grain in pounds) indicate that there is a significant difference between the two seeds?

Plot	Rust-resistant	Ordinary	Plot	Rust-resistant	Ordinary
1	49	47	6	49	44
2	58	57	7	66	67
3	53	49	8	55	52
4	60	57	9	44	42
5	45	44	10	52	53

25. In a paired feeding experiment, the gains in weight (in pounds) of hogs fed on two different diets are as listed below. Can either diet be considered superior?

Pair	Diet A	Diet B
1	25	19
2	30	32
3	28	21
4	34	34
5	23	19

26. A newly discovered apple is delicious in flavor. It is decided to test its yielding capacities by planting the apple adjacent to a standard apple in 8 orchards scattered about a region suitable for production of both varieties. When the trees are mature, the yields in bushels are measured. The data are as follows:

Orchard	New apple	Standard apple
1	13	12
2	14	16
3	19	17
4	10	9
5	15	16
6	14	12
7	12	10
8	11	8

Do these results indicate a greater yield for the new apple than for the standard apple?

27. A company wishes to study the effectiveness of a coffee break on the productivity of its workers. It selects 6 workers and measures their productivity on a day without a coffee break and later measures the productivity of the same 6 workers on a day when they are given a coffee break. The scores measuring the productivity are as follows:

Worker	Without coffee break	With coffee break
1	23	28
2	35	38
3	29	29
4	33	37
5	43	42
6	32	30

Do these results indicate that a coffee break raises productivity?

28. Some believe that the ability of students to perform arithmetic improves after they have taken this course in statistics. A sample of 8 students is given an aptitude test on arithmetic before and after taking this course, with the following results, based on a scale of 10.

Student	Before	After
1	7	8
2	6	5
3	7	9
4	8	8
5	6	3
6	8	9
7	5	8
8	2	9

Do these results confirm this belief (2.5% level)?

29. A teacher claims that his students perform similarly on the midterm and the final examinations. To test his claim he randomly chose to examine the records of 9 students from his last semester and found the following:

Student	Midterm	Final	Student	Midterm	Final
1	67	81	6	54	68
2	77	82	7	89	97
3	80	73	8	67	76
4	83	84	9	85	91
5	92	87			

At the 5% level of significance, do these data confirm the teacher's claim?

30. Two varieties of wheat are grown in adjacent plots at 9 different locations. The yields in pounds are as follows:

Location	Variety A	Variety B	Location	Variety A	Variety B
1	125	121	6	112	115
2	110	102	7	100	100
3	98	92	8	105	101
4	105	107	9	110	105
5	116	111			

(a) Are the variety means significantly different at the 5% level of significance?
(b) Set up 95% confidence limits for the mean difference in yields of the two varieties.

31. To test two promising new corn hybrids under normal farming conditions, a seed company selects 8 farms at random in Iowa and plants both hybrids in experimental plots on each farm. The yields, in bushels per acre, for the 8 locations are:

Farm	Hybrid A	Hybrid B
1	86	80
2	87	79
3	56	58
4	93	91
5	84	77
6	93	82
7	73	74
8	79	66

Set up the 95% confidence limits for the mean difference in yield of the two hybrids.

32. For a sample of 4 variates, the 90% confidence limits for the population mean were calculated to be 16.28 and 23.72. What is the value of ΣX^2 for this sample?

33. The estimate of the standard deviation of a population as determined from a sample of 9 variates is 1.20. One of the 95% confidence limits for the mean of the population from which the sample is drawn is known to be 10.20. What is the best information one can give about the other confidence limit?

34. At the University of California in Davis an experiment was carried out to determine the effect of light on the time of day at which eggs are laid. A group of

8 hens was kept in total darkness, and the number of eggs laid each hour was recorded over a 29-day period. Another group of 8 hens was exposed to 1 minute of light every hour for a 29-day period, and again the number of eggs laid each hour was recorded. The results are listed in the table below.

| | Number of eggs laid | |
Time	Total darkness	One minute of light per hour
Midnight– 1 A.M.	0	1
1– 2	0	1
2– 3	0	0
3– 4	6	2
4– 5	8	5
5– 6	5	2
6– 7	8	9
7– 8	5	9
8– 9	15	4
9–10	9	19
10–11	12	7
11–Noon	9	9
Noon– 1 P.M.	8	10
1– 2	8	9
2– 3	6	3
3– 4	2	4
4– 5	1	5
5– 6	1	2
6– 7	0	2
7– 8	0	0
8– 9	0	0
9–10	0	0
10–11	0	2
11–Midnight	0	0

Can we conclude from these data that 1 minute of light each hour had an effect on the average time at which eggs were laid?

11

Nonparametric Statistics

11.1 INTRODUCTION

The theorems of Chapters 8 and 10 are true only if the populations involved are normally distributed. For example, the sample mean satisfies a normal distribution only if the population is normally distributed, although for populations that are not normal the distribution of the sample mean approaches that of a normal distribution as the size of the sample increases. In general, the distributions of sample statistics depend upon the type of distribution that is satisfied by the population. Since in many cases little or no knowledge is available about the distribution of the population, it becomes desirable to employ methods that make no assumptions regarding the distribution and the parameters of the population.

Definition 11-1. Statistical analyses that do not depend upon the knowledge of the distribution and parameters of the population are called *nonparametric* or *distribution-free methods.*

There was relatively little interest in nonparametric statistics until about 1945, when Frank Wilcoxon proposed a test, now referred to as the Wilcoxon two-sample test, which is distinguished by both its simplicity and the very excellent results it gives, even in the case where the distribution of the population is known to be normal. For these reasons, the use of nonparametric techniques has become widespread. As in the use of Student's t-distribution, the case where the variates of the two samples are from two independent populations must be distinguished from the case where the variates are paired. For the first case, the Wilcoxon two-sample test will be discussed; for the second, the sign test and the Wilcoxon test for paired variates will be considered.

11.2 THE WILCOXON TWO-SAMPLE TEST FOR THE UNPAIRED CASE

Assume that two samples of n_1 and n_2 variates, respectively, are given and that we wish to test the hypothesis of equal population means. Then the two samples together have $n = n_1 + n_2$ variates. In Wilcoxon's test, these n variates are graded (or ranked) according to increasing size, that is, we substitute the scores (or ranks) 1, 2, 3, ..., n for the actual numerical data. If there is no significant difference between the two sample means, then the total of the ranks corresponding to the first and those corresponding to the second sample should be about the same. If, however, the total of the ranks for one sample is appreciably less than that of the other, we calculate—under the hypothesis of equal population means—the probability of obtaining by chance alone a sum of ranks less than or equal to that obtained in the given experiment. If this probability is less than the significance level, we reject the hypothesis; otherwise, we accept it.

The procedure can be better understood in terms of an example.

Example 11-1. A new method for making concrete has been proposed. To test whether the new method has increased the compressive strength, five sample blocks are made by each method. The compressive strengths in pounds per square inch are as follows:

New method: 148, 143, 138, 145, 141
Standard method: 139, 136, 142, 133, 140

Does this indicate that the new method has increased the compressive strength (5% level)?

Solution. The combined data of the two samples should be arranged in increasing order of magnitude with corresponding ranks indicated, and in order to distinguish between the two samples, the variates belonging to the sample with the smaller mean are underlined. In this example, the underlined variates belong to the sample made by the standard method.

Original data: 133, 136, 138, 139, 140, 141, 142, 143, 145, 148

Ranks: 1,   2,   3,   4,   5,   6,   7,   8,   9,   10

We now calculate the sum of the ranks corresponding to the standard method and find

$$W_1 = 1 + 2 + 4 + 5 + 7 = 19,$$

whereas the sum of the ranks corresponding to the new method is

$$W_2 = 3 + 6 + 8 + 9 + 10 = 36.$$

The total of all ranks is

$$W_1 + W_2 = 19 + 36 = 55.$$

The sum 55 is, of course, independent of the outcome of this experiment, since it represents the sum of the 10 ranks, that is, $1 + 2 + 3 + \cdots + 10$. If there is no effect of the new method of making concrete, then the total of the ranks for the standard method, W_1, should be about the same as that for the new method, W_2. Indeed, under the hypothesis of no difference in the two methods, any sum of 5 ranks, such as $3 + 4 + 6 + 8 + 9$ would be just as likely to occur for a value of W_1 as some other sum of 5 ranks, such as $2 + 4 + 5 + 6 + 8$.

Consequently, assuming a one-tailed test, we shall now determine the probability of obtaining, among all possible sums of ranks, one that is less than or equal to the value of $W_1 = 19$ obtained in this case. This calculation is easily made: The total number of possible ways of selecting 5 ranks from 10 is the number of combinations of 10 objects taken 5 at a time, that is, $C_{10,\,5} = 252$. Of these, the following have a sum less than or equal to 19 (note that we start by listing first the sum of ranks giving the smallest total and then proceed in a systematic way):

$$1 + 2 + 3 + 4 + 5 = 15$$
$$1 + 2 + 3 + 4 + 6 = 16$$
$$1 + 2 + 3 + 4 + 7 = 17$$
$$1 + 2 + 3 + 4 + 8 = 18$$
$$1 + 2 + 3 + 4 + 9 = 19$$

$$1 + 2 + 3 + 5 + 6 = 17$$
$$1 + 2 + 3 + 5 + 7 = 18$$
$$1 + 2 + 3 + 5 + 8 = 19$$

$$1 + 2 + 3 + 6 + 7 = 19$$

$$1 + 2 + 4 + 5 + 6 = 18$$
$$1 + 2 + 4 + 5 + 7 = 19$$

$$1 + 3 + 4 + 5 + 6 = 19$$

Consequently, since there are 12 cases, the probability of obtaining a sum of ranks less than or equal to 19 is given by

$$P = \frac{12}{252} = 0.0477.$$

Calculation of probabilities is facilitated by use of Appendix Table IV, giving, for various sample sizes n_1 and n_2 listed in the first two columns, the corresponding values of C_{n, n_1}, where $n = n_1 + n_2$, and the number of cases for which the sum of the ranks is less than or equal to that obtained in the experiment. The column headings 0, 1, 2, ..., 20 of Table IV represent values of the quantity

$$U = W_1 - \frac{1}{2} n_1(n_1 + 1).$$

Thus, if Table IV is used in the solution of Example 11-1, we calculate

$$U = 19 - \frac{1}{2} \cdot 5 \cdot 6 = 4.$$

Then $C_{10, 5} = 252$, and the number of cases for which the sum of the ranks is less than or equal to 19 is 12, the entry in the column headed 4.

It might be of interest to note that, if Example 11-1 had been solved by use of Student's t-distribution, using a one-tailed test, the probability would have been found to be $P = 0.033$. This would be the correct probability if the distribution of the population were normal; if it were not, the probability obtained by the Wilcoxon test might well be more accurate.

In the data of Example 11-1, none of the 10 given variates occurred more than once, that is, there were no ties. The question arises: How do we apply Wilcoxon's test to data with ties? The answer is supplied by Wilcoxon himself, who, when introducing his test in 1945 for the first time, gave an example in which there were several ties. This example is reproduced here as Example 11-2. In case of ties, the procedure is the same as in the case without ties, except for the assignment of ranks.

Example 11-2.   The following data give the results of tests on two preparations of a fly spray, in terms of the percentage of mortality.

    Sample A: 68, 68, 59, 72, 64, 67, 70, 74
    Sample B: 60, 67, 61, 62, 67, 63, 56, 58

Is there a significant difference (5 % level) between the two preparations?

Solution. As before, the data are arranged in increasing order of magnitude; those underlined correspond to the sample with the smaller mean, in this case, sample B. Three of the variates are 67; these variates correspond to ranks 9, 10, and 11. It seems reasonable, therefore, to assign to all three of these values the mean of these ranks, namely, $\frac{1}{3}$ of 30, or 10. The same procedure is followed for the two variates that are 68, corresponding to ranks 12 and 13. Again, since $12 + 13 = 25$, we assign to both variates the rank $\frac{1}{2}(25) = 12.5$. This, then, is the procedure generally followed in cases with tied data: equal ranks are assigned to tied data.

For this example, then, the ranks are determined as follows:

Original data: 56, 58, 59, 60, 61, 62, 63, 64, 67, 67, 67, 68,   68,   70, 72, 74

Ranks: 1, 2, 3, 4, 5, 6, 7, 8, 10, 10, 10, 12.5, 12.5, 14, 15, 16

Now, since

$$W_1 = 1 + 2 + 4 + 5 + 6 + 7 + 10 + 10 = 45,$$

and

$$U = 45 - \frac{1}{2} \cdot 8 \cdot 9 = 9,$$

we consult Appendix Table IV to find $C_{16,8} = 12{,}870$; the number of cases for which the sum of the ranks is less than or equal to 45 is 95, so that

$$P = \frac{95}{12{,}870} = 0.00738,$$

or, since a two-tailed test is needed here, the probability of obtaining a difference as great as or greater than that found between the means of the given two samples would equal

$$2P = 0.01476,$$

from which we conclude that there is a significant difference between the two preparations.

It should be noted that Appendix Table IV was constructed under the assumption that there are no ties. The reader may, therefore, question whether it is permissible to use Table IV also in the case in which there are ties, as was done in Example 11-2. It can be shown that, under appropriate assumptions concerning the probability of occurrence of ties, use of Table IV is justifiable also for that case. If ties occur, it is possible that the value of W_1 may not be an integer. In that case, the appropriate value to be taken from Table IV should be obtained by linear interpolation. Thus, if in Example 11-2 U had been 11.5, we would have calculated

$$P = \frac{212.5}{12{,}870}.$$

Wilcoxon's test is also applicable when the two samples are not of the same size. The procedure is essentially the same. In Table IV, n_1 and W_1 refer to the sample with the fewer observations. If the sample with n_1 observations has the larger mean, the observations should be ranked in order of decreasing size since Table IV deals only with the left tail of the distribution of U.

Example 11-3.   Solve Exercise 17, Chapter 10, by use of Wilcoxon's test.

Solution.   Since the sample for District 1 has the smaller mean, we have $n_1 = 5$, $n_2 = 7$, and we list the original data in increasing order of magnitude together with the ranks as follows:

Data: 11.9, 12.4, 12.6, 12.7, 12.8, 12.8, 13.0, 13.1, 13.3, 13.4, 13.4, 13.5

Ranks: 1,   2,   3,   4,   5.5, 5.5, 7,   8,   9,   10.5, 10.5, 12

Now, since

$$W_1 = 1 + 3 + 5.5 + 7 + 10.5 = 27,$$

and

$$U = 27 - \frac{1}{2} \cdot 5 \cdot 6 = 12,$$

from Table IV, we find $C_{12, 5} = 792$, and the number of cases for which the sum of the ranks is less than or equal to 27 is 171, so that

$$P = \frac{171}{792} = 0.216$$

and

$$2P = 0.432.$$

Therefore, there is no significant difference between the average proteins for the two districts.

Table IV can, of course, be used only for values of $n = n_1 + n_2$ not exceeding those listed in the first two columns. Fortunately, it has been proved that as n becomes larger and larger, the distribution of the sum of the ranks, W_1, approaches a normal distribution, so that we have the following theorem.

Theorem 11-1.   Consider two samples of sizes n_1 and n_2, respectively, from two identical populations, and let $n = n_1 + n_2$. If the sums of the ranks of the two samples are W_1 and W_2, respectively, and if there are no ties, then, for large n, W_1 satisfies an approximately normal distribution with

$$m_w = \frac{n_1(n + 1)}{2} \tag{11-1}$$

and standard deviation

$$\sigma_w = \sqrt{\frac{n_1 n_2 (n + 1)}{12}}. \tag{11-2}$$

Theorem 11-1 will not be proved, but the formula for m_w will be verified.

Proof of Formula (11-1). Before proving the formula for m_w in general, let us consider the special case of $n_1 = 5$, $n_2 = 5$, so that $n = 10$. Then m_w equals the mean of all possible sums of 5 ranks taken from the set of ranks 1, 2, 3, ..., 10. The smallest possible such sum is $1 + 2 + 3 + 4 + 5 = 15$, the next larger, $1 + 2 + 3 + 4 + 6 = 16$, and so on. Listing these sums in a systematic way, as in the solution of Example 11-1, we find that the rank 1 occurs in $C_{9,4}$ of these sums, that is, in the number of ways we can select from the 9 numbers 2, 3, 4, ..., 10 any 4 of them. By the same reasoning, the number 2 will also appear in $C_{9,4}$ of these sums. Thus the sum of all possible rank sums is equal to

$$S = C_{9,4} (1 + 2 + 3 + \cdots + 10) = C_{9,4} \cdot \frac{1}{2} \cdot 10 \cdot 11,$$

since

$$1 + 2 + 3 + \cdots + n = \frac{1}{2} n(n + 1).$$

The total number of possible ways of forming sums of 5 ranks from the 10 numbers 1, 2, 3, ..., 10 is, of course, given by $C_{10,5}$, so that

$$m_w = \frac{S}{C_{10,5}} = \frac{C_{9,4} \cdot \dfrac{1}{2} \cdot 10 \cdot 11}{C_{10,5}} = \frac{\dfrac{9!}{4!\,5!} \cdot \dfrac{1}{2} \cdot 10 \cdot 11}{\dfrac{10!}{5!\,5!}}$$

$$= \frac{5}{10} \cdot \frac{1}{2} \cdot 10 \cdot 11 = \frac{5 \cdot 11}{2} = 27.5,$$

which agrees with the value found by substituting $n_1 = 5$ and $n = 10$ in the formula for m_w, given by (11-1).

To prove Formula (11-1) in general, note that m_w equals the mean of all possible sums of n_1 ranks taken from the set of ranks 1, 2, 3, ..., n. The smallest possible such sum is $1 + 2 + 3 + \cdots + n_1$, the next larger is $1 + 2 + 3 + \cdots + (n_1 + 1)$, and so on. Listing these sums again in a systematic way, we find that the rank 1 appears in C_{n-1, n_1-1} of the sums, that is, in the number of ways we can select from the $n - 1$ numbers 2, 3, 4, ..., n any $n_1 - 1$

of them. Similarly, the number 2 will also appear in C_{n-1, n_1-1} of these sums. Thus, the sum of all possible rank sums is equal to

$$S = C_{n-1, n_1-1}(1 + 2 + 3 + \cdots + n) = C_{n-1, n_1-1} \frac{1}{2} n(n + 1).$$

The total number of possible ways of forming sums of n_1 ranks from the n numbers 1, 2, 3, ..., n is given by C_{n, n_1}, so that

$$m_w = \frac{S}{C_{n, n_1}} = \frac{C_{n-1, n_1-1} \frac{1}{2} n(n+1)}{C_{n, n_1}} = \frac{\frac{(n-1)!}{(n_1-1)!(n-n_1)!} \frac{1}{2} n(n+1)}{\frac{n!}{n_1!(n-n_1)!}}$$

$$= \frac{n_1}{n} \frac{1}{2} n(n+1) = \frac{n_1(n+1)}{2}.$$

By a similar procedure, Formula (11-2) can be verified.

Example 11-4.   Solve Example 11-1 by use of Theorem 11-1.

Solution.   Although the sizes of the samples of Example 11-1 are sufficiently small to permit use of Appendix Table IV, the normal curve approximation will be used in order to compare the results obtained by the exact method (use of Table IV) and the approximate method furnished by the normal curve approximation. The sizes of the samples are so small in this case that, in practice, the normal curve approximation should not be used.

Since $n_1 = 5$, $n_2 = 5$, and $n = 10$,

$$m_w = \frac{5 \cdot 11}{2} = 27.5,$$

and

$$\sigma_w = \sqrt{\frac{5 \cdot 5 \cdot 11}{12}} = \sqrt{22.92} = 4.79,$$

so that the probability of finding a sum of ranks less than or equal to 19 is the area under the normal probability curve to the left of

$$z = \frac{19.5 - 27.5}{4.79} = -1.67,$$

which equals 0.0475. This compares with 0.0477 obtained in Example 11-1.

For the case of tied ranks, Theorem 11-1 needs to be modified as follows: For each group of tied observations, let t_i be the number of tied observations in the ith group. Then let

$$T_i = (t_i - 1)t_i(t_i + 1)$$

for the ith group of ties. Now Theorem 11-1 is applicable with the formula for m_w remaining unchanged, but the formula for σ_w replaced by

$$\sigma_w = \sqrt{\frac{n_1 n_2 \left[n(n^2 - 1) - \sum\limits_{i=1}^{k} T_i \right]}{12n(n - 1)}}, \qquad (11\text{-}3)$$

where k is the number of groups containing ties. Note that when there are no ties, that is,

$$\sum_{i=1}^{k} T_i = 0,$$

Formula (11-3) reduces to (11-2), so that Formula (11-3) may be regarded as the general expression for σ_w, whether or not there are ties. Consequently, Theorem 11-1 can be stated in the more general form given by Theorem 11-2:

Theorem 11-2.   Consider two samples of sizes n_1 and n_2, respectively, from two identical populations, and let $n = n_1 + n_2$. If the sums of the ranks of the two samples are W_1 and W_2, respectively, then, for large n, W_1 satisfies an approximately normal distribution with m_w and σ_w given by (11-1) and (11-3), respectively.

Example 11-5.   Solve Example 11-2 by use of Theorem 11-2.

Solution.   Since $n_1 = 8$, $n_2 = 8$, and $n = 16$, we find

$$m_w = \frac{8 \cdot 17}{2} = 68,$$

and, since there are two groups of tied ranks (that is, those tied at 10 and those tied at 12.5), we have $t_1 = 3$, $t_2 = 2$, and $T_1 = 2 \cdot 3 \cdot 4 = 24$, $T_2 = 1 \cdot 2 \cdot 3 = 6$, and

$$\sum_{i=1}^{2} T_i = 24 + 6 = 30,$$

so that

$$\sigma_w = \sqrt{\frac{8 \cdot 8 \cdot [16(256 - 1) - 30]}{12 \cdot 16(16 - 1)}} = \sqrt{90} = 9.49.$$

Therefore, the probability of finding a sum of ranks less than or equal to 45 is the area under the normal probability curve to the left of

$$z = \frac{45.5 - 68}{9.49} = -2.37,$$

which equals 0.0089, so that, for a two-tailed test, twice this area equals 0.0178, which compares with 0.01476 obtained in Example 11-2. That there is not too close an agreement between the two values is not surprising in view of the relatively small value of n and the fact that, at the tail end of the distribution, the approximation is obviously worse than near its center. For larger values of n, however, the approximation will be much better.

Intuitively it would appear that Wilcoxon's test should be inferior to tests that use the actual numerical values of the samples. It would seem that replacement of the observations by their ranks, which merely indicate the order relationship between the observations, would disregard so much information that the test would not be an appropriate one to use in many cases. It is no surprise, then, that Wilcoxon's test and other nonparametric tests were regarded rather skeptically at first. Recent results, however, have shown this skepticism to be unwarranted; the reasons follow.

In order to determine how good a particular statistical test is, a precise definition of what is meant by a " good " statistical test is required. Clearly, one desirable property of a test is that, when the hypothesis is false, there is a high probability that the test will give a significant result, or, to put it differently, in a given experiment, the investigator would like to use the statistical test that gives the highest probability of detecting a false hypothesis. This probability is referred to as the *power* of the test. For a given statistical test, for example, the Student's *t*-test, the power of the test can always be increased by increasing the sample size. Clearly, with larger sample sizes, and assuming the hypothesis of equal population means to be false, the probability of rejecting the hypothesis is increased as the sample size increases.

A common method of comparing various statistical tests is to examine them for their *efficiency*; roughly speaking, the method compares the sample sizes for two tests that have the same power, i.e. that detect a false hypothesis with the same probability. If a certain nonparametric test (for example, Wilcoxon's test) is said to have an efficiency of 100 % when compared to another test (for example, Student's *t*-test), this means that, if the hypothesis is false, it will be detected by both tests with equal probability for the same size samples. If, however, this efficiency is 90 % for the Wilcoxon test, such a false hypothesis will be detected by means of the t-test with a sample only $\frac{9}{10}$ as large as that needed in the Wilcoxon test.

We would, of course, expect that the Wilcoxon test would have generally a much lower efficiency relative to Student's t-test. Surprisingly, however, it has recently been shown that the Wilcoxon test is about 95.5% efficient for large samples if the population is normally distributed and that its efficiency is more than 100% if the distribution deviates considerably from a normal distribution. Most surprising, however, is the result proved in December, 1963, by E. L. Lehmann: the efficiency of the Wilcoxon test relative to the t-test is never less than 86.4%, and it can be arbitrarily high and even infinite. Lehmann furthermore proved that this lowest efficiency of 86.4% is true even when the Wilcoxon test is used to compare the means of more than two samples; this, however, will not be discussed here.

Consequently, at the very best, Student's t-test requires a sample only 13.6% smaller than the Wilcoxon test to detect with the same probability a significant difference, but in many situations Student's t-test may actually require a larger sample.

For these reasons, use of Wilcoxon's test would appear nearly always to be preferable to the use of Student's t-test.

Many times it is not possible to obtain actual measurements for the samples, as in taste tests, where it is desired to determine preferences for certain foods. Clearly, in such tests it is not always possible to assign a meaningful numerical score to an individual's like or dislike for a particular food, yet it is possible to rank the foods according to increasing order of preference. Consequently, use of a nonparametric test, such as the Wilcoxon test, is imperative.

Further improvement in the efficiency is possible by means of recently developed nonparametric tests; in particular, by the use of Normal Scores. In this method, the observations are again first replaced by ranks, for example, in the case of 2 samples of 5 variates each, by 1, 2, 3, ..., 10. In the Normal Scores test, the ranks are assumed to follow a normal distribution and are therefore replaced by Normal Scores fitted to a normal distribution. This test will not be discussed, since, in most situations, the Wilcoxon test is entirely adequate.

11.3   THE SIGN TEST FOR THE PAIRED CASE

The simplest, by far, of all nonparametric methods is the sign test, which is used to test the significance of the difference between two means in a paired experiment. It is particularly suitable when the various pairs are observed under different conditions, a case in which the assumption of normality may not hold. However, because of its simplicity, the sign test is often used even though the populations are normally distributed. As is implied by its name, in this test only the sign of the difference between the paired variates is used. The method of applying the test will now be described.

First we determine for each pair the sign of the difference D. Under the hypothesis that the two population means are identical, the number of plus signs should approximately equal the number of minus signs. If p denotes the probability of a difference D being positive and q the probability of its being negative, we have as hypothesis $p = \frac{1}{2}$. Suppose we are given a sample from this population and wish to determine whether it differs significantly from this hypothesis. For this purpose, we count the number of plus signs (call it S) and the number of minus signs (pairs of observations for which $D = 0$ are disregarded) and calculate the probability of obtaining, from a population for which $p = \frac{1}{2}$, a sample in which the number of plus signs differs as much as S or more from the expected value.

Example 11-6.   In a paired experiment, the gains n weight, in pounds, of hogs fed on two different diets are as given in Table 11-1. Is either diet superior at the 5% level of significance?

Table 11-1
Gains in weight (in pounds) of hogs fed on two different diets.

Pair	Diet A	Diet B	Sign of $D = X - Y$	Pair	Diet A	Diet B	Sign of $D = X - Y$
1	25	19	+	10	28	26	+
2	30	32	−	11	32	30	+
3	28	21	+	12	29	25	+
4	34	34	0	13	30	29	+
5	23	19	+	14	30	31	−
6	25	25	0	15	31	25	+
7	27	25	+	16	29	25	+
8	35	31	+	17	23	20	+
9	30	31	−	18	26	25	+

Solution.   Disregarding the two cases where $D = 0$, we find that in the remaining sample of 16 differences 13 are positive. By means of the binomial distribution we can find the probability (under the hypothesis of equal chance for plus and minus signs) that 13 or more plus signs (or minus signs) are obtained.

Since $n = 16$ is fairly large, it is convenient to use the normal curve approximation of the binomial distribution, for which

$$m = np = 8, \qquad \sigma = \sqrt{npq} = \sqrt{4} = 2.$$

We need twice the area under the normal probability curve to the right of
$X = 12.5$. Since

$$z = \frac{12.5 - 8}{2} = 2.25,$$

we find from Appendix Table I

$$P = 2(0.0122) = 0.0244,$$

which is less than 0.05. Thus the first diet can be considered superior.

It should be noted that the sign test is applicable only if n, the number
of pairs, is not too small. For example, in an experiment where $n = 5$, the
probability of the most extreme case, that of all signs being either plus or
minus, is $2(\frac{1}{2})^5 = 0.0625$, which is more than 5%; thus, from a sample of 5 differences, it is never possible to conclude that the hypothesis of equal population means is incorrect. Clearly, then, use of the sign test at the 5% level of
significance with a two-tailed test requires that n be at least 6. It should
preferably be larger, and must be so if the 1% level of significance is used.

Clearly, even for large n, we expect the sign test to have a lower efficiency
than the t-test, which means that in many cases the sign test will not indicate
a significant difference when the t-test applied to the same problem actually
shows the result to be significant. Such a situation arises if the differences
corresponding to the more frequently occurring sign are large in absolute
value compared with those corresponding to the other sign. Another nonparametric test that does take into consideration the magnitude of the differences and, therefore, has a higher efficiency than the sign test, is given in
Section 11.4. On the other hand, if the sign test gives a significant result,
it is a valuable statistical technique requiring little calculation and, of course,
no knowledge of the distribution of the population.

11.4   THE WILCOXON TEST FOR THE PAIRED CASE

Assume that, as in Section 10.5, there are given two samples whose variates
are paired. In Wilcoxon's test for this paired experiment, the differences, D,
for each pair are calculated and their absolute values are ranked. If there is

no significant difference between the means of the two samples, then the total of the ranks corresponding to positive values of D and those corresponding to negative values of D should be about the same. If, however, the total of the ranks corresponding to one sign is appreciably less than that corresponding to the other, then, under the hypothesis of equal population means, the probability of obtaining by chance alone a sum of ranks less than or equal to W_1 is calculated, W_1 being the smaller of the rank totals. If this probability is less than the significance level, the hypothesis is rejected; otherwise, it is accepted.

Example 11-7.   The data below represent the stands of wheat on eight pairs of plots, one of each pair planted to seed having received treatment A and the other treatment B.

Pair	A	B	Pair	A	B
1	209	151	5	159	166
2	200	168	6	169	163
3	177	147	7	187	176
4	169	164	8	198	188

Using the 5% level of significance, is there a significant difference between the two seed treatments?

Solution.   Denoting the variates of the first sample by X and those of the second sample by Y, we find

X	Y	$D = X - Y$	Rank
209	151	58	8
200	168	32	7
177	147	30	6
169	164	5	1
159	166	−7	3
169	163	6	2
187	176	11	5
198	188	10	4

Note that ranks corresponding to a negative value of D are underlined; in this case, there is only one such rank, namely the rank 3, so that $W_1 = 3$. Now, under the hypothesis that there is no difference between the two treatments A and B, the probability of each difference D being positive is $\frac{1}{2}$. Consequently, the probability of obtaining any given sequence of signs for the eight differences (for example, the sequence $+ + - + - - - +$) is $(\frac{1}{2})^8 = \frac{1}{256}$, so that there are a total of 256 equally likely possibilities for the sequence of signs.

Assuming a two-tailed test, we determine the probability of obtaining among all possible sums of ranks (corresponding to negative D-values) one that is less than or equal to W_1 (in this case, 3) and then multiply this probability by 2. Now the following cases are the only ones for which the sum of the ranks (corresponding to negative D-values) is less than or equal to 3:

No negative D-value:	1 case
One negative D-value with rank either 1, 2, or 3:	3 cases
Two negative D-values with rank $1 + 2 = 3$:	1 case
Total:	5 cases

Consequently, the probability of obtaining a sum of ranks less than or equal to 3 is given by

$$P = \frac{5}{256} = 0.0195$$

and the desired probability equals

$$2P = 0.0390,$$

which leads to the conclusion that the two treatments are significantly different.

Calculation of probabilities for the Wilcoxon distribution in the paired case is facilitated by use of Appendix Tables Va and Vb. Two tables are given for the following reason: If it is desired to find the number of cases for which the sum of ranks is less than or equal to a given value of W_1, where W_1 is less than or equal to n, then this number of cases is independent of n and Table Va can be used. This was true in Example 11-7, where $W_1 = 3$ and $n = 8$. For this example, Table Va shows the number of cases for which W_1 is less than or equal to 3 to be 5.

If, however, it is desired to find the number of cases for which the sum of ranks is less than or equal to a given value of W_1, where W_1 is greater than n, then this number of cases depends on n and Table Vb must be used.

Example 11-8. Solve Example 11-7, assuming that in the data the values of 209 and 151 in pair 1 are interchanged, everything else remaining the same.

Solution. Interchanging the values of 209 and 151 in the first pair results in the value $D_1 = -58$, and the corresponding rank 8 must be underlined as corresponding to a negative D-value. Then $W_1 = 3 + 8 = 11$, and the number of cases for which the sum of the ranks is less than or equal to 11 (found in the row of Table Vb for which $W_1 - n = 11 - 8 = 3$ and the column for which $n = 8$) is 49, and, for a two-tailed test,

$$P = 2\left(\frac{49}{256}\right) = 0.383,$$

so that this would then not be a significant result.

Although Tables Va and Vb were constructed under the assumption that there are no ties, it can be shown that, under appropriate assumptions concerning the probability of occurrence of ties, use of these two tables is justifiable also when there are ties. When so, we rank the differences that remain after disregarding those that are zero in the same way we did in Section 11.2, that is, assign to differences that are the same in absolute value the mean of the ranks they would have been assigned if they had all been different. We underline the ranks corresponding to the sign for which the sum of the ranks is smaller and proceed as before, using linear interpolation in Tables Va and Vb if the value of W_1 thus obtained is not an integer.

Example 11-9.   Solve Example 10-5 by the use of Wilcoxon's test.

Solution.   After disregarding the difference that is zero, we rank the remaining 7 differences as indicated below

X	Y	$D = X - Y$	$Rank$
83	78	5	7
74	74	0	—
67	63	4	5
64	66	−2	2.5
70	68	2	2.5
67	63	4	5
81	77	4	5
64	65	−1	1

Since the sum of the ranks corresponding to negative values of D is smaller than that corresponding to the positive ones, we underline them and find

$$W_1 = 1 + 2.5 = 3.5.$$

From Appendix Table Va, we find, by linear interpolation, the entry corresponding to $W_1 = 3.5$ to be 6, so that, for a one-tailed test

$$P = \frac{6}{128} = 0.0469.$$

Consequently, we conclude from these data that kindergarten is of benefit in preparing for the first grade, a result that agrees with the one obtained in the solution of Example 10-5.

Appendix Tables Va and Vb can be used only for values of n not exceeding those listed. However, as n becomes larger and larger, the distribution for W_1 approaches a normal distribution, so that we have the following theorem.

Theorem 11-3. If, for a sample of n differences $X - Y$, where X and Y are the variables of two identical populations, the sum of the ranks corresponding to the negative differences is W_1 and that corresponding to the positive differences is W_2, and if there are no ties, then, for large n, W_1 satisfies an approximately normal distribution with mean

$$m_w = \frac{n(n + 1)}{4} \tag{11-4}$$

and standard deviation

$$\sigma_w = \sqrt{\frac{n(n + 1)(2n + 1)}{24}}. \tag{11-5}$$

No proof will be given for Theorem 11-3. The derivation of the formula for m_w is analogous to that for Formula (11-1) and is left as an exercise for the reader (see Exercise 25).

Example 11-10. Solve Example 11-7 by use of Theorem 11-3.

Solution. Since the size of the sample of Example 11-7 is sufficiently small to permit use of Appendix Table Va, this example should, as a rule, not be solved by means of the normal distribution. Nevertheless, this is done here to compare the result obtained by the exact method (Table V) with the approximation obtained by use of the normal curve.

Since $n = 8$,

$$m_w = \frac{8 \cdot 9}{4} = 18,$$

and

$$\sigma_w = \sqrt{\frac{8 \cdot 9 \cdot 17}{24}} = \sqrt{51} = 7.14,$$

so that the probability of finding a sum of ranks less than or equal to 3 is the area under the normal probability curve to the left of

$$z = \frac{3.5 - 18}{7.14} = -2.03,$$

which equals 0.0212. For a two-tailed test, the probability is therefore equal to

$$P = 2(0.0212) = 0.0424,$$

which compares with 0.0390 obtained in Example 11-7. The agreement is not too close, of course, because the value of n was smaller than that which should be used in applying Theorem 11-3.

EXERCISES

1. The measurements of the heights (in inches) of six adults of each of two different nationalities are given below:

$$\text{Nationality A:} \quad 64, 67, 68, 65, 62, 61$$
$$\text{Nationality B:} \quad 69, 70, 66, 63, 71, 72$$

   Use Wilcoxon's two-sample test to determine whether there is a significant difference in mean heights for these two nationalities (5% level).

2. Wire cable is being manufactured by two processes. To determine if they have a different effect on the mean breaking strength of the cable, a laboratory test is performed, and six pieces of wire from one process are put under tension and the critical values of the load recorded for each piece. A similar process is followed for the seven pieces from the other process. The results are as follows:

$$\text{Process I:} \quad 9, 4, 10, 7, 8, 10$$
$$\text{Process II:} \quad 14, 9, 13, 12, 11, 8, 10$$

   Use Wilcoxon's two-sample test to determine whether there is a significant difference (5% level) between the mean breaking strengths of cable manufactured by the two processes.

3. Solve Example 10-4 by the methods of this chapter.

4. Solve Exercise 13, Chapter 10, by the methods of this chapter.

5. Solve Exercise 16, Chapter 10, by the methods of this chapter.

6. Although in Exercise 1 of this chapter the samples are small, solve this exercise by use of Theorem 11-1 to compare the result with the result obtained in Exercise 1.

7. Solve Exercise 2 of this chapter by use of Theorem 11-2.

8. Solve Example 11-3 by use of Theorem 11-2.

9. The following scores are obtained in a personality test in random samples of married and unmarried women:

Unmarried: 85, 69, 76, 80, 61, 79

Married: 73, 76, 66, 73, 73, 65

Use the appropriate nonparametric test to determine if these data indicate that married and unmarried women have the same average scores on this personality test (5% level).

10. Two manufacturing processes are compared by making six experiments for each of the two processes. If each of the results obtained under one process is lower than those of any of the other and if there are no ties, do these results indicate a significant difference between the two processes at the 1% level?

11. The yields of two inbred lines of corn obtained from several different experiments are listed below. Is there a significant difference between the two lines of corn on the basis of the 1% level of significance? (Use the sign test).

Experiment	A Yield	B Yield	Experiment	A Yield	B Yield
1	48.6	50.1	15	39.8	40.8
2	43.0	48.6	16	37.5	37.3
3	41.0	42.9	17	28.1	27.5
4	39.1	42.6	18	33.3	32.4
5	36.8	37.5	19	43.0	48.6
6	40.8	41.3	20	29.0	31.1
7	28.9	38.6	21	27.4	28.0
8	35.9	37.3	22	28.0	28.7
9	39.2	42.6	23	28.3	28.8
10	47.8	46.1	24	26.8	26.8
11	42.1	43.4	25	42.2	42.2
12	47.6	48.2	26	38.9	39.1
13	26.4	26.3	27	39.0	39.4
14	30.6	31.7	28	33.6	34.0

12. A paired feeding experiment is carried out with pigs to determine the relative value of limestone and bonemeal for bone development. For each pair the ash content in percentage of scapulas is recorded first for pigs fed limestone (X) and then for pigs fed bonemeal (Y). If 35 such experiments are carried out, and if in 27 of them X is greater than Y and in 3 of them X is equal to Y, does this indicate that limestone yields higher ash content than bonemeal (5% level)?

13. In an experiment consisting of 25 pairs, the number of plus signs exceeds the number of minus signs. If a two-tailed test is applied, how many minus signs can there be, at most, to allow us to consider the results significant on the basis of the 5% level of significance?

14. In an experiment consisting of 30 pairs, most of the signs are minus signs. If our interest is in minus signs only (so that a one-tailed test would be applied), what is the largest possible number of plus signs for which the results would be considered significant on the basis of the 5% level of significance?

15. If in an experiment consisting of n pairs, 3 plus signs are obtained (all other signs being negative), no conclusions can be drawn if n is small, since in that case, 3 or fewer plus signs in n trials would by chance alone occur with probability greater than 5% (assuming a population in which half the signs are positive). Find the largest value of n for which, by chance alone, the probability of obtaining 3 or fewer plus signs in n trials is still larger than 5%. Use a one-tailed test.

16. Find the largest value of n such that, by chance alone, the probability of obtaining 4 or fewer plus signs (or minus signs) in n paired experiments (where no differences equal zero) is larger than 1%.

17. In a paired feeding experiment, the gains in weight (in pounds) of hogs fed on two different diets are as listed below. Can either diet be considered superior at the 5% level of significance? (Use the Wilcoxon test.)

Pair	Diet A	Diet B	Pair	Diet A	Diet B
1	27	19	6	31	34
2	30	32	7	39	34
3	31	21	8	34	28
4	34	33	9	33	26
5	23	27	10	35	26

18. Solve Exercise 17 of this chapter, assuming that in the data the values of 39 and 34 in pair 7 are interchanged, everything else remaining the same.

19. Solve Exercise 27, Chapter 10, assuming that the numbers 29 and 37 for the third and fourth workers in the last column (with coffee break) are replaced by 33 and 39, everything else remaining the same. Use the Wilcoxon test.

20. The manufacturer of a suntan lotion wishes to know whether a new ingredient increases the protection his lotion gives against sunburn. Seven volunteers have their backs exposed to the sun with the old lotion on one side and the new lotion on the other side of the spine. The experimental data on the degree of sunburn (as measured on some scale) for the seven volunteers are as follows:

Volunteer	Burn on side with old lotion	Burn on side with new lotion
1	42	38
2	51	53
3	31	36
4	61	52
5	44	33
6	55	49
7	48	36

Does this new ingredient improve the effectiveness of the lotion? (Use the Wilcoxon test.)

21. A random sample of six women agrees to be subjected to an experimental reducing diet. Their respective weights before and after the experiment are given as follows:

	Before	After
Mrs. Black	172	173
Mrs. White	138	130
Mrs. Brown	130	121
Mrs. Blue	166	169
Mrs. Green	139	132
Mrs. Grey	124	126

Does this indicate that the diet is effective at the 5% level of significance? (Use the Wilcoxon test.)

22. Solve Exercise 24, Chapter 10, by use of the Wilcoxon test.

23. Solve Exercise 26, Chapter 10, by use of the Wilcoxon test.

24. Although in Exercise 17 of this chapter the sample is small, solve this exercise by use of Theorem 11-3 to compare the result with the one obtained in Exercise 17.

25. Prove Formula (11-4).

12

Regression
and
Correlation

12.1 INTRODUCTION

In the preceding chapters we discussed only populations consisting of measurements of a single variable. We shall now consider populations made up of pairs of measurements.

Definition 12-1. If for every measurement of a variable X we know a corresponding value of a second variable Y, the resulting set of pairs of variates is called a *bivariate population*.

Two main questions will be discussed in this chapter:

1. Is there a relationship between the two variables of the bivariate population?
2. If such a relationship exists, how can it be expressed by means of an equation?

Let us consider the following example. The determination of the number of red blood cells by use of a microscope is time consuming and subject to considerable inaccuracies. On the other hand, the so-called packed cell volume of the blood is much easier to obtain. To find a possible relationship between these two variables, blood samples are taken from 10 dogs. For each of these samples the packed cell volume (X) and the corresponding red blood cell count (Y) are measured. The resultant data are presented in Table 12-1.

Table 12-1
Packed cell volume and red blood cell count of
10 dogs.

Packed cell volume (mm)	Red blood cell count (millions)
X	Y
45	6.53
42	6.30
56	9.52
48	7.50
42	6.99
35	5.90
58	9.49
40	6.20
39	6.55
50	8.72

Here we wish to decide whether the red blood cell count is related to the packed cell volume and, if such is the case, to find an equation that will express the red blood cell count in terms of the packed cell volume. The value of such an equation is obvious: It will allow us to estimate (or predict) the number of red blood cells from measurements of the packed cell volume.

Unless otherwise designated, the variable from whose values predictions are to be made will be denoted by X, and the variable whose values are to be estimated (or predicted) will be denoted by Y. In other words, X is considered the independent variable and Y the dependent variable. In the following discussion we shall assume that X can be measured exactly and that Y may be subject to fluctuations for a given value of X. (This situation would certainly obtain in the foregoing example, where for a given value of X—for example, 42—there may be several values of Y, owing to the inaccuracies inherent in a red blood cell count.) More precisely, we shall assume that the set of values of Y that can be obtained by performing the experiment repeatedly for the same value of X satisfies a normal distribution.

Let us suppose now that we are given a sample of *n* pairs of values, the first variate of each pair being the *X*-value, the second the *Y*-value.

To begin the study of a possible relationship between *X* and *Y*, let us represent the *n* pairs of values as *n* points on a graph.

Definition 12-2.   The graphical representation of a set of *n* pairs of values of *X* and *Y* in a coordinate system is called a *scatter diagram*.

Example 12-1.   Draw the scatter diagram for the data of Table 12-1.

Solution.   The desired scatter diagram is shown in Figure 12-1.

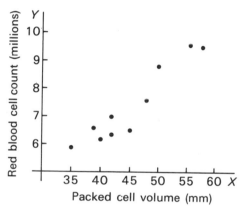

Figure 12-1
Scatter diagram for the data of Table 12-1.

Note that the 10 points of the scatter diagram lie nearly on a straight line. How do we find the equation of such a line? We shall answer this question and discuss the case where the scatter diagram indicates a linear relationship.

12.2   THE LINEAR REGRESSION EQUATION

In order to find the equation of the line that is a " best fit " to a scatter diagram of *n* points, we employ a theorem from elementary mathematics.

Theorem 12-1.   The equation of any nonvertical line (Figure 12-2) can be

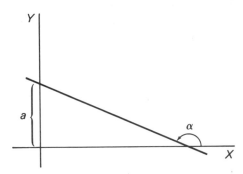

Figure 12-2
Graph of straight line showing Y-intercept a
and angle of inclination α.

written in the form $Y = a + bX$, where a is the Y-intercept and b the slope of
the line. The slope is given by $b = \tan \alpha$, where α is the angle measured from
the positive X-axis to the line, and is positive or negative as α is acute or
obtuse. Conversely, the graph of an equation of the form $Y = a + bX$ for any
values of the constants a and b is always a straight line.

Proof of Theorem 12-1.   Let us suppose we are given a line, such as that
in Figure 12-3, for which we know a, the Y-intercept, and α, the angle of

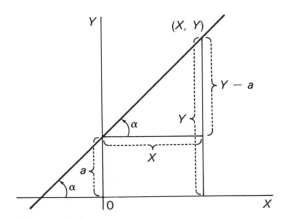

Figure 12-3
The relations between the coordinates of an
arbitrary point on a straight line and the Y-intercept
and angle of inclination.

inclination. The equation of a line is the relationship that the coordinates of any point must satisfy in order that the point lies on the line. Let the point (X, Y) be a point on the line; then from the upper right triangle in Figure 12-3 we have $b = \tan \alpha = (Y - a)/X$ or $bX = Y - a$, from which we obtain

$$Y = a + bX.$$

Conversely, if $Y = a + bX$, then its graph goes through the point $(0, a)$ and since $Y = a + bX$ can be written as $b = Y - a/X$, all points (X, Y) satisfying the equation are situated so that the ratio $Y - a/X$ is a constant, which means that all points must lie on a straight line through the point $(0, a)$ forming an angle α with the positive X-axis, where $\tan \alpha = b$.

Thus the graph of the equation $Y = -2 + X$ is a straight line, crossing the Y-axis 2 units below the X-axis and having a slope $b = \tan \alpha = 1$; since $\tan \alpha = 1$ when $\alpha = 45°$, the line forms an angle of $45°$ with the positive X-axis.

A particular line, therefore, is completely determined if the values of the constants a and b in its equation are given. Returning to the problem of finding the line of best fit to a scatter diagram of n points, we must determine a and b in such a way that the n points lie "as close to the line as possible" (the precise meaning of the phrase in quotes will be explained in Definition 12-3). Note that after the equation of the line of best fit has been determined, it will yield for each X-value a certain Y-value, which will be an estimate of the actual Y-value. To distinguish between these two Y-values, we shall denote the estimate, which is the value obtained from the line, by Y_e, and the actual Y-value, which is obtained by measurement, by Y. We can then write the equation of the line of best fit in the form

$$Y_e = a + bX. \tag{12-1}$$

In order to determine the line of best fit we must decide what is meant by this term, or more precisely, we need a definition of the term. At first thought we might define such a line as being one for which half the points of the scatter diagram fall below the line, the other half above it. This, however, would not yield a single line, since there could be many such lines. Instead, a definition of the term "line of best fit" should give a unique line. Let us proceed by considering, for each value of X, the absolute value $|Y - Y_e|$ of the difference between the actual Y-value and the estimated Y-value (see Figure 12-4). This difference represents the error committed by using the estimated Y-value instead of the actual value. One way to determine the line of best fit

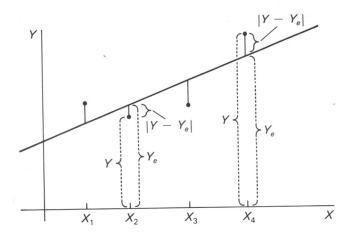

Figure 12-4
The differences between observed values, Y, and estimated
values Y_e.

might be to find that line for which the sum of these errors for all the given
values of X, that is, $\Sigma| Y - Y_e|$, has the smallest possible value. However,
this procedure, while yielding a unique line, leads to the mathematical difficulties customarily associated with the occurrence of absolute values. A
better method, and one which accomplishes the same aims, defines the line
in such a way that $\Sigma(Y - Y_e)^2$ has the smallest value. This method, called
the method of least squares, is the one most generally used in statistics for
obtaining the line of best fit.

Definition 12-3.   The *method of least squares* is the method of fitting a line
to a set of n points in such a way that

$$D = \Sigma(Y - Y_e)^2 \tag{12-2}$$

has its smallest possible value, where the sum is calculated for the given n
pairs of values of X and Y. The line thus obtained is called the *line of best
fit* or *the line of regression of Y on X*.

For the purposes of determining this line and in order to be able to discuss
how good a fit to a given set of n points the line of regression is, we need to
introduce some notation and another definition (Definition 12-4).

For the given sample of n values of (X, Y), we shall denote the standard
deviation of the X's by $\hat{s}_x$, so that

$$\hat{s}_x^2 = \frac{\Sigma(X - \bar{X})^2}{n} = \frac{\Sigma x^2}{n}, \tag{12-3}$$

where $x = X - \overline{X}$. Similarly, the standard deviation of the Y's shall be denoted by $\hat{s}_y$, so that

$$\hat{s}_y^2 = \frac{\Sigma(Y - \overline{Y})^2}{n} = \frac{\Sigma y^2}{n}. \tag{12-4}$$

The computing form for $\hat{s}_x^2$ is given by

$$\hat{s}_x^2 = \frac{\Sigma X^2}{n} - \overline{X}^2. \tag{12-5}$$

Proof of Formula (12-5). Using Formula (12-3) and the fact that $\overline{X} = (\Sigma X)/n$ or $\Sigma X = n\overline{X}$, we have

$$\hat{s}_x^2 = \frac{\Sigma(X - \overline{X})^2}{n} = \frac{\Sigma(X^2 - 2X\overline{X} + \overline{X}^2)}{n}$$

$$= \frac{\Sigma X^2 - 2\overline{X}\Sigma X + n\overline{X}^2}{n} = \frac{\Sigma X^2 - 2\overline{X}n\overline{X} + n\overline{X}^2}{n}$$

$$= \frac{\Sigma X^2 - 2n\overline{X}^2 + n\overline{X}^2}{n} = \frac{\Sigma X^2 - n\overline{X}^2}{n} = \frac{\Sigma X^2}{n} - \overline{X}^2.$$

Similarly the computing form for $\hat{s}_y$ is given by

$$\hat{s}_y^2 = \frac{\Sigma Y^2}{n} - \overline{Y}^2. \tag{12-6}$$

Definition 12-4. The *covariance* for a sample of n pairs of X, Y-values is defined as

$$\hat{s}_{xy} = \frac{\Sigma xy}{n}, \tag{12-7}$$

where, as before, $x = X - \overline{X}$ and $y = Y - \overline{Y}$.

The computing form for the covariance is

$$\hat{s}_{xy} = \frac{\Sigma XY}{n} - \overline{X}\,\overline{Y}. \tag{12-8}$$

Proof of Formula (12-8).   From (12-7), we have

$$\hat{s}_{xy} = \frac{\Sigma xy}{n} = \frac{\Sigma(X - \bar{X})(Y - \bar{Y})}{n}$$

$$= \frac{\Sigma(XY - X\bar{Y} - \bar{X}Y + \bar{X}\bar{Y})}{n} = \frac{\Sigma XY - \bar{Y}\Sigma X - \bar{X}\Sigma Y + n\bar{X}\bar{Y}}{n}$$

$$= \frac{\Sigma XY - \bar{Y}n\bar{X} - \bar{X}n\bar{Y} + n\bar{X}\bar{Y}}{n} = \frac{\Sigma XY - n\bar{X}\bar{Y}}{n}$$

$$= \frac{\Sigma XY}{n} - \bar{X}\bar{Y}.$$

We are now ready to find, for a given sample of n pairs of values of X and Y, the line of regression $Y_e = a + bX$, which means we need to determine the constants a and b in such a way that $\Sigma(Y - Y_e)^2$ is minimized. The answer is furnished by the following theorem:

Theorem 12-2.   The equation of the line of regression of Y on X fitted to a set of n points (X, Y) by the method of least squares is given by

$$Y_e = a + bX,$$

where

$$b = \frac{\Sigma XY - n\bar{X}\bar{Y}}{\Sigma X^2 - n\bar{X}^2} \tag{12-9}$$

and

$$a = \bar{Y} - b\bar{X}. \tag{12-10}$$

Proofs of Theorem 12-2 will follow Example 12-4.

Example 12-2.   Find the line of regression of Y on X for the data of Table 12-1.

Solution.   From Formula (12-3) it is evident that, for the calculation of b, it is necessary to find the values of XY and their sum, as well as the values of X^2 and their sum. These quantities are shown in the third and fourth columns, respectively, of Table 12-2. (The other columns of the table are not needed for the present calculations, but will be used later.)

Table 12-2
Calculation of regression line and related quantities for the data of Table 12-1.

1	2	3	4	5	6	7	8
X	Y	XY	X^2	Y^2	Y_e	$Y-Y_e$	$(Y-Y_e)^2$
45	6.53	293.85	2025	42.6409	7.28	−0.75	0.5625
42	6.30	264.60	1764	39.6900	6.75	−0.45	0.2025
56	9.52	533.12	3136	90.6304	9.22	0.30	0.0900
48	7.50	360.00	2304	56.2500	7.81	−0.31	0.0961
42	6.99	293.58	1764	48.8601	6.75	0.24	0.0576
35	5.90	206.50	1225	34.8100	5.52	0.38	0.1444
58	9.49	550.42	3364	90.0601	9.57	−0.08	0.0064
40	6.20	248.00	1600	38.4400	6.40	−0.20	0.0400
39	6.55	255.45	1521	42.9025	6.22	0.33	0.1089
50	8.72	436.00	2500	76.0384	8.16	0.56	0.3136
455	73.70	3441.52	21,203	560.3224			1.6220

Since

$$\bar{X}=\frac{455}{10}=45.50 \quad \text{and} \quad \bar{Y}=\frac{73.70}{10}=7.37,$$

using Formula (12-3), we obtain

$$b=\frac{3441.52-10(45.50)(7.37)}{21,203-10(45.50)^2}=\frac{88.17}{500.50}=0.176,$$

and from (12-4)

$$a=7.37-(0.176)(45.50)=7.37-8.01=-0.64;$$

thus the equation of the line of regression of Y on X takes the form

$$Y_e=-0.64+0.176X.$$

This equation can now be used in our original example to estimate the red blood cell count for various values of the packed cell volume. However, care should be taken to use the estimating equation only for such values of X as lie within the range of the X-values of the given sample; in our original example the estimating equation could be safely applied only for values of X between 35 and 58.

Example 12-3.   From the data of Table 21-1 estimate the red blood cell count for a packed cell volume reading of 52 millimeters.

Solution.   Using the estimating equation obtained in Example 12-2 for $X=52$, we find

$$Y_e=-0.64+(0.176)52=8.51.$$

In order to graph the line of regression we need determine only two points on the line. To facilitate drawing, the two points should be fairly far apart.

Example 12-4.   Draw the line of regression for the data of Table 12-1.

Solution.   Using the equation of the line of regression obtained in Example 12-2, we find for the two extreme X-values:

$$X = 35: \qquad Y_e = -0.64 + (0.176)35 = 5.52,$$
$$X = 58: \qquad Y_e = -0.64 + (0.176)58 = 9.57.$$

Connecting these two points by a straight line, we obtain the graph shown in Figure 12-5.

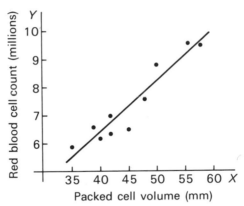

Figure 12-5
Regression line fitted to the data of Table 12-1.

Proof of Theorem 12-2.   Readers who are familiar with the methods of differential calculus will find it fairly easy to prove—and are urged to do so by use of partial differentiation—that a and b must be the solution of the following two linear equations.

$$\left.\begin{array}{l} an + b\Sigma X = \Sigma Y \\ a\Sigma X + b\Sigma X^2 = \Sigma XY \end{array}\right\} \qquad (12\text{-}11)$$

Definition 12-5.   Equations (12-11) are called *normal equations.*

Observe that for a given sample of n pairs of values of X and Y, all of the quantities occurring in (12-11), except a and b, will be known numbers (or numbers that can be calculated from the given data). The sums appearing in (12-11) extend over all pairs in the sample.

Solving Equations (12-11) for b gives (12-9); then, using this value of b in the first of the two Equations (12-11) and solving for a gives (12-10).

Since readers of this text are, however, not expected to be familiar with calculus, we shall now prove Theorem 12-2 by a method that does not require knowledge of calculus.

Alternate Proof of Theorem 12-2.   For any line, the difference between an observed Y-value and the corresponding Y-value on the line can be written as

$$
\begin{aligned}
Y - Y_e &= Y - (a + bX) \\
&= (Y - \bar{Y} + \bar{Y}) - a - b(X - \bar{X} + \bar{X}) \\
&= (y + \bar{Y}) - a - b(x + \bar{X}) \\
&= y - bx + (\bar{Y} - a - b\bar{X}).
\end{aligned}
$$

If, for purposes of simplification, we let $K = \bar{Y} - a - b\bar{X}$, we have

$$
\begin{aligned}
D = \Sigma(Y - Y_e)^2 &= \Sigma(y - bx + K)^2 \\
&= \Sigma(y - bx)^2 + 2K\,\Sigma(y - bx) + nK^2 \\
&= \Sigma(y - bx)^2 + nK^2,
\end{aligned} \tag{12-12}
$$

where the last step follows from the fact that $\Sigma(y - bx) = \Sigma y - b\Sigma x$, which equals 0, since both $\Sigma y = 0$ and $\Sigma x = 0$.

Consequently, D is the sum of two nonnegative terms. The minimum value of D is obtained, therefore, by making each of these two terms as small as possible. Let us first make the first term, $\Sigma(y - bx)^2$, as small as possible; that is, find the value of b that minimizes this term. To do this, we rewrite this term, by use of (12-4), (12-7), and (12-3) in that order, as follows:

$$
\begin{aligned}
\Sigma(y - bx)^2 &= \Sigma y^2 - 2b\Sigma xy + b^2 \Sigma x^2 \\
&= n\hat{s}_y^2 - 2bn\hat{s}_{xy} + b^2 n\hat{s}_x^2 \\
&= n\hat{s}_y^2 - 2bn\hat{s}_{xy} + b^2 n\hat{s}_x^2 + n\frac{\hat{s}_{xy}^2}{\hat{s}_x^2} - n\frac{\hat{s}_{xy}^2}{\hat{s}_x^2} \\
&= n\hat{s}_x^2 \left(b^2 - 2b\frac{\hat{s}_{xy}}{\hat{s}_x^2} + \frac{\hat{s}_{xy}^2}{\hat{s}_x^4} \right) + n\hat{s}_y^2 - n\frac{\hat{s}_{xy}^2}{\hat{s}_x^2} \\
&= n\hat{s}_x^2 \left(b - \frac{\hat{s}_{xy}}{\hat{s}_x^2} \right)^2 + n\hat{s}_y^2 - n\frac{\hat{s}_{xy}^2}{\hat{s}_x^2}.
\end{aligned} \tag{12-13}
$$

Only the first of these terms contains b and, since it is nonnegative, it will be as small as possible when it is 0; that is,

$$b = \frac{\hat{s}_{xy}}{\hat{s}_x^2}. \tag{12-14}$$

But Formula (12-14) can be rewritten by use of (12-8) and (12-5) as

$$b = \frac{\hat{s}_{xy}}{\hat{s}_x^2} = \frac{\dfrac{\Sigma XY}{n} - \bar{X}\,\bar{Y}}{\dfrac{\Sigma X^2}{n} - \bar{X}^2} = \frac{\Sigma XY - n\bar{X}\,\bar{Y}}{\Sigma X^2 - n\bar{X}^2},$$

which proves (12-9).

Now, for this value of b, nK^2 can be made as small as possible by setting $K = \bar{Y} - a - b\bar{X} = 0$; that is,

$$a = \bar{Y} - b\bar{X},$$

which proves (12-10).

The minimum value of D, which we shall denote by $D_{\min}$, is obtained by substituting the above values of a and b into (12-12); note that $K = 0$ then, and that the first term on the right-hand side is given by (12-13), so that

$$D_{\min} = n\hat{s}_y^2 - n\frac{\hat{s}_{xy}^2}{\hat{s}_x^2} = n\hat{s}_y^2\left(1 - \frac{\hat{s}_{xy}^2}{\hat{s}_x^2\,\hat{s}_y^2}\right). \tag{12-15}$$

If we let

$$r = \frac{\hat{s}_{xy}}{\hat{s}_x\,\hat{s}_y}, \tag{12-16}$$

the expression for $D_{\min}$ can be written in the simple form

$$D_{\min} = n\hat{s}_y^2(1 - r^2). \tag{12-17}$$

One of the important properties of the line of regression is that it goes through the point for which $X = \bar{X}$ and $Y = \bar{Y}$. This is seen to be true for the line of regression obtained in Example 12-2, since for $X = \bar{X} = 45.50$, we find

$$Y_e = -0.64 + (0.176)(45.50) = 7.37 = \bar{Y}.$$

We can easily prove that the point $(\bar{X}, \bar{Y})$ will always lie on the regression line. Substituting the value of a from (12-10) into the equation of the line of regression (12-1), we can write

$$Y_e = (\bar{Y} - b\bar{X}) + bX; \tag{12-18}$$

thus for $X = \bar{X}$ we obtain $Y_e = \bar{Y}$, which can also be seen directly from (12-10).

We shall now establish a formula that is of fundamental importance in regression analysis. To do this we need the following definitions.

Definition 12-6.   The *sum of squares for regression* is defined as

$$S.S.R. = \Sigma(Y_e - \bar{Y})^2. \tag{12-19}$$

It is a measure of the dispersion of the predicted (or estimated) Y-values, Y_e, on the regression line about the mean, $\bar{Y}$, of the sample of Y-values. It can also be written in the form

$$S.S.R. = n\hat{s}_y^2 r^2. \tag{12-20}$$

Proof of Formula (12-20).   From (12-18), it follows that $Y_e = \bar{Y} + b(X - \bar{X})$ $= \bar{Y} + bx$, so that we can write

$$S.S.R. = \Sigma(Y_e - \bar{Y})^2 = \Sigma(\bar{Y} + bx - \bar{Y})^2 = \Sigma(bx)^2 = b^2\Sigma x^2 = b^2 n\hat{s}_x^2.$$

From (12-16), it follows that

$$r\hat{s}_y = \frac{\hat{s}_{xy}}{\hat{s}_x} = b\hat{s}_x; \tag{12-21}$$

thus, substituting this expression for $b\hat{s}_x$ into the one just obtained for $S.S.R.$ proves (12-20).

Definition 12-7.   The *sum of squares for error* for a regression line is defined as

$$S.S.E. = \Sigma(Y - Y_e)^2. \tag{12-22}$$

It is a measure of the dispersion of the observed values, Y, about their corresponding values on the regression line, Y_e. Its value is clearly the same as $D_{\min}$ as given by (12-15) or (12-17), so that we can also write

$$S.S.E. = n\hat{s}_y^2(1 - r^2). \tag{12-23}$$

Let us now introduce the additional notation

$$S.S. = \Sigma(Y - \bar{Y})^2, \tag{12-24}$$

which is a measure of the dispersion of the observed values, Y, about their mean. By use of (12-4) we can write (12-24) as

$$S.S. = n\hat{s}_y^2 . \tag{12-25}$$

Adding Equations (12-20) and (12-23), we obtain

$$S.S.R. + S.S.E. = n\hat{s}_y^2 r^2 + n\hat{s}_y^2(1 - r^2) = n\hat{s}_y^2 = S.S.,$$

so that

$$S.S. = S.S.R. + S.S.E. \tag{12-26}$$

This relationship shows that the dispersion of the Y-values about their mean is equal to the sum of the dispersion of the predicted Y-values about that mean and the dispersion of the actual Y-values about their corresponding values on the regression line.

12.3 THE STANDARD ERROR OF ESTIMATE

We shall now determine whether or not a linear relationship can be said to exist between two variables. Sometimes, as we saw in Example 12-1, the scatter diagram shows fairly clearly the existence of a linear relationship; but if the points are widely scattered about the line of regression, there may be considerable doubt that the assumption of a linear relationship is warranted. We must therefore define a quantity that measures the spread of a set of points about the line of regression. Two such measures are in common use: the standard error of estimate, and the correlation coefficient. In practice, the latter is more widely employed.

For a population or a sample consisting of measurements of a single variable, the standard deviation is used in measuring the spread of the variates around the mean. A similar measure, which indicates the spread of the given points around the fitted line of regression, is called the standard error of estimate.

Definition 12-8. The *standard error of estimate* for a regression line fitted to a sample of n pairs of X, Y-values is defined as

$$\hat{s}_e = \sqrt{\frac{\Sigma(Y - Y_e)^2}{n}} . \tag{12-27}$$

It is evident that the nearer the points lie to a straight line, the smaller is the value of the standard error of estimate. A value of $\hat{s}_e = 0$ means that all points lie on a straight line. Although the standard error of estimate ordinarily is defined only for the regression line, i.e. the line obtained by the method of least squares, Definition 12-8 could be used to define the standard error of estimate for other lines fitting the given n points. From Equation (12-27) it is evident that the regression line is that line among all lines fitting the given n points for which the standard error of estimate has its smallest possible value, since the regression line is determined by minimizing $\Sigma(Y - Y_e)^2$ and consequently $\hat{s}_e$.

Example 12-5.   Find the standard error of estimate for the data of Table 12-1.

Solution.   Using the equation of the line of regression obtained in Example 12-2, we calculate for each of the given X-values the corresponding value of Y_e; see the sixth column of Table 12-2. The values of $Y - Y_e$ and $(Y - Y_e)^2$ are listed in the seventh and eighth columns, respectively, of that table. Then

$$\hat{s}_e = \sqrt{\frac{1.6220}{10}} = \sqrt{0.162} = 0.40.$$

Since the standard error of estimate is the standard deviation of the errors of estimation, it has the property that, provided certain conditions (which we shall not state here) are satisfied, about two-thirds of the errors $|Y - Y_e|$

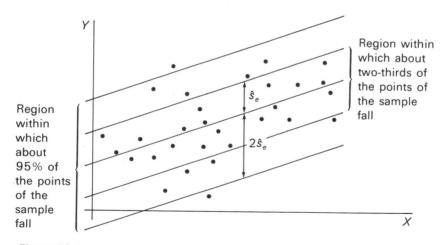

Figure 12-6

Relation between regression line, points of scatter diagram, and standard error of estimate for large samples.

will be less than the standard error of estimate, and about one-third of the errors will exceed it. Thus two lines drawn parallel to the line of regression and at a vertical distance of $\hat{s}_e$ from it will define a region within which about two-thirds of the points of the given sample will fall. This statement is valid only if n is fairly large. Furthermore, under the same conditions about 95% of the points will fall within a region bounded by two lines drawn parallel the line of regression at a vertical distance of $2\hat{s}_e$ from it.

These regions are illustrated in Figure 12-6, the points representing a sample of 30 pairs of X, Y-values.

12.4   THE CORRELATION COEFFICIENT

In studying the possibility of a relationship between X and Y, if a sample of n pairs is given, construction of the scatter diagram for the data is useful in suggesting the existence and nature of any such relationship. Such a scatter diagram is shown in Figure 12-7. If new axes are drawn through the point $O'(\overline{X}, \overline{Y})$ as origin, parallel to the original axes, the points of the scatter diagram will be distributed fairly uniformly below and above and to the right and left of the new origin. In Figure 12-7 we see that the coordinates X and

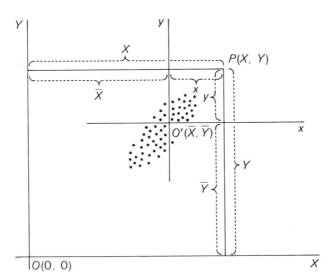

Figure 12-7
Scatter diagram with X, Y and x, y coordinate systems.

Y of any point P, when related to the new axes, become x and y, where $x = X - \overline{X}$ and $y = Y - \overline{Y}$.

Figure 12-8 shows the signs of the product xy in the four quadrants. If in

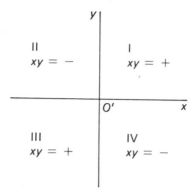

Figure 12-8

Signs of xy in quadrants of x, y coordinate system.

each quadrant the number of points of the scatter diagram is approximately the same, no linear relationship is indicated, and the sum of the products Σxy will be numerically small. If, however, the points follow a linear trend and occur mostly in quadrants I and III, the positive products will predominate, and Σxy will be positive; if most of the points fall in quadrants II and IV, Σxy will be negative. It is evident therefore that Σxy and likewise the mean product $(1/n)\Sigma xy$ are indicators of a linear relationship between the values of X and Y, and that the larger Σxy is in absolute value the closer the points lie to a straight line and the stronger is the evidence of a linear relationship. However, the quantities Σxy and $(1/n)\Sigma xy$ have units, and their magnitudes will depend upon the units in which X and Y are measured. If we want to use these quantities in making comparisons, we must free them of this disadvantage. To free them of units we shall proceed as we did for the z-value used in applications of the normal curve where deviations from the mean were expressed in terms of the standard deviation. If we take the unit of measurement in the x direction as $\hat{s}_x$ and that in the y direction as $\hat{s}_y$, the expression for the average product becomes

$$\frac{1}{n} \Sigma \frac{x}{\hat{s}_x} \cdot \frac{y}{\hat{s}_y} = \frac{\Sigma xy}{n\hat{s}_x \hat{s}_y} = \frac{\Sigma xy}{\sqrt{\Sigma x^2 \Sigma y^2}},$$

which is called the coefficient of correlation. This is the most frequently used measure for determining the usefulness of a regression line for estimating purposes.

Definition 12-9. The *correlation coefficient* for a sample of n pairs of X, Y-values is defined as

$$r = \frac{\Sigma xy}{\sqrt{\Sigma x^2 \Sigma y^2}}. \tag{12-28}$$

Dividing numerator and denominator of the formula for r by n, we see that Formula (12-28) is equivalent to the one previously given for r, namely (12-16),

$$r = \frac{\hat{s}_{xy}}{\hat{s}_x \hat{s}_y}.$$

For purposes of computation (in order to avoid the use of deviations) it is best to write r in either of the two forms:

$$r = \frac{\Sigma XY - \dfrac{\Sigma X \Sigma Y}{n}}{\sqrt{\left[\Sigma X^2 - \dfrac{(\Sigma X)^2}{n}\right]\left[\Sigma Y^2 - \dfrac{(\Sigma Y)^2}{n}\right]}}; \tag{12-29}$$

$$r = \frac{\Sigma XY - n\bar{X}\,\bar{Y}}{\sqrt{(\Sigma X^2 - n\bar{X}^2)(\Sigma Y^2 - n\bar{Y}^2)}}.$$

The proof that both forms of (12-29) are equivalent to (12-16) follows immediately from use of the computing forms for $\hat{s}_{xy}$, $\hat{s}_x$, and $\hat{s}_y$, as given by (12-8), (12-5), and (12-6), respectively.

Example 12-6. Calculate the correlation coefficient for the data of Table 12-1.

Solution. Using the second of the two alternate forms of r given by (12-29), we observe that the only values needed, in addition to those already calculated in finding the slope of the regression line b (see Example 12-2), are the values of Y^2 and their sum, which are given in the fifth column of Table 12-2. Then

$$r = \frac{3441.52 - 10(45.50)(7.37)}{\sqrt{[21{,}203 - 10(45.50)^2][560.3224 - 10(7.37)^2]}}$$

$$= \frac{88.17}{\sqrt{(500.50)(17.153)}} = \frac{88.17}{\sqrt{8585.08}} = \frac{88.17}{92.66} = 0.952.$$

In order to interpret the meaning of the correlation coefficient, we shall show that there exist three formulae relating the correlation coefficient to, respectively, the standard error of estimate, the slope of the line of regression, and the sum of squares of regression; namely

$$\hat{s}_e = \hat{s}_y\sqrt{1 - r^2}; \tag{12-30}$$

$$r = b\frac{\hat{s}_x}{\hat{s}_y}; \tag{12-31}$$

$$r^2 = \frac{S.S.R.}{S.S.}. \tag{12-32}$$

Proof of Formula (12-30). Dividing both sides of (12-17) by n and recalling that D_{min} is equal to the value of $\Sigma(Y - Y_e)^2$ for the regression line obtained by the method of least squares, we have

$$\frac{D_{min}}{n} = \frac{\Sigma(Y - Y_e)^2}{n} = \hat{s}_e^2 = \hat{s}_y^2(1 - r^2),$$

from which (12-30) follows by taking the square root of both sides.

Formula (12-31) follows directly from (12-21) by dividing the first and last terms of the equation by $\hat{s}_y$.

The proof of Formula (12-32) is left as an exercise (see Exercise 3).

Since r^2 appears in many other formulae, as well as in Formula (12-32), it is sometimes given a special name.

Definition 12-10. The square of the correlation coefficient is defined as the *coefficient of determination.*

It can be seen from Formula (12-32) that the coefficient of determination is the proportion of the total variation (or variance) in Y that can be explained by the linear relationship existing between X and Y. When multiplied by 100, the proportion is converted to a percent. Thus, the correlation coefficient of 0.952 found in Example 12-6 indicates that $(0.952)^2 \cdot 100 = 90.6\%$ of the variation in red blood cell counts is due to the linear relationship between the packed cell volume and the red blood cell count; the rest of the variation is due to unexplained factors, and is called the experimental error. Note that this interpretation in terms of percentages applies only to the variance of the Y's, not to the standard deviation of the Y's.

From (12-30), (12-31), and (12-32), we can conclude that the correlation coefficient satisfies the following properties:

1. If $r = +1$ or $r = -1$, then all points of the scatter diagram lie on a straight line. This follows from (12-30) since, if $r = +1$, then $\hat{s}_e = 0$, which, as we saw earlier, means that all points lie on a straight line. Conversely, if all points lie on a straight line, that is, $\hat{s}_e = 0$, then $r^2 = 1$ and r must be $+1$ or -1. The same conclusion can also be drawn from (12-32), since, if $r = \pm 1$, it follows that $S.S.R. = S.S.$, so that $S.S.E. = 0$, and all points must lie on a straight line; and conversely, if all points lie on a straight line, then $S.S.E = 0$.

2. If $r = 0$, then the regression line becomes the horizontal line $Y = \overline{Y}$, since it follows from (12-31) that b must be 0, as $\hat{s}_x$ is never 0. This means no linear relationship exists between the X- and Y-values, and for *any* value of X, the same value of Y, namely $\overline{Y}$, is the estimated Y-value. Conversely, if there is no linear relationship, that is, $Y = \overline{Y}$ is the line of best fit, then $b = 0$ and consequently also $r = 0$. The same conclusion can also be drawn from (12-32), since, for $r = 0$, $S.S.R. = 0$, hence no linear relationship exists.

3. In any case r is always between -1 and $+1$. This follows from (12-30), since the number under the square root sign, namely, $1 - r^2$, can never be negative; that is, r^2 can never exceed 1, so r itself is always between -1 and $+1$. The same conclusion also follows from (12-32), since the ratio given in that formula can never exceed 1 or be less than 0, so r^2 is always between 0 and 1.

4. From the foregoing it also follows that the closer r is to $+1$ or -1, the better the correlation is. There remains the question of what conclusion can be drawn concerning a possible linear correlation between X and Y if r is some value between 0 and $+1$ or between 0 and -1. This will be answered in Section 12.5.

5. From (12-31) it follows that r is positive if and only if b is positive, since $\hat{s}_x$ and $\hat{s}_y$, being standard deviations, are always nonnegative. Similarly, r will be negative if and only if b is negative. Thus, a positive r indicates that, since $b = \tan \alpha$ (see Figure 12-2), the regression line forms an acute angle with the positive X-axis, or that Y increases with increasing X; a negative r indicates that the regression line forms an obtuse angle with the positive X-axis, or that Y decreases with increasing X.

Figures 12-9, 12-10, and 12-11 show examples of $r = -1$, r slightly less than 1, and $r \cong 0$.

Figure 12-9
Diagram of $r = -1$.

Figure 12-10
Diagram of r slightly less
than 1.

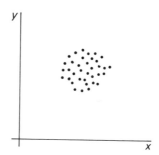

Figure 12-11
Diagram of r approximately
equal to 0.

12.5 SIGNIFICANCE OF A CORRELATION COEFFICIENT

For values of r close to zero, there is no linear relationship; for values of r near ± 1, a very strong linear relationship exists. We must yet formulate precise conclusions for intermediate values of r. We may do this by considering the bivariate population from which the sample of n pairs is taken. Let us denote the correlation coefficient of the population by ρ and assume the population to be normal; i.e. for each X all values of Y that are obtained satisfy a normal distribution. If we take all possible samples of n pairs from this population and consider the distribution of the correlation coefficients of these samples, we find that this distribution is not one of the well-known distributions we have studied. To illustrate this, the distributions of r for various sample sizes when $\rho = 0$, 0.50, and 0.90 are shown in Figures 12-12, 12-13, and 12-14.

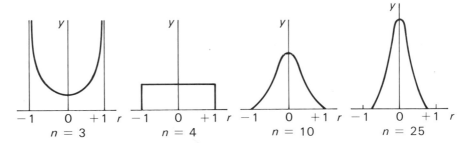

Figure 12-12
The distribution of r for $\rho = 0$.

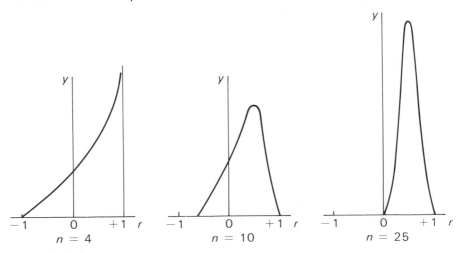

Figure 12-13
The distribution of r for $\rho = 0.50$.

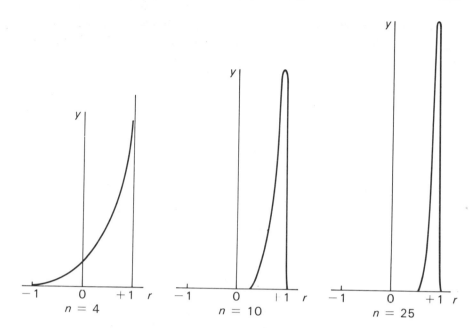

Figure 12-14
The distribution of r for $\rho = 0.90$.

These figures show that the distribution of r assumes widely different forms for various values of ρ and n. As we shall see, the study of the case where $\rho = 0$ (Figure 12-12) is the most important study for practical purposes. Figure 12-12 shows what at first might seem paradoxical: for $n = 3$ the curve is U-shaped. This shows that when a sample of size 3 is selected from a normal bivariate population in which there is no linear correlation the chance of obtaining a fairly high correlation coefficient for the sample is greater than the chance of obtaining a small correlation coefficient. For samples of size 4, the fact that the curve is a horizontal straight line shows that the chance of obtaining a particular correlation coefficient is equal to that of obtaining any other.

Considering now the special case where $\rho = 0$, it can be shown that, while r itself does not satisfy one of the distributions studied so far, the quantity

$$\frac{r}{\sqrt{\dfrac{1-r^2}{n-2}}}$$

satisfies a Student's t-distribution. This results in the following theorem.

Theorem 12-2.   If from a normal bivariate population with correlation coefficient $\rho = 0$ all samples of size n are taken and their correlation coefficients are denoted by r, then

$$t = \frac{r}{\sqrt{\dfrac{1 - r^2}{n - 2}}} \qquad\qquad (12\text{-}33)$$

satisfies a Student's t-distribution with $n - 2$ degrees of freedom.

The number of degrees of freedom is taken as $n - 2$, since for a regression line the constants a and b are fixed; i.e. there exist two independent relationships involving the n pairs of values of X and Y.

In order to decide whether, for a sample of size n, a given correlation coefficient indicates a linear relationship, we test the hypothesis that the sample is chosen from a population for which $\rho = 0$ and, therefore, determine the probability that from such a population a sample of size n is taken for which the correlation coefficient equals or exceeds the absolute value of r calculated for the given sample. Here, we use a two-tailed test. If the probability is less than 5% (or whatever other level of significance is chosen), we reject the hypothesis that the sample is taken from a population in which there is no linear relationship. The value of r then indicates that the X- and Y-values can be assumed to be linearly related.

Example 12-7.   In a sample of 42 pairs of X, Y-values the correlation coefficient is found to be 0.220. Does this indicate a significant correlation on the basis of the 5% level of significance?

Solution.   We are testing here the hypothesis that $\rho = 0$. Using Theorem 12-2, we find

$$t = \frac{0.220}{\sqrt{\dfrac{1 - (0.220)^2}{40}}} = \frac{0.220}{0.154} = 1.43.$$

For $\nu = 40$, $t_{.05} = 2.021$.

Since $t = 1.43 < 2.021$, the correlation coefficient of 0.220 cannot be considered significant. This means that from a population with $\rho = 0$, samples of 42 pairs with a correlation coefficient of 0.220 or larger or -0.220 or less are obtained rather frequently—more than 5% or the time (actually, as we see from Appendix Table III, more than 10% of the time).

Since Theorem 12-2 is applicable only where $\rho = 0$, it can be used only to decide whether in a particular sample the correlation coefficient is significantly different from zero; it cannot be used to establish confidence limits for ρ.

Where ρ does not equal zero, even for larger values of n the distributions do not assume a bell-shaped form but are decidedly skewed, as shown in Figures 12-13 and 12-14. Indeed, the equation of the distribution curve of r where ρ is different from zero is so complicated it will not be given here. On the other hand, R. A. Fisher found that if instead of r the quantity Z is considered, Z being defined in terms of r as

$$Z = \frac{1}{2} \ln \frac{1+r}{1-r},\qquad(12\text{-}34)$$

where "ln" means "logarithm to the base e," its distribution will be approximately normal. Thus we have the following theorem.

Theorem 12-3. If from a bivariate population with correlation coefficient ρ all samples of size n are taken, then

$$z = \frac{Z - m_Z}{\sigma_Z}\qquad(12\text{-}35)$$

satisfies an approximately normal distribution with

$$m_Z = \frac{1}{2} \ln \frac{1+\rho}{1-\rho},\qquad \sigma_Z = \frac{1}{\sqrt{n-3}}.\qquad(12\text{-}36)$$

The values of Z corresponding to various positive values of r, as calculated from (12-34), are given in Appendix Table VI. To find the values of Z corresponding to negative values of r, we note that for $r = -|r|$, where $|r|$ is the absolute value of r and, therefore, positive, we have

$$Z = \frac{1}{2} \ln \frac{1+r}{1-r} = \frac{1}{2} \ln \frac{1-|r|}{1+|r|}$$

$$= \frac{1}{2} [\ln(1-|r|) - \ln(1+|r|)]$$

$$= -\frac{1}{2} [\ln(1+|r|) - \ln(1-|r|)]$$

$$= -\frac{1}{2} \ln \frac{1+|r|}{1-|r|},$$

so that the value of Z corresponding to $r = -|r|$ is the negative of the value of Z corresponding to the absolute value of r.

Example 12-8.   In a sample of 35 pairs the correlation coefficient is found to be 0.80. Use the 5% level of significance to test the hypothesis that $\rho = 0.90$.

Solution.   In order to test this hypothesis, we determine the probability that from a population with $\rho = 0.90$ a sample with correlation coefficient $r = 0.80$ or less is taken (more precisely, we take twice this probability, since we apply a two-tailed test); if the probability exceeds 5%, we accept the hypothesis; otherwise, we reject it.

From Appendix Table VI, we find $Z = 1.099$ for $r = 0.80$. Appendix Table VI can also be used to find the value of m_z, since the procedure for finding this value, as given by (12-36), is identical to that for finding Z from r, as given by (12-34). From Appendix Table VI, we find $m_z = 1.472$ for $\rho = 0.90$. Furthermore,

$$\sigma_z = \frac{1}{\sqrt{32}} = 0.177;$$

thus,

$$z = \frac{1.099 - 1.472}{0.177} = -2.11.$$

From Appendix Table I we find that the area under the normal probability curve to the left of $z = -2.11$ is

$$A = 0.0174;$$

hence, the desired probability has a value of

$$P = 2(0.0174) = 0.0348,$$

which is less than 5%. Therefore, at the 5% level of significance, we reject the hypothesis that the sample with correlation coefficient $r = 0.80$ is taken from a population with $\rho = 0.90$.

Note that the probability found here refers to the chance of obtaining a value of z less than or equal to -2.11 or greater than or equal to 2.11. The value of r corresponding to $z = -2.11$ is, of course, 0.80; however, the value of r corresponding to $z = 2.11$ is not 1, as a superficial inspection might indicate, but is that value of r for which the corresponding Z satisfies the equation

$$\frac{Z - 1.472}{0.177} = 2.11,$$

from which we find $Z = 1.845$ and $r = 0.95$. Thus, the value of $P = 0.0348$ obtained above represents the probability that from a population with correlation coefficient 0.90 a sample of 35 pairs is taken for which the correlation coefficient is 0.80 or less or 0.95 or more.

Theorem 12-3 now enables us to set up confidence limits for ρ when a sample correlation coefficient is known.

Example 12-9.   In a sample of 52 pairs the correlation coefficient is found to be 0.53. Find the 95% confidence limits for the correlation coefficient of the population.

Solution.   From Appendix Table VI we find $Z = 0.590$ for $r = 0.53$. Furthermore,

$$\sigma_z = \frac{1}{7} = 0.143.$$

Since, for the 95% confidence limits, $z = \pm 1.96$, we have

$$\frac{Z - m_z}{0.143} = \pm 1.96.$$

Hence,

$$Z - m_z = \pm 0.280,$$

and

$$m_z = Z \pm 0.280 = 0.590 \pm 0.280.$$

Thus, m_z is between 0.31 and 0.87, and, correspondingly, ρ is between 0.30 and 0.70, the latter two values being obtained from Appendix Table VI for Z-values of 0.31 and 0.87, respectively.

Why are the confidence limits for ρ, 0.30 and 0.70, not symmetric with respect to the given value of $r = 0.53$? The answer is that the distribution of r, shown in Figures 12-13 and 12-14, is not symmetric but skewed, except when $\rho = 0$. Thus, whenever confidence limits for ρ are determined from a given positive value of r, the upper confidence limit will be closer to r than will be the lower one. If confidence limits are determined from a negative value of r, the lower confidence limit will be closer to r than will be the upper one.

Clearly, Theorem 12-3 can be applied even when $\rho = 0$. But since Theorem 12-3 involves a distribution that is only approximately normal, whenever the hypothesis $\rho = 0$ is to be tested Theorem 12-2 should be used.

In Chapter 8 we stated without proof the formula for the standard deviation of the differences between sample means from two independent populations—see Formula (8-11). We are now in a position to state a much more general formula.

The standard deviation of the difference between sample means from two populations, i.e. the differences shown in Figure 8-3, is given by

$$\sigma_{\bar{x}-\bar{y}} = \sqrt{\sigma_{\bar{x}}^2 - 2\rho\sigma_{\bar{x}}\sigma_{\bar{y}} + \sigma_{\bar{y}}^2}, \qquad (12\text{-}37)$$

where ρ is the correlation coefficient between the $\bar{X}$ and $\bar{Y}$; that is,

$$\rho = \frac{\Sigma(\bar{X} - m_{\bar{x}})(\bar{Y} - m_{\bar{y}})}{N\sigma_{\bar{x}}\sigma_{\bar{y}}.} \qquad (12\text{-}38)$$

Proof of Formula (12-37).

$$\sigma_{\bar{x}-\bar{y}}^2 = \frac{1}{N}\{\Sigma[(\bar{X} - \bar{Y}) - m_{\bar{x}-\bar{y}}]^2\} = \frac{1}{N}\{\Sigma[(\bar{X} - \bar{Y}) - (m_{\bar{x}} - m_{\bar{y}})]^2\}$$

$$= \frac{1}{N}\{\Sigma[(\bar{X} - m_x) - (\bar{Y} - m_{\bar{y}})]^2\}$$

$$= \frac{1}{N}\Sigma(\bar{X} - m_{\bar{x}})^2 - \frac{2}{N}\Sigma(\bar{X} - m_{\bar{x}})(\bar{Y} - m_{\bar{y}}) + \frac{1}{N}\Sigma(\bar{Y} - m_{\bar{y}})^2$$

$$= \sigma_{\bar{x}}^2 - 2\frac{\Sigma(\bar{X} - m_{\bar{x}})(\bar{Y} - m_{\bar{y}})}{N\sigma_{\bar{x}}\sigma_{\bar{y}}}\sigma_{\bar{x}}\sigma_{\bar{y}} + \sigma_{\bar{y}}^2 = \sigma_{\bar{x}}^2 - 2\rho\sigma_{\bar{x}}\sigma_{\bar{y}} + \sigma_{\bar{y}}^2.$$

Taking the square root of both sides of this equation results in Formula (12-37).

If two populations are independent, there exists no relationship between X and Y and consequently none between $\bar{X}$ and $\bar{Y}$; thus $\rho = 0$, and (12-37) becomes

$$\sigma_{\bar{x}-\bar{y}} = \sqrt{\sigma_{\bar{x}}^2 + \sigma_{\bar{y}}^2}, \qquad (12\text{-}39)$$

which is identical to Formula (8-11).

12.6. THE LINEAR REGRESSION EQUATION WHEN VARIABLES ARE CHANGED

In many cases the calculations required to find a regression line can be simplified by a change of variable; for example, if we introduce new variables x and y given by

$$x = X - \bar{X}, \qquad y = Y - \bar{Y},$$

each point (X, Y) in the X, Y-coordinate system becomes a point (x, y) in a new x, y-coordinate system (see Figure 12-7).

The equation of the regression line (12-1) then becomes

$$y_e = bx, \tag{12-40}$$

since for the new variables x and y, we have $\bar{x} = 0$ and $\bar{y} = 0$, so that $a = 0$.

The formula for b, as obtained from (12-9), becomes

$$b = \frac{\Sigma xy}{\Sigma x^2}. \tag{12-41}$$

Sometimes, it is desirable just to introduce x as a new variable, but to keep Y. Such a change of variables is the most convenient whenever all consecutive X-values of the given sample differ by the same amount, in which case the mean $\bar{X}$ may be found by inspection. In practice, this situation arises oftenest when the X's are consecutive years, months, or other periods of time. Since again $\bar{x} = 0$, the values of a and b in the regression line

$$Y_e = a + bX$$

are given by

$$b = \frac{\Sigma xY}{\Sigma x^2}, \qquad a = \bar{Y}. \tag{12-42}$$

If, for example, the X-values are the years 1950, 1951, 1952, 1953, 1954, 1955, 1956, letting $x = X - 1953$ replaces these values by the much simpler numbers $-3, -2, -1, 0, 1, 2, 3$. In the corresponding regression line the x's are measured with 1953 as origin. Whenever the X-values are an odd number of consecutive years (or other consecutive time periods), x will be measured from the year in the center; for an even number of years, x will be measured from the midpoint between the two years in the center. This procedure of replacing the X's by x's is advantageous only in the situations described above; in other types of problems such substitution frequently leads to unnecessary complications.

12.7   COMPARISON OF REGRESSION LINE OF *Y* ON *X* WITH THAT OF *X* ON *Y*

If a sample of n pairs of X, Y-values is given, there are occasions when either X or Y can be considered the independent variable. For example, from the data of Table 12-1 we may wish to determine a regression line from which to predict packed cell volumes from given red blood cell counts—in Example 12-2 we estimated red blood cells counts from given packed cell volumes. Also, it would be of interest to compare these two regression lines. When Y is considered the independent variable, the regression line can be written in the form

$$X_e = c + dY, \tag{12-43}$$

where c and d play the roles of a and b in the original line and are obtained by interchanging the X's and Y's in the formulae previously given for a and b. Formula (12-43), then, is the result of minimizing $\Sigma(X - X_e)^2$ rather than $\Sigma(Y - Y_e)^2$.

When the two lines of regression of Y on X and X on Y are graphed, will the graphs of these two lines ever coincide? This question is answered by Theorem 12-4.

Theorem 12-4.   The graphs of the regression lines of Y on X and X on Y are identical if and only if all points of the given sample lie on a straight line.

Proof of Theorem 12-4.   If the graphs of the two lines are identical, the equations of the regression lines given by Formulae (12-43) and (12-1) must be the same. Solving $Y_e = a + bX$ for X, we obtain

$$X = -\frac{a}{b} + \frac{1}{b}Y_e.$$

This equation will be identical to $X_e = c + dY$ if corresponding coefficients are the same; thus

$$\frac{1}{b} = d.$$

Substituting for b and d in this equation the value of b given by (12-14) and the value of d obtained from (12-14) by interchanging the roles of x and y, we obtain

$$\frac{\hat{s}_x^2}{\hat{s}_{xy}} = \frac{\hat{s}_{xy}}{\hat{s}_y^2},$$

from which it follows that

$$\frac{\hat{s}_{xy}^2}{\hat{s}_x^2 \hat{s}_y^2} = 1$$

or $r^2 = 1$.

This means that all points lie on a straight line.

12.8 LINES OF REGRESSION THROUGH THE ORIGIN

Often the conditions of a practical problem require that the line of regression go through the origin. For example, if X represents time measured from the germination of a plant, and Y its height, then the value of Y for $X = 0$ must equal zero. This is true also for the data of Table 12-1, since we know that for a packed cell volume of zero the red blood cell count must equal zero.

If, then, the hypothesis is made that the line of regression goes through the origin, it can be shown either by differential calculus or algebraically as in the proof of Theorem 12-2 that the normal equations reduce to a single equation that is identical to the second of the Equations (12-11) with $a = 0$. Thus we obtain immediately

$$b = \frac{\Sigma XY}{\Sigma X^2}. \tag{12-44}$$

The line of regression then takes the form

$$Y_e = bX. \tag{12-45}$$

Example 12-10. Find the line of regression for the data of Table 12-1, assuming that the line goes through the origin.

Solution. Using the results of Table 12-2 we find

$$b = \frac{3441.52}{21,203} = 0.162;$$

hence the regression line takes the form

$$Y_e = 0.162X.$$

In order to establish a possibly more accurate formula for the relationship between packed cell volume and red blood cell count, 562 pairs of observations on canine blood were made at the clinic of the School of Veterinary Medicine at the University of California at Davis, and the corresponding regression line was calculated. Its equation was precisely the same as that calculated above.

12.9 EXPONENTIAL AND POWER CURVES

It may be evident from the scatter diagram, or it may be known from theoretical considerations, that a straight line is not a good fit to a given set of observed points. The correlation coefficients may indicate that a line is practically useless for estimating purposes. In such cases, we try to determine whether there is another curve that might be a "good" fit to the observed points. In this and in Section 12-10, we shall consider the curves most frequently used to fit a set of points if a line is not appropriate.

It may be that when we graph the given points on "semilog paper" we find that the points lie close to a straight line. This means that the coordinate lines on the graph paper are drawn so that we are graphing (X, Y'), where $Y' = \ln Y$, where "ln" means "logarithm to the base e." It suggests that the equation of the curve that is a good fit to the data is of the form

$$Y_e = c \cdot d^X, \tag{12-46}$$

which is called an *exponential function*.

If we take the logarithm to the base e of both sides, we obtain

$$\ln Y_e = \ln c + (\ln d)X.$$

Letting $a = \ln c$, $b = \ln d$ and $Y'_e = \ln Y_e$, we have

$$Y'_e = a + bX. \tag{12-47}$$

This is the equation of a straight line. To find it, we calculate $Y' = \ln Y$ for each of the observed values Y and, using (X, Y') as the observed points, fit a straight line of the form (12-47) to these points; that is, we find the line (12-47) for which $\Sigma(Y' - Y'_e)^2$ is a minimum. Then, solving $a = \ln c$ and $b = \ln d$, for c and d, respectively, we find $c = e^a$ and $d = e^b$, which are the constants needed for Equation (12-46).

Another situation arises if, when we graph the points on "double log paper," we find that the points lie close to a straight line. This means that the coordinate lines on the graph paper are drawn so that we are graphing (X', Y'), where $X' = \ln X$ and $Y' = \ln Y$. It suggests that the equation of the curve that is a good fit to the data is of the form

$$Y_e = c \cdot X^b, \tag{12-48}$$

which is called a *power function*.

If we take the logarithm to the base e of both sides, we obtain

$$\ln Y_e = \ln c + b \ln X.$$

Letting $Y'_e = \ln Y_e$, $a = \ln c$, and $X' = \ln X$, we have

$$Y'_e = a + bX'. \tag{12-49}$$

This is the equation of a straight line. To find it, we calculate $X' = \ln X$ for each of the given values of X and $Y' = \ln Y$ for each of the observed values Y and, using (X', Y') as the observed points, fit a straight line of the form (12-49) to these points. Then since again $c = e^a$, we have the constants needed for equation (12-48).

12.10 POLYNOMIAL REGRESSION

The scatter diagram might indicate the existence of a polynomial relationship, i.e. a relationship that, for the case of a polynomial of degree 3, would be of the form

$$Y_e = a + bX + cX^2 + dX^3. \tag{12-50}$$

Here our problem is that of determining from a given sample of n pairs of X, Y-values the constants a, b, c, and d by the method of least squares so that again $\Sigma(Y - Y_e)^2$ has its smallest possible value. Using the differential calculus it is fairly easy to show that a, b, c, and d must be the solution of the following system of four linear equations.

$$\left.\begin{array}{l} an \ \ + b\Sigma X \ + c\Sigma X^2 + d\Sigma X^3 = \Sigma Y \\ a\Sigma X \ + b\Sigma X^2 + c\Sigma X^3 + d\Sigma X^4 = \Sigma XY \\ a\Sigma X^2 + b\Sigma X^3 + c\Sigma X^4 + d\Sigma X^5 = \Sigma X^2 Y \\ a\Sigma X^3 + b\Sigma X^4 + c\Sigma X^5 + d\Sigma X^6 = \Sigma X^3 Y \end{array}\right\} \tag{12-51}$$

These equations are rather obvious generalizations of the corresponding equations (12-11) found for the linear case, i.e. for a polynomial of degree 1, and are also called normal equations.

The "polynomial of best fit," as obtained by the method of least squares, would then be found by solving Equations (12-51) for a, b, c, and d and substituting these values into (12-50).

12.11 MULTIPLE REGRESSION

We have confined our attention to the case of one independent variable. Now let us assume that a variable Z is dependent upon the two independent variables X and Y, and, to take the simplest possible case, let us suppose that there exists a linear relationship relating Z to X and Y such as

$$Z_e = a + bX + cY. \tag{12-52}$$

The reader familiar with solid analytic geometry will recall that, graphically, Equation (12-52) represents a plane in three-dimensional space.

If we are now given n sets of values for X, Y, Z, our problem is that of determining the constants a, b, and c in such a way that the sum of squares $\Sigma(Z - Z_e)^2$ has its smallest possible value. It can be shown, again by use of the differential calculus, that a, b, and c must be the solution of the following system of three linear equations.

$$\left.\begin{array}{l} an \ \ + b\Sigma X \ \ + \ c\Sigma Y \ = \Sigma Z \\ a\Sigma X + b\Sigma X^2 \ + c\Sigma XY = \Sigma XZ \\ a\Sigma Y + b\Sigma XY + \ c\Sigma Y^2 \ = \Sigma YZ \end{array}\right\} \qquad (12\text{-}53)$$

The " plane of best fit," as obtained by the method of least squares, then would be the result of substituting into (12-52) the values of a, b, and c obtained by solving (12-53).

EXERCISES

1. Given the following values of X and Y:

X:	0	2	3	5	6
Y:	6	-1	-3	-10	-16

   (a) Find the equation of the regression line of Y on X.
   (b) Draw the scatter diagram, and graph the line.
   (c) Use the line to estimate Y when $X = 4$.
   (d) Calculate the standard error of estimate.
   (e) Calculate the correlation coefficient.

2. Given the following pairs of values of X and Y:

X:	-1	0	1	2	3
Y:	7	6	2	-2	-3

   (a) Find the equation of the regression line of Y on X.
   (b) Draw the scatter diagram, and graph the line.
   (c) Calculate the standard error of estimate.
   (d) Calculate the coefficient of correlation.

3. Derive Formula (12-32) from (12-20) and (12-25).

4. In the table below, S is the weight of potassium bromide that will dissolve in 100 grams of water at $T°C$.

T:	0	20	40	60	80
S:	54	65	75	85	96

(a) Find the regression line for predicting S.
(b) Use this equation to estimate S when $T = 50°$.
(c) Find the standard error of estimate.
(d) Find the correlation coefficient.

5. A manufacturer of optical equipment has the following data on the unit cost (in dollars) of certain custom-made lenses and the number of units made in each order.

Number of units (X)	Cost per unit (Y)
1	58
3	55
5	40
10	37
12	22

(a) Use the regression equation to predict the unit cost in an order of 8 of these lenses.
(b) Find the correlation coefficient.

6. A study is made relating aptitude test scores to productivity in a factory after three months of personnel training. The 6 pairs of scores shown below are obtained by testing 6 randomly selected applicants and later measuring their productivity.

Applicant	Aptitude score (X)	Productivity (Y)
1	9	23
2	17	35
3	20	29
4	19	33
5	20	43
6	23	32

(a) Find the equation of the line that can be used to predict the productivity from the aptitude score.
(b) What is the expected productivity of a worker who has an aptitude score of 16?
(c) Find the line of regression that will permit prediction of aptitude scores from productivity.
(d) Find the correlation coefficient.
(e) What percent of the variance in Y is explained by the linear relationship existing between X and Y?

7. Given the following data on age in weeks, X, and mean height in centimeters, Y, of soybean plants:

$$
\begin{array}{cccccc}
X: & 1 & 2 & 3 & 4 & 5 \\
Y: & 5 & 17 & 24 & 33 & 41
\end{array}
$$

Find the equation of the line of regression of Y on X. (*Note*: This line must go through the origin.)

8. The regression line of wormy fruits, Y, on the size of crop on the tree in hundreds of fruits, X, is

$$Y_e = 64 - 1.4X.$$

If it is known that

$$\bar{Y} = 45, \qquad \hat{s}_x = 7.2, \qquad \hat{s}_y = 10.5,$$

find
(a) $\bar{X}$;
(b) the correlation coefficient;
(c) the regression line of X on Y.

9. A correlation coefficient of 0.50 is found in a sample of 27 pairs. Can it be regarded as significantly different from zero on the basis of the 5% level of significance?

10. Suppose that a sample of 11 pairs taken from an approximately normal bivariate population yields a correlation coefficient of -0.65. Can this be considered a significant correlation at the 1% level?

11. How large a correlation coefficient is needed for a sample of size 25 in order to justify the claim that the variables are linearly related? Use the 5% level of significance.

12. How large a correlation coefficient is needed for a sample of size 11 so that the variables can be considered linearly correlated at the 1% level of significance?

13. For what range of values can the coefficient of correlation of a sample of 62 pairs be considered significantly different from zero at the 5% level of significance?

14. Test at the 5% level the hypothesis that $\rho = 0.60$, if in a sample of 103 pairs the correlation coefficient is found to be equal to 0.90.

15. A machine, which can be run at different speeds, produces articles of which a certain number are defective. The number of defective items produced per hour seems to depend on machine speed, as indicated in the following results of an experimental run.

Speed in r.p.s.	Number of defectives per hour
8	5
12	8
14	9
16	11

(a) Find the regression line of the number of defectives per hour on the machine speed in r.p.s.

(b) Find the correlation coefficient.

(c) Would you use the line obtained in part (a) to estimate machine speed for a given number of defectives per hour? Explain.

(d) Which of the following two hypotheses is more reasonable? (i) The population correlation coefficient is 0. (ii) It is 0.7.

16. In a sample of 63 pairs the correlation coefficient is found to be 0.31. Find the 95% confidence limits for the correlation coefficient of the population.

17. In a sample of 103 pairs the coefficient of correlation is found to be 0.63. Find the 95% confidence limits for the coefficient of correlation of the population from which the sample was selected.

18. Find the 98% confidence limits for the correlation coefficient of a population from which a sample of 50 pairs with correlation coefficient -0.48 is taken.

19. A sample is found to have a correlation coefficient of 0.69. If the hypothesis that this sample was taken from a population with correlation coefficient 0.57 is rejected at the 5% level of significance, find the smallest possible size of the sample.

20. In a sample of 45 pairs the correlation coefficient is found to be -0.18.

(a) Find the 99% confidence limits for the correlation coefficient of the population.

(b) Use the result of part (a) to determine whether there is a significant correlation in this sample at the 1% level of significance.

21. In a sample of 25 pairs of parent and child the correlation in a certain character is found to be 0.60. Is this value consistent with the view that the true correlation is 0.46, using the 5% level of significance?

22. A correlation coefficient of 0.56 is said to be highly significant. Assuming that this refers to the 1% level of significance, what is the least number of observations that must have been made in order to warrant this statement?

23. At most, how many observations could have been made in a sample for which a correlation coefficient of 0.25 was not considered significant at the 5% level?

24. A sample of size 52 is taken from a population in which the correlation coefficient is 0.59. Find the probability that the correlation coefficient of the sample is 0.40 or less.

25. Find the probability that a sample of size 67 taken from a population with correlation coefficient 0.45 has a correlation coefficient between 0.55 and 0.65.

26. From a population with correlation coefficient 0.65, a large number of samples of equal size is taken. If 15% of these samples have correlation coefficients of 0.75 or more, what is the expected size of the sample?

27. The 98% confidence limits for the correlation coefficient of a normal bivariate population, as determined from a sample, are 0.29 and 0.71. What is the expected size of the sample?

28. In a bivariate population, the correlation coefficient is 0.60. If 3000 samples of 39 pairs are taken from this population, how many would you expect to have a correlation coefficient of at most 0.55?

29. From a population with correlation coefficient 0.80, an experimenter selects 500 samples, all of equal size. If 115 of these samples show a correlation coefficient greater than 0.85, what is the expected sample size?

30. The 95% confidence limits for the correlation coefficient of a bivariate population are found to be 0.20 and 0.58. What would be the 99% confidence limits for the correlation coefficient of this population?

31. The 99% confidence limits for the correlation coefficient of a bivariate population are found to be -0.22 and 0.51. Find the number of pairs in the sample from which these confidence limits were calculated.

32. For a set of n pairs of values of X, Y, the correlation coefficient is found to be 0.80. If, for this sample, $\Sigma(Y - Y_e)^2 = 9$, find $\Sigma(Y - \bar{Y})^2$.

33. If b is the slope of the regression line of Y on X, and d the slope of the regression line of X on Y, prove that

$$bd = r^2,$$

where r is the correlation coefficient between X and Y.

34. If, from a sample, the regression line of Y on X is found to be

$$Y_e = 10 - \frac{4}{3} X$$

and the one of X on Y

$$X_e = 5 - \frac{1}{3} Y,$$

find
(a) the correlation coefficient;
(b) the values of $\bar{X}$ and $\bar{Y}$ for the sample.

35. Prove that the formula for the sum of squares for regression—Formula (12-20) —can be rewritten as

$$S.S.R. = b\Sigma xy,$$

where b is the slope of the regression line.

36. Prove that for a bivariate population of X, Y-values the variance of the $X + Y$ values, that is, σ_{x+y}^2, is given by

$$\sigma_{x+y}^2 = \sigma_x^2 + 2\rho\sigma_x\sigma_y + \sigma_y^2.$$

37. Prove that linear transformations of X and Y into W and Z, respectively, i.e.

$$W = aX + b$$
$$Z = cY + d,$$

where a, b, c, d are constants, do not change the absolute value of the correlation coefficient. That is, if r_{xy} is the correlation coefficient between the X-values and Y-values, and r_{wz} is the correlation coefficient between the W-values and Z-values, then show that

$$|r_{wz}| = |r_{xy}|.$$

13

Chi-Square
Distribution

13.1   DEFINITION

If an experiment has only two possible outcomes, such as the appearance
of a head or a tail in the tossing of a coin, the normal distribution can be used
to determine whether the observed frequencies of these two events depart
significantly from the expected frequencies. Whenever more than two events,
say k events, can occur, the normal distribution no longer can be applied to
test for a possible significant difference between the observed and expected
frequencies.

If we are considering more than two events, we must first define a quantity
that measures the discrepancy between the k observed frequencies $o_1, o_2, \ldots,$
o_k and their corresponding expected frequencies $e_1, e_2, \ldots, e_k$. This statistic,
called chi-square, is defined as

$$\chi^2 = \sum_{i=1}^{k} \frac{(o_i - e_i)^2}{e_i}. \tag{13-1}$$

Clearly, the closer the agreement between the expected and observed frequencies, the smaller will be the value of χ^2. If $\chi^2 = 0$, each of the terms of

the sum in Formula (13-1) must be zero, and there is perfect agreement between observed and expected frequencies for all k events.

Example 13-1. A die is tossed 120 times, and the results are listed in the second column of Table 13-1 (a 1 is obtained 13 times, a 2 is obtained 28 times, and so on). Assuming that the die is honest, find the value of χ^2.

Table 13-1
Observed and expected results of 120 tosses of a die and calculation of chi-square.

Face	o	e	$o-e$	$(o-e)^2$	$\dfrac{(o-e)^2}{e}$
1	13	20	−7	49	2.45
2	28	20	8	64	3.20
3	16	20	−4	16	0.80
4	10	20	−10	100	5.00
5	32	20	12	144	7.20
6	21	20	1	1	0.05
	120	120	0		18.70

Solution. Since each face of an honest die is expected to show one sixth of the time, in this example each expected value equals 20 and χ^2 is the sum of the numbers in the last column of Table 13-1. Consequently,

$$\chi^2 = 18.70.$$

13.2   DISTRIBUTION OF CHI-SQUARE

If an experiment, such as the preceding one, is repeated many times, always under the hypothesis that the expected values remain unchanged, and each time the value of χ^2 is calculated, a set of values of χ^2 is obtained for which the corresponding histogram can be drawn. It is important, for later discussion, to realize that χ^2 is a discrete variable; that is, χ^2 can only assume certain, but not all, nonnegative values. The reason for this may be clarified if we consider again the tossing of a die 120 times. For perfect agreement of observed and expected frequencies, χ^2 equals 0; the next larger value of χ^2 is obtained when one of the observed values is 21, another 19, and all the rest

20, in which case, as can be readily verified, $\chi^2 = 0.10$. Thus, in this example, χ^2 cannot assume any value between 0 and 0.10.

Fortunately, as was true for the binomial distribution, there is a frequency curve that is an excellent approximation to the histogram of χ^2, and, correspondingly, a probability curve that approximates the distribution of χ^2.

Theorem 13-1.   The distribution of χ^2 is well approximated by the curve

$$y = c(\chi^2)^{(v-2)/2}e^{-\chi^2/2}, \tag{13-2}$$

where v is the number of degrees of freedom, and c is a constant depending on v and determined in such a way that the total area under the probability curve (13-2) is equal to 1.

This means that the function $P(\chi^2) = c(\chi^2)^{(v-2)/2}e^{-\chi^2/2}$ qualifies as a probability density function (or distribution), as defined in Definition 7-1, and the probability that χ^2 is between two given numbers a and b, with $b > a$, is the area under the curve (13-2) between a and b.

In graphing the curves given by (13-2) for various values of v we note that for $v \geq 3$ the curve starts at the origin, reaches its highest point for $\chi^2 = v - 2$, and then decreases, approaching the horizontal axis asymptotically without, however, being symmetric about any vertical line. For $v = 2$, the curve has its highest point on the y-axis, and for $v = 1$, the curve has no maximum value for $\chi^2 \geq 0$; i.e. it is infinite for $\chi^2 = 0$, decreases continuously as χ^2 increases, and approaches the horizontal axis. The graphs of (13-2) for $v = 1$, 3, 5, and 15 are shown in Figure 13-1.

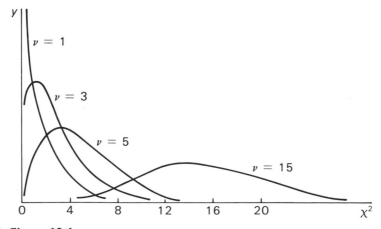

Figure 13-1
The χ^2-distributions for $v = 1, 3, 5, 15$.

Appendix Table VII gives for various v-values (first column) the values of χ^2 for which the right-tail areas have the values listed at the heads of the other columns: e.g., for $v = 3$ the value of χ^2 for a right-tail area of 0.30 is 3.66, which means that for 3 degrees of freedom the probability of obtaining a value of χ^2 greater than or equal to 3.66 is 0.30.

Example 13-2.   On the basis of the 5% level of significance, do the results of Example 13-1 indicate that the die is dishonest?

Solution.   The hypothesis to be tested is that of the honesty of the die. This means we wish to test whether the expected frequencies are those given in the third column of Table 13-1. Under this hypothesis, the value of χ^2 is the one already calculated in Example 13-1. Since the total of the observed values must equal 120, there is one relationship between them, and consequently the number of degrees of freedom is 1 less than the number of events, or $6 - 1 = 5$. At the 5% level the tabular value of χ^2 corresponding to $v = 5$ is $\chi^2_{.05} = 11.07$. Inasmuch as the calculated χ^2 of 18.70 exceeds this, the result is significant and the hypothesis of an honest die has to be rejected. Thus the indication is that the die is dishonest. Since $\chi^2 = 18.70$ falls between the values 15.09 and 20.52 (in Appendix Table VII) corresponding to $A = 0.01$ and $A = 0.001$, respectively, we can say that, for an honest die, the result obtained in Table 13-1, or one for which the observed values differ even more widely from the expected (i.e. one for which χ^2 is equal to or larger than 18.70), can be expected by chance alone to occur with a probability of less than 0.01.

13.3   APPLICATION TO GENETICS

The χ^2 distribution is particularly useful in genetics, where in breeding experiments it is desired to determine whether observed results are consistent with expected segregations, the latter usually being given in the form of genetic ratios.

Example 13-3.   According to Mendelian inheritance theory, in crossing two kinds of peas, four types of seeds, A, B, C, and D, are expected to occur in the ratio $9 : 3 : 3 : 1$. In such an experiment an experimenter obtains 102 seeds of type A, 30 of type B, 42 of type C, and 15 of type D. Are these results consistent with the theory on the basis of the 5% level of significance?

Solution.   Since a total of 189 seeds is used in this experiment, the number of type A seeds expected under the hypothesis of a $9 : 3 : 3 : 1$ segregation ratio is

$\frac{9}{16}$ of 189, or 106.3. (Although expressing expected values to one decimal place when dealing with numbers of seeds may seem strange, it is considered advisable in order to assure greater precision in the results.) Expected frequencies of occurrence for the other types of seeds are calculated similarly, and they are listed in the second column of Table 13-2.

Table 13-2
Observed segregation in four types of seeds with calculation of chi-square.

o	e	$o-e$	$(o-e)^2$	$\dfrac{(o-e)^2}{e}$
102	106.3	−4.3	18.49	0.17
30	35.4	−5.4	29.16	0.82
42	35.4	6.6	43.56	1.23
15	11.8	3.2	10.24	0.87
189	188.9	0		3.09

From Table 13-2, $\chi^2 = 3.09$, which has to be compared with $\chi^2_{.05} = 7.82$, the tabular value for $\nu = 3$ at the 5% level. This means that by chance alone we would expect to obtain more than 5% of the time (according to Appendix Table VII, about 37%) a value of χ^2 as large as or larger than 3.09 when sampling from a population in which the 9 : 3 : 3 : 1 ratio holds. Therefore, the result is not significant, and there is good indication that observation agrees with expectation.

13.4 APPLICATION TO CONTINGENCY TABLES

Definition 13-1. A *contingency table* is an arrangement in which a set of objects is classified according to two criteria of classification, one criterion being entered in rows, the other in columns.

The number of rows of a contingency table is usually denoted by j, and the number of columns by k; such a table is also referred to as a $j \times k$ table. Our main interest is in determining whether a relationship exists between the two criteria of classification or if they are independent. Therefore, we are testing the hypothesis of independence of the two criteria, and we calculate expected values in accordance with this hypothesis.

The application of χ^2 to contingency tables is illustrated in Example 13-4 for a 2×2 table.

Example 13-4.   Table 13-3 gives data on the number of men and women who became airsick on an airplane trip in extremely rough weather. Do these data indicate that, in general, women are less susceptible to air sickness than are men (5 % level)?

Table 13-3
Number of men and women affected by airsickness on an airplane trip.

	Number of cases of airsickness	Number of cases free of airsickness	Totals
Men	24	31	55
Women	8	26	34
	32	57	89

Solution.   Since a quick examination of the table reveals that the percentage of women who became airsick is considerably smaller than the percentage of men, we might be led to conclude that this is also true for the population at large. This conclusion, however, would be completely unjustified without a statistical test.

As in any contingency table, we set up the hypothesis that the two criteria of classification (in this case, sex and susceptibility to airsickness) are independent. Under this hypothesis, we would expect the ratio of airsick men to all men on the plane to be the same as the ratio of airsick women to all women on the plane, and also the same as the ratio of all airsick people to the total number of people on the plane. Denoting by x the expected number of airsick men on the plane, we have

$$x : 55 = 32 : 89$$

and $x = 19.8$. The shortest way of finding the expected number of men free of airsickness is to subtract the expected number of airsick men from the total number of men on the plane, which in this case gives $55 - 19.8 = 35.2$. The other expected numbers are found similarly by subtraction, and are listed in the second column of Table 13-4.

Table 13-4
Calculation of chi-square for data of Table 13-3.

o	e	$o - e$	$(o-e)^2$	$\dfrac{(o-e)^2}{e}$
24	19.8	4.2	17.64	0.89
31	35.2	−4.2	17.64	0.50
8	12.2	−4.2	17.64	1.45
26	21.8	4.2	17.64	0.81
89	89.0	0		3.65

Since for a 2×2 table the number of degrees of freedom ν is equal to 1 (for an explanation, see the end of this section), and since the tabular value at the 5% level for $\nu = 1$ is $\chi^2_{.05} = 3.84$, a value larger than the calculated value, we conclude that under the hypothesis of independence of sex and susceptibility to airsickness a deviation from expectation as large as or larger than that which occurred in this sample is found more than 5% of the time. Therefore, this result is not significant, and these data do not indicate that, in general, women are less susceptible to airsickness than men.

You may have noted that in Table 13-4 all of the absolute values of $o - e$ are the same. It is easy to prove that for any 2×2 table all the values of $|o - e|$ are identical; however, this is not true for a $j \times k$ table in general. The proof of this is left as an exercise (see Exercise 26).

Whenever the χ^2-value in a contingency table gives a significant result, it indicates that the two criteria of classification are not independent but related. However, such a result does not indicate a causal relationship. For example, although a contingency table classifying a large number of people according to smoking habits and incidence of lung cancer might yield a significant χ^2-value that would indicate that lung cancer occurs more frequently among smokers than among nonsmokers, this does not prove by any means that smoking increases the incidence of lung cancer. The relationship between lung cancer and smoking might be due to some factor that causes people to have a desire to smoke and also causes them to have lung cancer. Much misinterpretation of data is due to the erroneous assumption that a relationship between two criteria means that one causes the other.

Why is the number of degrees of freedom taken as 1 for a 2×2 table? Let us consider the general 2×2 arrangement shown in Table 13-5.

Table 13-5
General 2×2 contingency table.

			Totals
	a	b	A
	c	d	$N-A$
Totals	B	$N-B$	N

For the observed frequencies a, b, c, d, the following relationships hold:

$$a + b = A;$$
$$c + d = N - A;$$
$$a + c = B;$$
$$b + d = N - B.$$

But these four equations are not independent; indeed, the last equation can be obtained by subtracting the third from the sum of the first two. There are, therefore, only three independent relationships; thus, according to Definition 10-2, $v = 4 - 3 = 1$, since v is the total number of variates minus the number of independent relationships existing between them. An alternative approach would be to note that for fixed totals only one of the numbers a, b, c, d has to be given to fix the remaining numbers.

For a general $j \times k$ table $v = (j - 1)(k - 1)$, since if all entries except those in one row and one column are given, the rest can be determined from the marginal totals. The methods applied to any $j \times k$ tables are completely analogous to those discussed for the special case of 2×2 tables and are illustrated in Example 13-5.

Example 13-5.   Some recent studies have indicated the possibility that there might be a relationship between month of birth and intelligence. To determine whether there is such a relationship, J. E. Orme reported in the British Journal of Medical Psychology, (Volume 35, Part 3, 1962, page 233) the results of intelligence tests conducted on two groups of intellectual subnormals (defined as those scoring more than 2 standard deviations below the mean—that is, less than 70 IQ). His results are given in Table 13-6. What conclusions can you draw from these results (5% level)?

Table 13-6
The season of birth and IQ incidence.

	Summer	Autumn	Winter	Spring	Totals
55–69 IQ	29	19	12	18	78
40–54 IQ	13	17	20	20	70
Totals	42	36	32	38	148

Solution.   This is a 2×4 contingency table. We test here the hypothesis that the two criteria of classification (IQ and season of birth) are independent. Under this hypothesis, we would expect all of the following ratios to be the same: (1) the ratio of persons born in the summer who have 55–69 IQ to all persons having 55–69 IQ; (2) the ratio of persons born in the summer who have 40–54 IQ to all persons having 40–54 IQ; and (3) the ratio of all persons born in the summer to the total number of persons involved in this investigation. Denoting by x the expected number of persons born in the summer who have 55–69 IQ, we have

$$x : 78 = 42 : 148$$

and $x = 22.1$. Denoting now by y the expected number of persons born in the autumn with 55–69 IQ, we have

$$y : 78 = 36 : 148$$

and $y = 19.0$. Note that we could not have found this value of $y = 19.0$ by subtracting some number from the marginal total, as we were able to in the case of a 2×2 table (see Example 13–4). Denoting next by z the expected number of persons born in the winter with 55–69 IQ, we have

$$z : 78 = 32 : 148$$

and $z = 16.9$. Now the expected number of people born in the spring can be obtained by subtracting from the total number of people with 55–69 IQ those expected to be born in the summer, autumn, and winter, which gives $78 - (22.1 + 19.0 + 16.9) = 20.0$. The other expected numbers are found similarly by subtraction, and are listed in the second column of Table 13-7.

Table 13-7
Calculation of chi-square for data of Table 13-6.

o	e	$o - e$	$(o - e)^2$	$\dfrac{(o - e)^2}{e}$
29	22.1	6.9	47.61	2.15
19	19.0	0.0	0.00	0.00
12	16.9	−4.9	24.01	1.42
18	20.0	−2.0	4.00	0.20
13	19.9	−6.9	47.61	2.39
17	17.0	0.0	0.00	0.00
20	15.1	4.9	24.01	1.59
20	18.0	2.0	4.00	0.22
				7.97

Note that, in this case, we had to set up three proportions for the purpose of calculating expected numbers (rather than only one as in the case of a 2×2 table) before we could determine the remaining expected numbers by subtraction from the marginal totals. It is readily seen, therefore, that, in a $j \times k$ table in general, the number of proportions that have to be set up for the calculation of expected numbers before the remaining ones can be determined by subtraction is precisely the number of degrees of freedom for the $j \times k$ table, namely $(j - 1)(k - 1)$.

Since for a 2×4 table the number of degrees of freedom ν is equal to $1 \cdot 3 = 3$, and since the tabular value at the 5% level for $\nu = 3$ is $\chi^2_{.05} = 7.82$, we conclude that the result is significant, although only barely so, and that, for intellectual subnormals, these data indicate a relationship between season of birth and intelligence, although not very convincingly.

To determine what this relationship is, we note that those born in the summer and autumn appear to have a higher IQ than those born in the winter and spring. In order to determine how significant this latter relationship is, we might investigate it by analyzing the 2 × 2 table shown as Table 13-8. Analyzing

Table 13-8
The season of birth and IQ incidence.

	Summer and Autumn	Winter and Spring
55–69 IQ	48	30
40–54 IQ	30	40

this 2 × 2 table by the method shown previously for such tables, we find (see Exercise 30) $\chi^2 = 5.18$, which is more substantially significant, when compared with $\chi^2_{.05} = 3.84$, than was the comparison between 7.97 and 7.82. An even more revealing procedure is to analyze the data for summer and winter only as a 2 × 2 table. This analysis (see Exercise 29) results in a value of $\chi^2 = 7.24$, which is significant even at the 1 % level. Our conclusion, then, is that, for intellectual subnormals, there is a strong indication of a relationship between season of birth and IQ, and that, in particular, those born in the summer tend to have a higher IQ than those born in the winter.

Note that from Table 13-6 we can conclude a relationship between season of birth and IQ only for intellectual subnormals, or, more precisely, for the population from which the sample of 148 individuals of Example 13-5 was selected. If, for example, these individuals all happened to have been white males, then clearly our conclusion should be restricted to this population; if, however, also white females in approximately the same number as males had been included in the study, then we could extend our conclusion to both sexes. In any case, however, we must restrict our conclusion to intellectual subnormals; it is indeed conceivable that, for intellectual normals or superiors, there may not exist a relationship between season of birth and intelligence.

In general, it is essential when analyzing experiments to draw conclusions concerning only the populations from which the random samples of the experiments were selected. Thus, for example, if in Example 13-4 it had turned out that there had been a significant relationship between sex and susceptibility to airsickness (which did not happen to be the case), that conclusion could only have been applied to the particular population of which those on the plane were a random sample. This might, for example, have been residents of the Southern United States. On the other hand, it might have been an international plane, in which case the 89 individuals travelling on it might have included people from all over the world representing most races; then the conclusion could properly have been extended to the population of the world at large.

13.5   APPLICATION TO TESTING FOR NORMALITY

One of the important applications of χ^2 is in determining whether or not a population satisfies a normal distribution. Here, we must have a fairly large sample from the population. First, we construct a frequency table for the data of the sample, and then calculate its mean $\overline{X}$ and standard deviation $\hat{s}$, which, since the sample is large, we use for the mean m and standard deviation σ, respectively, of the population. Finally, we calculate, under the hypothesis that the population is normally distributed, the expected frequencies corresponding to each of the classes of the frequency table and compare them by means of a χ^2-test with the observed frequencies.

This procedure requires that each of the expected frequencies be at least 5. If the expected frequencies of some classes at the beginning or end of the frequency table are less than 5, which they often are, such classes should be pooled with adjoining ones in order that the expected frequency of the pooled class will be at least 5.

In testing for normality, since for fixed total number, mean, and standard deviation three relationships exist between the variates of the sample, the number of degrees of freedom equals $v = h - 3$, where h is the number of classes in the frequency table used in the calculation of χ^2.

Example 13-6.   Consider the data of Table 2-3 for the total seasonal rainfall at Sacramento, California, for the 120-year period 1855–1975. Using the 5% level of significance, is there reason to doubt that these data are normally distributed?

Solution.   In Table 13-9 the data of Table 2-3 are listed in a frequency table that is similar to Table 2-4 except that the last three classes are pooled into a single class in order that no expected frequency is below 5 (in this case 4.9 is so close to 5 no further pooling will be made). Using the mean and standard deviation of this sample as estimates of their corresponding parameters, we have $m = 17.92$ and $\sigma = 6.09$ (see Examples 4-2 and 4-13). This allows us to calculate

Table 13-9

Observed and expected frequencies under the assumption of normality and the calculation of chi-square.

Class boundaries	o	e	$o - e$	$(o - e)^2$	$\dfrac{(o - e)^2}{e}$
7.5–10.5	14	13.3	0.7	0.49	0.04
10.5–13.5	16	14.6	1.4	1.96	0.13
13.5–16.5	23	21.2	1.8	3.24	0.15
16.5–19.5	25	23.2	1.8	3.24	0.14
19.5–22.5	13	20.5	−7.5	56.25	2.74
22.5–25.5	18	14.3	3.7	13.69	0.96
25.5–28.5	5	8.0	−3.0	9.00	1.12
28.5–37.5	6	4.9	1.1	1.21	0.25
	120	120.0	−0.2		5.53

the expected frequencies for each of the classes of Table 13-9; thus, for the second class we find

$$z_1 = \frac{10.5 - 17.92}{6.09} = -1.22, \qquad z_2 = \frac{13.5 - 17.92}{6.09} = -0.73.$$

From Appendix Table I we find the area between z_1 and z_2 to be

$$A = 0.3888 - 0.2673 = 0.1215,$$

and the expected frequency for that class is

$$e_2 = (0.1215)(120) = 14.6.$$

In determining the expected frequency for the first class, we include the total area to the left of 10.5 under the normal probability curve, and for the last class we include the total area to the right of 28.5. This is done to make the total of all expected frequencies agree with the total of all observed frequencies (in this example the total is 120).

Since $\chi^2 = 5.53$ and the tabular value for $\nu = 8 - 3 = 5$ is $\chi^2_{.05} = 11.07$, there is no reason to doubt that the rainfall data as given in Table 2-3 are normally distributed.

13.6 ADJUSTED CHI-SQUARE

The necessity for introducing adjusted chi-square can best be demonstrated by an example.

Example 13-7. According to the Mendelian theory of inheritance, certain crosses of peas should give yellow and green peas in the ratio $3 : 1$. If an experimenter obtains 200 yellow peas and 48 green peas, are these results consistent with the theory on the basis of the 5% level of significance?

Solution.   Calculating χ^2 in the usual fashion (which, as we shall see later, is not very accurate), we obtain, from Table 13-10, $\chi^2 = 4.21$.

Table 13-10
Calculation of chi-square for the data of Example 13-7.

o	e	$o-e$	$(o-e)^2$	$\dfrac{(o-e)^2}{e}$
200	186	14	196	1.05
48	62	-14	196	3.16
248	248	0		4.21

Comparing this value of χ^2 with the tabular value $\chi^2_{.05} = 3.84$ at the 5% level for $\nu = 1$, we conclude that these results deviate significantly from the theory.

For the data of Example 13-7, we find, by linear interpolation, that since for $\chi^2 = 3.84$, $A = 0.05$, and for $\chi^2 = 5.41$, $A = 0.02$, the probability of getting χ^2 greater than or equal to 4.21, assuming that the Mendelian theory is valid, is 0.043. Note that this value represents the probability of obtaining 200 or more or 172 or fewer yellow peas, since also for 172 yellow peas and 76 green peas the value of χ^2 is 4.21.

This probability can also be calculated using the normal distribution, approximating the binomial distribution involved, since in this example only two events can occur. Recalculating this probability, we find

$$m = 248 \cdot \frac{3}{4} = 186, \qquad \sigma = \sqrt{248 \cdot \frac{3}{4} \cdot \frac{1}{4}} = 6.82.$$

Thus the probability of obtaining 200 or more or 172 or fewer yellow peas is equal to twice the area to the right of

$$z = \frac{199.5 - 186}{6.82} = 1.98$$

under the normal probability curve; that is, $P = 0.048$, which is a larger value than 0.043 obtained by the χ^2-method. The reason for this discrepancy immediately becomes evident if we recall that χ^2 is a discrete variable. We can readily verify that the next lower value of χ^2 that is obtainable here (when $o_1 = 199$, $o_2 = 49$) is 3.64 (see Figure 13-2, which also shows that the next higher value of χ^2 is 4.84). By using the χ^2-distribution we have found the

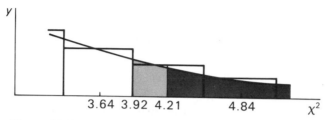

Figure 13-2
Adjusted chi-square.

area under the curve to the right of $\chi^2 = 4.21$ (shaded area in Figure 13-2). Actually, the probability that χ^2 is greater than or equal to 4.21 includes the entire area of the rectangle of which 4.21 is the midpoint; thus we have omitted the dotted area. Therefore, in finding the desired probability, the area under the chi-square curve to the right of the left border of the rectangle with midpoint 4.21 should be taken. This left border is located at approximately

$$\frac{1}{2}(3.64 + 4.21) = 3.92.$$

An alternate, and quicker, method of obtaining this smaller value of χ^2, corresponding to the left boundary of the rectangle (instead of its midpoint), and called *adjusted chi-square*, was given for 1 degree of freedom by Frank Yates:

$$\text{adj. } \chi^2 = \sum_{i=1}^{k} \frac{(|o_i - e_i| - \frac{1}{2})^2}{e_i}. \tag{13-3}$$

Note that the absolute value of each of the differences of the observed and expected frequencies is reduced by $\frac{1}{2}$. If the number of degrees of freedom is greater than 1, the correction to be applied is more complicated but also less important in that it will not alter the results materially; for these reasons, it will not be discussed.

Definition 13-2. The adjustment made on χ^2 and given by (13-3) is called *Yates' correction for continuity.*

From the preceding discussion it is clear that, for 1 degree of freedom, adjusted χ^2 should always be used. If all expected values, however, are large, the difference between unadjusted and adjusted χ^2 is small. This situation is analogous to that where the binomial distribution is approximated by the normal curve: when it is desired to find from the normal probability curve the probability that a variate taken at random will be greater than or equal to a given number A, the smaller the number of trials, the more important it becomes to subtract $\frac{1}{2}$ from A (i.e. to find the area to the right of $X = A - \frac{1}{2}$). On the other hand, note that in Example 13-7, where certainly all expected values are large, the failure to adjust χ^2 results in an error of about 10% in the probability. Such an error in a particular problem (see Exercise 5) may make a result appear significant whereas it really is not significant. Therefore, consistent use of adjusted χ^2 for all cases involving 1 degree of freedom is recommended. The reader should note that adjusted χ^2 should have been used in Example 13-4 if it had been discussed sooner than Section 13.4.

Example 13-8.   Solve Example 13-7 by using adjusted χ^2 and find the probability of the event discussed in that example.

Solution.   Using (13-3), we find, from Table 13-11, adjusted $\chi^2 = 3.92$.

Table 13-11
Calculation of adjusted chi-square for the data of Table 13-10.

o	e	$|o-e|$	$|o-e| - \frac{1}{2}$	$(|o-e| - \frac{1}{2})^2$	$\dfrac{(|o-e| - \frac{1}{2})^2}{e}$
200	186	14	13.5	182.25	0.98
48	62	14	13.5	182.25	2.94
248	248				3.92

Since adjusted $\chi^2 = 3.92$ (which agrees with the value found previously from Figure 13-2), we find by interpolation that the area under the chi-square curve to the right of $\chi^2 = 3.92$ equals 0.048, which agrees exactly with the value obtained by use of the normal curve.

EXERCISES

The reader should refer also to the mixed set of exercises in the Appendix entitled " Review Exercises," which covers the material from Chapters 3 to 13.

In the following exercises, if no statement to the contrary is made, use the 5% level of significance. Also use one-decimal accuracy for expected values.

1. A die is tossed 100 times with the following observed results.

Face	o
1	12
2	17
3	20
4	22
5	13
6	16

Determine whether this die could still be considered honest on the basis of the 1% level of significance.

2. The number of books borrowed from a public library, which is open 5 days a week, is given below:

	Mon.	Tues.	Wed.	Thurs.	Fri.
Number of books borrowed	153	108	120	114	145

Do these data indicate that the number of books borrowed is dependent on the day of the week (5% level)?

3. In a cross between ivory and red snapdragons the following counts are observed in the F_2 generation.

Phenotype	Numbers of plants
Red	20
Pink	55
Ivory	25

On the basis of these data, can segregation be assumed to occur in the simple Mendelian ratio $1 : 2 : 1$?

4. In an experiment involving crossing two hybrids of a species of flower the results shown below are observed. Are these results consistent with the expected proportion $9:3:3:1$?

Magenta flower, Green stigma	Magenta flower, Red stigma	Red flower, Green stigma	Red flower, Red stigma
120	49	36	12

5. In a certain cross of two varieties of peas genetic theory leads us to expect half the seeds to be wrinkled and half of them to be smooth. In a sample of 800 seeds we find 440 wrinkled.
   (a) Find the value of unadjusted χ^2.
   (b) If in a sample of 400 seeds the count is 220 wrinkled, would unadjusted χ^2 also be half of its original value?
   (c) Using unadjusted χ^2, calculate the probability of obtaining, in a total of 400 seeds, a number of wrinkled seeds differing by 20 or more from the value expected by the theory (that is, 220 or more or 180 or fewer).
   (d) Solve part (c), using the normal distribution.
   (e) Solve part (c), using adjusted χ^2.

6. An honest coin is tossed 120 times.
   (a) Using adjusted χ^2, find the probability of obtaining 68 or more or 52 or fewer heads.
   (b) Find the same probability, using the normal curve.

7. A well-balanced die is tossed 180 times. Find the probability of obtaining 35 or more sixes,
   (a) using the normal curve approximation,
   (b) using adjusted χ^2.

8. It is known that blood groups are inherited in a simple Mendelian manner so that a cross of parents both of which have AO blood should give three-fourths type A children and one-fourth type O children. What is the probability that among 400 children from such parents, at least 312 are of type A,
   (a) using the normal distribution,
   (b) using adjusted χ^2?

9. If 8% of certain articles produced are defective, calculate the probability that in a sample of 200 such articles more than 20 are defective,
   (a) using the normal distribution,
   (b) using adjusted χ^2.

10. A manufacturer of automobile tires claims that 98% of his products are perfect. Assuming that this claim is correct, find the probability that in a sample of 500 tires he has 15 or more defective tires,
    (a) using the normal distribution,
    (b) using adjusted χ^2.

11. Of 64 offspring of a certain cross between guinea pigs, 34 are red, 10 are black, and 20 are white. According to the genetic model, these numbers should be in the ratio 9 : 3 : 4. Are the data consistent with the model?

12. According to the genetic model, offspring of a certain cross between guinea pigs are red, black, or white in the ratio 9 : 3 : 4. A worker wishing to disprove this theory for a particular experiment has reason to believe that the offspring will occur in the ratio 8 : 5 : 3. If in this experiment the results occur precisely in the ratio 8 : 5 : 3, how many offspring must there be in the sample in order that the worker can be 95% confident he has disproved the hypothesis of a 9 : 3 : 4 ratio?

13. In a random sample of 1000 housewives 55% state a preference for brand A and 45% for brand B. Is this result compatible with the hypothesis that 50% of all housewives prefer brand A?

14. During an epidemic of cholera the following data on the effectiveness of inoculation as a means of preventing the disease were obtained.

	Not attacked	Attacked
Inoculated	192	4
Not inoculated	113	34

Do these data indicate the effectiveness of the inoculation on the basis of the 1% level of significance?

15. Fifty individuals are classified according to eye color and shade of hair. Can we conclude from the data shown below that for these individuals there is a significant connection between eye color and hair shade?

	Light hair	Dark hair
Blue eyes	23	7
Brown eyes	4	16

16. A thousand individuals are classified according to sex and according to whether they are color-blind:

	Male	Female
Not color-blind	442	514
Color-blind	38	6

What conclusion regarding color-blindness can we draw from these data?

17. To determine whether an engine oil additive decreases the chances of needing an engine overhaul within two years, a truck company with 25 new trucks selects 10 of these at random to be given the additive. The results are shown below.

	No overhaul	Overhaul
Additive	5	5
No additive	3	12

Do these data confirm that the additive decreases the chances of needing an overhaul within two years?

18. A group of 100 men and 100 women is interviewed to determine if they listen to a certain radio program. Sixty-five of the men do not listen to the program, and 35 do listen. Eighty of the women are nonlisteners, and 20 are listeners. Is there a significant difference in the listening habits of this group of men and women?

19. In a preference test for soap, 25 men and 50 women participated. Of a total of 9 people preferring one brand, 6 were men. Do these data indicate that preference for a particular brand is related to sex (5% level)?

20. In an epidemic of a certain disease, of 950 children who contracted the disease, 380 received no treatment, and, of those, 105 suffered aftereffects. Of the remainder, who did receive treatment, 195 suffered aftereffects. Do these data support the conclusion that the treatment was not effective (5% level)?

21. The following data show the effect of a certain type of fumigation on fruit spoilage.

	Spoiled	Unspoiled
Unfumigated	8	16
Fumigated	2	14

Does the amount of fruit spoiled depend upon whether it has been fumigated (2% level of significance)?

22. A group of 30 men and 80 women is asked to express a preference for the aromas of two pipe tobaccos. The results are shown below.

	Brand A	Brand B
Men	9	21
Women	8	72

(a) Do these data indicate that preference for a particular brand is related to sex?

(b) What is the probability that in part (a) a type 1 error has been committed?

23. Twenty-two animals are suffering from a disease, the severity of which is about the same in each case. In order to test the therapeutic value of a serum, it is administered to 10 of the animals; 12 remain uninoculated as a control. The result is:

	Recovered	Died
Inoculated	7	3
Not inoculated	3	9

Determine the probability of obtaining this result or one more favorable to the treatment.

24. In an experiment conducted by W. E. Howard of the University of California, Davis, the survival of two groups of laboratory rats gavaged with the rodenticide sodium fluoroacetate was studied. One group, F_4, was chosen after 5 generations of selective breeding for tolerance to the toxic chemical and the other group was drawn from the same gene pool as the original parents. Do the data below indicate that selective breeding for tolerance was effective in reducing the death rate (1% level)?

	Survived	Died
F_4	93	19
Parent stock	13	37

25. In a 2×2 contingency table giving the preference for two brands of wine among 30 men and 30 women, half of them preferred one brand and half the other. If the value of adjusted χ^2 is 5.40, what are the observed values in this table?

26. Prove that for a 2×2 contingency table all values of $|o - e|$ are the same.

27. Show that for a 2×2 contingency table with observed frequencies a, b, c, d the value of χ^2 used to test for independence can be written as

$$\chi^2 = \frac{n(ad - bc)^2}{(a + b)(c + d)(b + d)(a + c)},$$

where $n = a + b + c + d$.

28. Show that for a 2×2 contingency table with observed frequencies a, b, c, d the value of adjusted χ^2 can be written as

$$\text{adj. } \chi^2 = \frac{n\left(|ad - bc| - \dfrac{n}{2}\right)^2}{(a+b)(c+d)(b+d)(a+c)},$$

where $n = a + b + c + d$.

29. For the data of Table 13-6 calculate the χ^2 value for comparing summer and winter births only.

30. Calculate the χ^2 value for the data of Table 13-8.

31. In three groups of people, chosen from different geographical regions, the distribution of hair color is as follows.

	Red hair	Light hair	Dark hair
Group A	2	9	9
Group B	3	6	21
Group C	15	15	20

Do these data indicate that hair color is dependent on geographical region? (Use the 2% level of significance.)

32. A gray female zebra finch mated to a gray-cinnamon male produced 3 gray, 2 cinnamon, and 1 gray-cinnamon progeny. A gray-cinnamon female mated to a gray male produced 4 gray, 3 cinnamon, and 3 gray-cinnamon progeny. On the basis of these results, is there a relation between color distribution of progeny and sex placement (color of the parents) in a cross of gray-cinnamon and gray parents?

33. Questionnaires were mailed to the graduates of a certain university with the following results:

	B.S.	M.S.	Ph.D.
Returned	78	44	18
Did not return	22	6	7

Do these data indicate that the rate of return of these questionnaires is independent of the degree earned?

34. A company carries two lines of products and has three salesmen. A three-month record shows the following number of units of the two lines sold by each salesman.

	Salesman		
	1	2	3
Line 1	20	8	15
Line 2	17	16	5

Does the record support the claim that each salesman's ability to sell depends on the line he is selling?

35. The following are tabulated data on 82 strains of oats divided into 2 groups according to the presence or absence of awns, and into 3 groups according to yield. Do these data permit the conclusion that more of the awned strains occur in the highest yielding classes than do the awnless strains?

	Yield class (weight in grams)		
	151–200	201–250	251–325
Awned	6	7	21
Awnless	18	21	9

36. The data of Exercise 5 of Chapter 2 giving the total seasonal rainfall at Davis, California, for the 70-year period 1905-1975, are represented in the following frequency table:

Class boundaries	Frequency	Class boundaries	Frequency
5– 8	1	20–23	7
8–11	9	23–26	6
11–14	15	26–29	5
14–17	13	29–32	1
17–20	13		

Can these data be considered as satisfying a normal distribution? Note that for the above frequency table $m = 16.96$, $\sigma = 5.60$.

37. Eighty sets of 10 seeds each are placed on damp filter paper. The frequencies of sets having various numbers of seeds germinating are recorded below. Can these data be assumed to follow a normal distribution?

Number germinating:	0	1	2	3	4	5	6	7	8	9	10
Frequency:	6	20	28	12	8	6	0	0	0	0	0

14

Index Numbers

14.1 INTRODUCTION

In many business and economics problems we are interested in reducing data to purely relative numbers amenable to comparison. Numbers that are all relative to the same quantity (or base) are called *index numbers*. Index numbers are used in comparing price, production, employment, or population changes over a certain period of time. They may also be used in studying differences between places or differences between comparable categories such as animals or persons or commodities. Index numbers play an important role today: The wages of many workers in the United States are tied by union contract to the consumer price index, which is an index number compiled by the Bureau of Labor Statistics. This type of agreement between a labor union and management became widespread after General Motors Corporation signed such a contract in May 1948 with the United Automobile Workers (at that time CIO) and in 1950 entered into a five-year pact on this basis.

14.2   SELECTION OF THE BASE PERIOD

Although many types of index numbers are in use, all of them initially require selection of a base period as a point of reference. For the base period, the index number is set at 100.

The choice of a particular base period is guided by the intended use of the index numbers. Often it is desirable that the base represent as closely as possible a "normal" or "typical" situation. However, because the base period represents the time from which changes are measured, selection of a period not too distant from the present is usually advisable.

Until 1950, many of the price indices in use in the United States had as their base the average of prices during the five-year period 1935–1939, a period that can be considered rather normal inasmuch as it was sufficiently far removed from the depression years and was not as yet affected by the economic changes that occurred with the beginning of World War II. Indeed, the Central Statistical Board of the United States Government suggested the selection of indices on this basis. Beginning in early 1950, the most common base for indices in the United States was the average of the three-year period 1947–1949. At present, 1967 is used as base year.

The base need not refer to a single year, month, day, or hour. Frequently it is the average of several consecutive periods. If data have been collected for relatively few years, and no knowledge of what represents a normal or typical result is available, the first year is usually selected as the base period.

14.3   SIMPLE INDEX NUMBERS

Definition 14-1. *Simple index numbers* are the percents of each of a set of variates relative to some quantity that is selected as a base.

Consequently, if the variate in the ith period is denoted by p_i and the quantity for the base period by p_o, the ith simple index number is given by

$$S_i = \frac{p_i}{p_o} \cdot 100. \qquad (14\text{-}1)$$

Example 14-1.   Table 14-1 lists the net income of Ford Motor Company for each of the years from 1965 to 1974. Using the year 1965 as the base year, express the net income for each year as an index number.

Table 14-1
Net income (in millions of dollars) of Ford
Motor Company.

Year	Net income	Index number
1965	703.0	100.0
1966	621.0	88.3
1967	84.1	12.0
1968	626.6	89.1
1969	546.5	77.7
1970	515.7	73.3
1971	656.7	93.4
1972	870.0	123.8
1973	906.5	128.9
1974	360.9	51.3

Solution.   Since 703.0 is the variate for the base period, we find for 1966

$$S_1 = \frac{621.0}{703.0} \cdot 100 = 88.3.$$

The index numbers for the net income in the other years are found similarly and are listed in the last column of Table 14-1. Note that these index numbers facilitate comparisons with the base year. For example, we see that from 1965 to 1966 the net income of Ford Motor Company declined by 11.7%, whereas from 1965 to 1973 the net income increased by 28.9%.

In Example 14-1 our interest was restricted to the study of a single quantity for each year. Most investigations involve a number of quantities corresponding to each period.

Definition 14-2.   If the price per unit of a number of items is given for a series of time intervals, the *simple aggregate price index* for a given time interval is the sum of the prices per one unit of each of these items at that time interval expressed as a percent relative to the corresponding price for the period selected as the base period.

Consequently, if the prices per unit of k items in the base period are denoted by $p_{o,1}, p_{o,2}, p_{o,3}, \ldots, p_{o,k}$, those in the first period by $p_{1,1}, p_{1,2}, \ldots, p_{1,k}$, and in general those in the ith period by $p_{i,1}, p_{i,2}, p_{i,3}, \ldots, p_{i,k}$, then the simple aggregate price index for the ith period is given by

$$S_i = \frac{p_{i,1} + p_{i,2} + \cdots + p_{i,k}}{p_{o,1} + p_{o,2} + \cdots + p_{o,k}} \cdot 100 = \frac{\sum p_i}{\sum p_o} \cdot 100. \qquad (14\text{-}2)$$

Example 14-2. In the third, fourth, and fifth columns of Table 14-2 are listed the prices per pound of three metals used by a certain factory in the years 1950, 1951, 1952 (the other columns of this table are not used in this example). Calculate the simple aggregate price index for these metals for the years 1951 and 1952, using 1950 as the base year.

Table 14-2

Prices and typical quantities of three metals used by a certain factory.

1	2	3	4	5	6	7	8
	Typical quantities used	Price per pound 1950	1951	1952	Price of quantities used 1950	1951	1952
Metal	(w)	p_0	p_1	p_2	$p_0 w$	$p_1 w$	$p_2 w$
Copper	20,000 lbs	$0.242	$0.242	$0.242	$4840	$4840	$4840
Lead	12,000 lbs	0.170	0.190	0.148	2040	2280	1776
Zinc	14,000 lbs	0.175	0.195	0.125	2450	2730	1750
		0.587	0.627	0.515	9330	9850	8366

Solution. For the base year 1950, the total price of 1 pound each of copper, lead, and zinc is $0.587. In 1951 the price of these metals is $0.627; thus, the simple aggregate price index for 1951 is

$$S_1 = \frac{0.627}{0.587} \cdot 100 = 106.8.$$

Similarly, the corresponding simple aggregate price index for 1952 is found to be 87.7.

14.4 WEIGHTED INDEX NUMBERS

As is clear from Example 14-2, simple index numbers do not take into account the relative importance (called the *weight*) of the various items considered. Therefore, simple index numbers are also referred to as *unweighted index numbers*.

Frequently, unweighted index numbers are rather meaningless. For example, if the consumer price index were based upon the unit price of each of the consumer items used by a typical family, and if no consideration were given the relative importance of these items, the price index would not give a true picture of the amount spent for such items by the typical family.

Definition 14-3. An index number that takes into account the weight (used here in the sense of importance) of each of the items considered is called a *weighted index number*.

Since the weights of items may fluctuate during the period under consideration, it is important that the weighted index number be a measure of the change in the price of the item and not a measure of the change in its weight. One method of constructing a weighted index number is to use the same weights for all periods, determining these weights not for any one period but rather as averages of several periods.

Let $w_1, w_2, \ldots, w_k$ be the typical (or average) weights of k items applicable to the period under study, and let $p_{o,1}, p_{o,2}, \ldots, p_{o,k}, p_{1,1}, p_{1,2}, \ldots, p_{1,k}, p_{i,1}, p_{i,2}, \ldots, p_{i,k}$ be defined as before. Then the cost of the k items with weights $w_1, w_2, \ldots, w_k$ in the base period is

$$p_{o,1}w_1 + p_{o,2}w_2 + \cdots + p_{o,k}w_k = \sum p_o w,$$

and in the ith period is

$$p_{i,1}w_1 + p_{i,2}w_2 + \cdots + p_{i,k}w_k = \sum p_i w,$$

and the weighted aggregate price index in the ith period is given by

$$W_i = \frac{\sum p_i w}{\sum p_o w} \cdot 100. \tag{14-3}$$

Example 14-3. The second column of Table 14-2 lists the average quantities of three metals used by a certain factory in the period 1950–1952. Calculate the weighted aggregate price index for these metals for 1951 and 1952, using 1950 as the base year.

Solution. As shown in the sixth column of Table 14-2, the 1950 prices of the average quantities of the three metals are $4840, $2040, and $2450. Therefore, the total cost of these metals for the base period is

$$\sum p_o w = 9330.$$

For 1951 the total cost for the same quantities is

$$\sum p_1 w = 9850.$$

Thus, the weighted aggregate price index for 1951 is

$$W_1 = \frac{9850}{9330} \cdot 100 = 105.6.$$

And the weighted aggregate price index for 1952 is 89.7.

It is rarely possible to use the same weights over long periods of time. Because the relative importance of the items under study changes from day to day or from year to year, and because new items often must be added, weights should be periodically revised. There have been several changes in the United States system of weights used in contructing the consumer price index; for example, one was made in early 1950 in accordance with new conditions following World War II that resulted in certain items, such as frozen foods, assuming a major role in the budget of the typical family unit. This change was made at the same time that the base period was changed to the years 1947–1949. At present, 1967 is used as base year for the consumer price index.

Although defining a weighted index number by Formula (14-3) is only one of many ways of constructing weighted index numbers, this method is the one most generally used. Irving Fisher, in his treatise, "The Making of Index Numbers" (1922), listed 134 different formulae which have been suggested for use in constructing weighted index numbers.

EXERCISES

1. The following figures show the production of crude oil and natural gas liquids in the United States by Phillips Petroleum Company (thousands of barrels per day).

1965	1966	1967	1968	1969	1970	1971	1972	1973	1974
251.1	262.7	277.0	288.7	280.2	286.0	281.1	282.1	272.2	270.7

Express these production figures as index numbers,
(a) using the year 1965 as the base year;
(b) using the year 1966 as the base year.

2. The sales of petroleum products by Phillips Petroleum Company in the United States are given by the following figures (thousands of barrels per day).

1965	1966	1967	1968	1969	1970	1971	1972	1973	1974
457	517	566	585	596	616	583	605	553	541

Express these sales figures as index numbers,
(a) using the year 1965 as the base year;
(b) using the year 1966 as the base year.

3. Residential building cost indices (base period 1947–1949) are shown by the following figures.

1915	1920	1925	1930	1935	1940	1945	1950	1955
26.7	59.3	47.9	48.7	40.3	50.5	70.1	107.7	123.9

If building a house in 1945 cost $7500, what would it have cost to build the same house in 1955?

4. A company buys three commodities in the typical quantities and at the prices shown in the following table.

			Price per unit		
Commodity	Units	Weights	1974	1975	1976
A	pounds	20,000	$0.62	$0.58	$0.71
B	dozen	50,000	2.05	2.05	2.09
C	pounds	60,000	0.83	0.75	0.85

Using 1974 as the base year, find for 1975 and 1976
(a) the simple aggregate price index;
(b) the weighted aggregate price index.

5. Every year a farmer purchases four items in the typical quantities and at the prices shown in the following table.

			Price per unit		
Commodity	Units	Weights	1974	1975	1976
A	pounds	120	$0.45	$0.46	$0.48
B	pounds	50	0.60	0.61	0.62
C	pounds	60	0.80	0.70	0.66
D	dozen	100	1.20	1.30	1.40

Using 1974 as the base year, find for 1975 and 1976

(a) the simple aggregate price index;

(b) the weighted aggregate price index.

(c) If from 1976 to 1977 the prices of commodities A and B are expected to increase by 1 cent per pound, and the price of commodity D is expected to increase by 10 cents per dozen, how much must the cost per pound of commodity C decrease during that period in order that the weighted aggregate price index for 1977 remain the same as for 1976?

15

Time Series

15.1. INTRODUCTION

In this chapter, attention is given to data that show variations with time.

Definition 15-1. Series of successive observations of the same phenomenon over a period of time are called *time series*.

Time series are of particular importance in economic and business studies, but their use is not restricted to such data. The data of Tables 14-1 and 15-1 are examples of time series. Ordinarily, the observed values of the variable studied, such as the price of a commodity, are the results of various influences. Discovering and measuring the effects of these influences are the primary purposes of a time series analysis. Although the effects cannot always be determined exactly, they can often be approximated if observations have been made over a sufficiently long period. Time series analyses are basic to understanding past behavior, evaluating current accomplishments, planning future operations, and comparing different time series.

It is customary to consider time series variations as being the result of four well-defined influences: the secular trend, the seasonal variation, the cyclical fluctuation, and the random or erratic variation.

Definition 15-2.   The *secular trend* is that characteristic of a time series which extends consistently throughout the entire period of time under consideration.

The secular trend is the basic long-term tendency of a particular activity to grow or to decline. It indicates the presence of factors that persist for a considerable duration—factors such as population changes, technological improvements, price level fluctuations, or various conditions that are peculiar to individual industries or establishments.

Frequent and sudden changes in a secular trend are inconsistent with the notion that it represent the regular, steady, long-run movement of a time series. Occasionally an important new element is introduced into the complex of forces affecting the time series, or an old factor is eliminated. Then an abrupt change in the trend is real and should be recognized.

Definition 15-3.   *Seasonal variations* are variations that occur in regular sequence at specific intervals of time.

Although the word "seasonal" might seem to imply a connection with the seasons of the year, the term is meant to include any kind of variation that is of a periodic nature, and whose repeating cycles are of relatively short duration. Seasonal variations are the result of such seasonal factors as climatic conditions, holidays, business operations, or human habits. Prices of perishable agricultural commodities show pronounced seasonal variation, being high early in the season, then declining sharply as supply is increased at the peak of the season, and finally rising again as supply is diminished. A time series giving the number of visitors per hour at a bank would show a seasonal variation since noon hours and pay days are always busy periods. The borrowing of books at college libraries shows seasonal variation; there is a surprising increase before each examination period and a sharp decline thereafter.

Definition 15-4.   *Cyclical fluctuations* are long-term movements that represent consistently recurring rises and declines in activity.

Cyclical fluctuations in time series of business activities are usually called business cycles. Although markedly less periodic than seasonal variations, cyclical fluctuations are characterized by considerable regularity. Most business establishments reflect the alternating periods of depression and prosperity experienced by the general economy. An example of this type of fluctuation would be the time series showing building activity in the United States between 1830 and 1940. A graph of this series would reveal more or less regular cycles, with an approximate interval of 18 years between successive tops and bottoms of the curve. In addition to business cycles there are certain other cyclical movements. For example, hog prices move in characteristic cycles averaging about four years in length; the cycles in hog prices are caused by, or at least lag behind, opposite cycles in hog production. Beef cattle prices move in cycles of about 15 years.

Definition 15-5. *Random or erratic variations* are variations that occur in a completely unpredictable fashion.

Random variations are due to unforeseeable causes and may be the result of wars, strikes, floods, fires, earthquakes, unusual weather, or some political events. More simply, random or erratic variations in a time series are variations that cannot be accounted for by the secular trend, seasonal variations, or cyclical fluctuations.

Most time series show a definite secular trend, upon which are superimposed seasonal variations, cyclical fluctuations, and random variations. Each observation of a time series can therefore be thought of as being the sum of these four (positive or negative) components. If we denote the secular trend component by T, the seasonal component by S, the cyclical component by C, and the random or erratic component by E, we can express each observation Y of a time series as

$$Y = T + S + C + E. \tag{15-1}$$

Isolating those parts of the over-all variation of a time series that are traceable to each of these four components and measuring each part independently are major aims of a time series analysis. In this chapter we shall discuss the various methods employed in isolating and measuring these four components.

15.2 THE SECULAR TREND

15.2.1 The Method of Least Squares

Since the secular trend is the long-term tendency of a time series (disregarding the usually minor seasonal variations and the major cyclical and random variations), we may think of the secular trend as being the main tendency of a time series. It is natural, therefore, to represent a time series graphically by a scatter diagram and to find the curve, preferably a line, that best fits the points of the scatter diagram. If a straight-line trend is indicated, the most widely used procedure of fitting the data is by the method of least squares (see Chapter 12). Inasmuch as the observations Y of a time series are usually measured at uniformly spaced intervals, such as successive hours, days, months, or years, the formulae given by (12-42) are applicable.

Since the secular trend disregards seasonal influences, the calculation can be further simplified by condensing the data in such a way that short-term variations are eliminated. For example, a monthly time series extending over several years is best expressed first as a yearly time series.

Example 15-1.   Express the data of Table 15-1 as a yearly time series, and find the corresponding equation of the trend line.

Table 15-1
Number of hired farm workers on Pacific Coast (in thousands).

Year	Jan.	Feb.	Mar.	Apr.	May	June	July	Aug.	Sept.	Oct.	Nov.	Dec.
1939	156	156	158	198	205	237	260	317	380	255	210	165
1940	146	157	159	202	227	232	283	260	345	340	195	151
1941	134	143	154	170	203	275	368	413	430	361	294	231
1942	160	162	172	171	210	244	374	394	476	339	332	236
1943	153	161	167	185	209	269	356	404	474	350	309	243
1944	165	178	184	185	207	285	359	385	460	349	315	236
1945	174	179	179	188	211	263	364	417	501	336	341	224
1946	176	174	172	188	240	290	388	422	468	357	343	248
1947	188	190	201	200	236	313	418	432	453	387	339	249
1948	196	187	183	205	239	306	388	436	484	436	373	250

Solution.   Determining the average number of hired workers for the twelve months of each year, we obtain the values of Y' listed in the second column of Table 15-2. (We use Y' to distinguish the calculated values from the observed values of Y given in Table 15-1.)

Since the values of Y' are averages of twelve observed values for the months January 1 to December 1, they should refer to the midpoint of the twelve-month period, i.e. they should be centered at July 1. The mean of the ten years occurs halfway between July 1, 1943 and July 1, 1944, at January 1, 1944, which therefore corresponds to time $x = 0$. July 1, 1944, half a year later, is represented by $x = 0.5$, July 1, 1945, by $x = 1.5$, and so on, as shown in the third column of Table 15-2. The values of Σx^2, $\Sigma x Y'$, required in the calculation of the values for b and a, are listed in the fourth and fifth columns of Table 15-2.

Table 15-2
Calculation of trend line for the data of Table 15-1 by
method of least squares

Year	Y'	x	x^2	xY'
1939	225	—4.5	20.25	—1012.5
1940	225	—3.5	12.25	—787.5
1941	265	—2.5	6.25	—662.5
1942	272	—1.5	2.25	—408.0
1943	273	—0.5	0.25	—136.5
1944	276	0.5	0.25	138.0
1945	281	1.5	2.25	421.5
1946	289	2.5	6.25	722.5
1947	300	3.5	12.25	1050.0
1948	307	4.5	20.25	1381.5
	2713	0	82.50	706.5

Thus,

$$b = \frac{706.5}{82.50} = 8.56, \qquad a = \frac{2713}{10} = 271.$$

The trend line is given by

$$Y'_e = 271 + 8.56\, x.$$

This line, together with the values of Y', is shown in Figure 15-1.

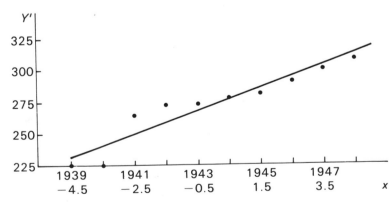

Figure 15-1
Trend line for data of Table 15-2 obtained by method of least squares.

15.2.2   The Method of Semiaverages

Another method of determining the linear trend is the method of semi-averages. Its chief advantage lies in the simplicity of calculation, its disadvantage in a somewhat greater inaccuracy than the method of least squares.

First, the time series is divided into two equal-length parts. Thus the time series in Table 15-2, covering the years from 1939 to 1948, would be divided into a period 1939–1943 and a period 1944–1948. If a time series consists of an odd number of periods, the middle period is omitted in order to obtain an even number of periods. The means of these two halves are then calculated and graphed as points referring to the midpoints of the two respective time intervals. The line joining these two points is the trend line, as obtained by the method of semiaverages.

Example 15-2.   Using the data of Table 15-2, graph the trend line by the method of semiaverages.

Solution.   The mean value of Y' for the five-year period 1939–1943 is 252, and the mean value for the five-year period 1944–1948 is 291. Since each half of the time series covers an odd number of years, the two means are centered at July 1, 1941, and July 1, 1946. These two means are plotted and connected by a straight line, denoted by (1), in Figure 15-2. For comparison, the line obtained by the method of least squares, line (2), is also included in this graph. There is very good agreement between the two lines in this case.

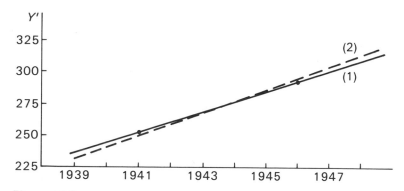

Figure 15-2

Trend lines for data of Table 15-2 obtained by method of semiaverages (1) and method of least squares (2).

15.2.3   Isolation of the Trend from a Time Series

After the trend line has been determined it is possible to determine the so-called trend value—the values of Y_e obtained from the trend line. From the equation of a straight line, $Y_e = a + bx$, it is evident that the amount of increase (or decrease) in Y_e between two consecutive values of x (i.e. between two consecutive periods) is equal to b, the slope of the line of regression. If the trend value for a particular period has been determined, the trend values for the succeeding periods can be obtained by the repeated addition of b.

Example 15-3.   Determine the trend values for the data of Table 15-2.

Solution.   Using the equation of the line obtained in Example 15-1, we find for $x = -4.5$ the trend value for 1939:

$$Y_e' = 271 + 8.56(-4.5) = 232.$$

The trend value for 1940 is $271 + 8.56(-3.5) = 241$. These and the other trend values are listed in the third column of Table 15-3.

Table 15-3
Trend values for the data of Table 15-2.

Year	Y′	Trend values
1939	225	232
1940	225	241
1941	265	250
1942	272	258
1943	273	267
1944	276	275
1945	281	284
1946	289	292
1947	300	301
1948	307	310

Trend values are the values that would be observed if the secular trend is the only factor influencing a time series; i.e. if there is no effect whatsoever due to a seasonal variation, a cyclical fluctuation, or a random disturbance.

In Example 15-3 we confined our attention to the annual trend increment, which was 8.56. In many cases, such as that represented by the data of Table

15-1, it is also of interest to consider the monthly trend increment, which is found by dividing the annual trend increment by 12. Monthly trend increments are of particular importance if it is desired to eliminate the trend for the given data—that is, if it is desired to study the time series that would be obtained in the absence of any trend.

Example 15-4. Eliminate the trend from the data of Table 15-1, considering the number of hired farm workers in January 1939 as being fixed (considering 156 as the base figure).

Solution. From Example 15-1 we find the monthly trend increment to be

$$\frac{8.56}{12} = 0.71.$$

Eliminating the trend from the observed value for February 1939 we obtain

$$156 - 0.71 = 155,$$

rounded off to the nearest integer, and for March 1939 we obtain

$$158 - 2(0.71) = 157.$$

These values, and those for the rest of 1939, are given in Table 15-4.

Table 15-4
Elimination of the trend for the data of Table 15-1 (year 1939).

Month	Y (observed values)	Values after elimination of the trend
Jan.	156	156
Feb.	156	155
Mar.	158	157
Apr.	198	196
May	205	202
June	237	233
July	260	256
Aug.	317	312
Sept.	380	374
Oct.	255	249
Nov.	210	203
Dec.	165	157

Clearly, not all trends can be represented by straight lines. The general question of what curves should be fitted to various types of nonlinear trends will not be discussed here. Instead, we shall mention briefly an example that is of particular importance in the study of various types of growth, such as those of bacteria, insects, and animals. Analysis of such growth has revealed that initially the rate of growth is proportional to the size of the population. Later, however, some factor—usually, though not always, attributable to the limitations of space and food—begins to decelerate growth. Sometimes this decelerating factor appears long before limitations of land, food, or other resources could conceivably influence the growth of the population. The point at which acceleration of growth has changed to deceleration is called the critical point of the growth curve, and plays an important role in the study of population trends. After the critical point has been determined, the saturation level can be estimated.

The curve that represents the growth pattern just described is called the *logistic curve*. Its equation is

$$Y_e = \frac{k}{1 + be^{-aX}},\qquad (15\text{-}2)$$

where a, b, and k are constants determined from the data of the time series and e is approximately $2.71828 \cdots$, the same number referred to in Chapter 7 in connection with Equation (7-1).

The study of the logistic curve originated with work by Raymond Pearl and Lowell J. Reed in 1920. Pearl, a zoologist, found the equation of the curve from an analysis of the growth of fruit flies enclosed in bottles with fixed amounts of food. Figure 15-3 shows the average number of fruit flies in two half-pint bottles, each started on the same day with two adult wild-type flies (one male and one female).

The observations (averages of bottles 1 and 2), indicated by dots in Figure 15-3, were fitted with a logistic curve, as given by Equation (15-2), for which it was found that $k = 212$, $a = -0.27$, and $b = 108.85$. It is clear that the observed growth of this fly population is, in general, accurately described by Equation (15-2) with the values of k, a, and b given above. The only serious discrepancies (April 21 and 23) are due to starting both bottles with only one pair of adult flies; we are thus dealing only with the children of the two original parents, not yet with any grandchildren—we are considering only a family, not population growth. Not until April 26 did overlapping of generations begin, and only from then on are we dealing with population growth.

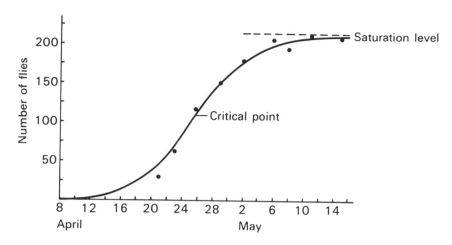

Figure 15-3
Growth of wild-type fruit flies in half-pint bottles.

Since for large values of X the denominator of the fraction in Equation (15-2) approaches 1, the constant k is seen to equal the estimated size of the population when it has reached the saturation level. It can also be shown from (15-2) that the critical point of the curve occurs at

$$X = \frac{1}{a} \ln b \tag{15-3}$$

and that the estimated size of the population at that time is $Y_e = \frac{1}{2}k$.

Thus, for the data in Figure 15-3, the saturation level is 212 flies, and the critical point was reached on April 26, when there were an estimated 106 flies.

Their success in using the logistic curve to estimate population growth in fruit flies led Pearl and Reed to apply the same curve to estimating human population growth in various countries, including the United States. A forecast made by them in 1920 on the basis of the logistic curve predicted the 1950 population of the United States with an error of only about 1%. Use of the same logistic curve, however, resulted in an error of more than 11% in the prediction of the 1960 population of the United States, and thus this logistic curve (which, incidentally, was shown in earlier editions of this text) can no longer be used to predict the growth of the United States population.

Attempts by Pearl and Reed to predict population growth of other countries, e.g., Germany and Algeria, resulted in similar experiences: very good estimates for earlier years, but disappointingly large errors for later years.

It has, therefore, become evident that the early successes of the use of the logistic curve to predict population growth were coincidental. The failure of the logistic curve to describe accurately growth of human populations is not surprising when it is recalled that Pearl and Reed obtained this curve from an experiment with fruit flies in which there was a fixed amount of space, food, and other resources. Their hypothesis is generally not applicable to human populations—except, possibly, in the very long run. Control of the environment has made it increasingly possible to overcome limitations of resources that otherwise might cause a deceleration of population growth. Attempts to determine maximum populations by extrapolation from present technology to future expected improvement appear to be just as hazardous as projecting the number of people directly. It must be concluded that the influences on human population growth are too varied and dependent on too many unpredictable factors to allow accurate description by any particular curve or curves.

15.3 THE SEASONAL VARIATION

To study the seasonal variation in a time series we first consider a purely hypothetical time series in which all variations are entirely due to seasonal influences. Of course, a time series such as this is hardly ever actually encountered. In such a time series, as shown in the second, third, and fourth columns of Table 15-5, the yearly totals are identical; if they were not, some influence other than seasonal would be present.

Table 15-5
Sales (in thousands of dollars) of a corporation in the period 1974–1976.

	Sales			Specific seasonals			Typical seasonals
Quarter	1974	1975	1976	1974	1975	1975	
1st	112	104	115	72	67	74	71
2nd	172	162	179	110	104	115	110
3rd	212	221	211	136	142	135	138
4th	128	137	119	82	88	76	82
	624	624	624	400	401	400	401

Definition 15-6. If the variations in a time series are entirely due to seasonal influences, the *specific seasonal* for a given period of the time series is an index number whose base is the mean quarterly or monthly variate (depending on whether the time series is a quarterly or monthly one).

Example 15-5. Calculate the specific seasonals for each quarter of the time series of Table 15-5.

Solution. The mean quarterly sale, in thousands of dollars, for each year of the time series is 156. The specific seasonal for the first quarter of 1974 is

$$\frac{112}{156} \cdot 100 = 72.$$

The specific seasonals for the other quarters are listed in the fifth, sixth, and seventh columns of Table 15-5.

To study the typical effect of any particular season in the time series, we introduce the following definition:

Definition 15-7. For a quarterly time series, the *typical seasonal* for the first quarter is the mean of all specific seasonals for all the first quarters of the time series. Typical seasonals for the other three quarters are defined analogously.

For a monthly time series, the typical seasonal for January is the mean of all specific seasonals for January, and so on.

For a time series showing only seasonal variation, the mean of all typical seasonals is always 100. The proof of this fact is left as an exercise.

Example 15-6. Calculate the typical seasonals for the four quarters of the time series of Table 15-5.

Solution. For the first quarter, the typical seasonal is

$$\frac{72 + 67 + 74}{3} = 71.$$

The typical seasonals for the other three quarters are calculated in like manner; their values are listed in the last column of Table 15-5.

If a time series is influenced by variations other than seasonal ones, the mean quarterly or monthly values are no longer constant. This might suggest that, in the calculation of specific seasonals, a different base be used for each year of the time series. But such a procedure might still give a distorted picture, especially if a seasonal low coincides with the bottom of a business cycle, in which case too much weight might be ascribed to the seasonal influence. In order to avoid this possibility, three methods are commonly used to calculate the seasonal variation in a time series: *the ratio-to-moving-average method, the ratio-to-trend method,* and *the method of link-relatives.* These methods differ only in the way in which they define the specific seasonals. The typical seasonals are defined in terms of the specific seasonals in the same way in all three methods. The first of these methods is used much more frequently than are the other two.

15.3.1   The Ratio-to-moving-average Method

As the name implies, this method makes use of a moving average.

Definition 15-8.   If the observations in a time series are denoted by Y_1, Y_2, ..., Y_n, then a *moving average* of length k is the series of successive means of k consecutive observations. That is, the series of moving averages is

$$\overline{Y}_1 = \frac{Y_1 + Y_2 + \cdots + Y_k}{k},$$

$$\overline{Y}_2 = \frac{Y_2 + Y_3 + \cdots + Y_{k+1}}{k},$$

$$\overline{Y}_3 = \frac{Y_3 + Y_4 + \cdots + Y_{k+2}}{k}, \cdots.$$

Since the values of these means, $\overline{Y}_1$, $\overline{Y}_2$, ..., are based upon k observations, they are centered in a graph at the midpoint of the set of observations for which the mean has been calculated. Thus, if $k = 7$, then $\overline{Y}_1$ is centered at Y_4; similarly, $\overline{Y}_2$ is centered at Y_5, and so on. If $k = 8$, then Y_1 is centered at the midpoint between Y_4 and Y_5; similarly, $\overline{Y}_2$ is centered at the midpoint between Y_5 and Y_6, and so on.

The ratio-to-moving-average method makes use of the fact that if a monthly time series is affected by the seasonal fluctuations due to the twelve months of the year, then a moving average of length 12 will remove these fluctuations.

Definition 15-9.   For a general time series, the *specific seasonal* for a given observation is an index number whose base is the moving average that is centered at that observation.

Example 15-7. Calculate the specific seasonals for the data of Table 15-1.

Solution. As explained on the next page we obtain the specific seasonals shown in Table 15-6.

Table 15-6

Calculation of specific seasonals for 1939 and 1940 for the data of Table 15-1 by the ratio-to-moving-average method.

Year and month	Y	12-month moving total	12-month moving average	Centered 12-month moving average	Specific seasonal
1939					
Jan.	156				
Feb.	156				
Mar.	158				
Apr.	198				
May	205				
June	237				
		2697	225		
July	260			224.5	116
		2687	224		
Aug.	317			224	142
		2688	224		
Sept.	380			224	170
		2689	224		
Oct.	225			224	114
		2693	224		
Nov.	210			225	93
		2715	226		
Dec.	165			226	73
1940		2710	226		
Jan.	146			227	64
		2733	228		
Feb.	157			225.5	70
		2676	223		
Mar.	159			221.5	72
		2641	220		
Apr.	202			223.5	90
		2726	227		
May	227			226.5	100
		2711	226		
June	232			225.5	103
		2697	225		
July	283				
Aug.	260				
Sept.	345				
Oct.	340				
Nov.	195				
Dec.	151				

The first twelve-month moving average,

$$\overline{Y}_1 = \frac{156 + 156 + 158 + \cdots + 165}{12} = \frac{2697}{12} = 225,$$

is centered between June 1 and July 1, 1939; the second twelve-month moving average,

$$\overline{Y}_2 = \frac{156 + 158 + 198 + \cdots + 146}{12} = \frac{2687}{12} = 224,$$

is centered between July 1 and August 1, 1939. In order to obtain an average centered at July 1, 1939, we use the mean of $\overline{Y}_1$ and $\overline{Y}_2$:

$$\frac{225 + 224}{2} = 224.5.$$

These centered twelve-month moving averages are listed in the fifth column of Table 15-6. They are used as the bases in the calculation of the specific seasonals. Thus, the specific seasonal for July 1939 is

$$\frac{260}{224.5} \cdot 100 = 116.$$

In a similar manner, specific seasonals for all months from July 1939 to June 1940 are calculated; they are listed in the last column of Table 15-6.

Note that each twelve-month moving total (except the first) can be calculated from the preceding one by adding the next following observation and subtracting the first one used to calculate the preceding total. Thus, the second moving total in Table 15-6 is obtained from the first as

$$2697 + 146 - 156 = 2687.$$

In a quarterly time series, centered four-quarter moving averages are used in determining the specific seasonals for each quarter.

The ratio-to-moving-average method does not allow the calculation of moving averages for the first and last six months of the time series under study. Similarly, for quarterly time series, calculation of the moving averages for the first two and last two quarters is impossible. This disadvantage of the ratio-to-moving-average method is not serious when we are dealing with a time series covering an extended period of time, since the omission of a few specific seasonals at the beginning and end of the series hardly affects the calculation of the typical seasonals.

We can now calculate typical seasonals from the specific seasonals. Recalling Definition 15-7, it would seem natural to obtain again the typical seasonal for January as the mean of all available specific seasonals for January. However, this might give undue weight to a few specific seasonals that might be excessively low or high, as compared to the rest. In order to avoid this, one of two procedures is used. Either the median of all specific seasonals is employed, or a certain fixed number of the exceptionally high and low specific seasonals is disregarded and the mean of the remaining specific seasonals is calculated. For simplicity, we shall adopt the first procedure. Usually the results of the two methods will differ little. However, the medians (or means) thus obtained for each season will not in general have a mean of 100. But since typical seasonals were previously defined in such a way that their mean is 100, the set of medians (or means) obtained must be expressed as index numbers whose base is the average of these medians (or means). Thus the index numbers will have a mean of 100 and will therefore represent the typical seasonals.

Example 15-8.   Calculate the typical seasonals for the data of Table 15-1.

Solution.   Using the results of Table 15-6 and extending the calculations to include all the data of Table 15-1, we obtain the specific seasonals listed in Table 15-7 (1939 and 1948 are omitted, since for the first and last six months of a monthly time series no specific seasonals can be calculated by the ratio-to-moving-average method).

Table 15-7
Specific seasonals for 1940–1947 for the data of Table 15-1.

Year	Jan.	Feb.	Mar.	Apr.	May	June	July	Aug.	Sept.	Oct.	Nov.	Dec.
1940	64	70	72	90	100	103	126	116	155	154	89	69
1941	60	61	63	68	80	105	138	154	160	135	109	86
1942	60	61	64	63	78	90	138	145	175	125	122	86
1943	56	59	61	67	76	99	130	147	171	126	112	88
1944	59	64	67	67	75	103	130	139	167	126	114	86
1945	63	65	64	67	75	93	129	148	178	120	121	79
1946	62	61	60	66	84	101	134	145	160	121	116	84
1947	63	63	67	67	79	104	139	144	151	129	113	83
Median	61	62	64	67	78.5	102	132	145	163.5	126	113.5	85
Typical seasonal	61	62	64	67	78.5	102	132	145	163.5	126	113.5	85

The median of the eight specific seasonals for January is found to be 61, that for February is 62, and so on, as listed in the next-to-last row of Table 15-7. Since the mean of the twelve medians is 100, i.e.

$$\frac{61 + 62 + 64 + \cdots + 85}{12} = \frac{1199.5}{12} = 100,$$

we find the typical seasonal for January to be

$$\frac{61}{100} \cdot 100 = 61.$$

In this special case the typical seasonals are identical to the corresponding medians.

15.3.2   The Ratio-to-trend Method

This method differs from the ratio-to-moving-average method in that the specific seasonals are not calculated by use of moving averages but are based upon the trend values. Thus, each observation is expressed as a percentage of the trend value corresponding to that particular observation. This method is particularly suitable where an abrupt change occurs in the trend: In 1941 fresh citrus shipments showed a sudden change in trend when there was a material decrease due to the introduction of and subsequent increase in processing, initially by canning, then by concentration of juice, and later by freezing (see Exercise 5). The ratio-to-trend method has the disadvantage that it does not differentiate between seasonal influences and cyclical and random influences.

15.3.3 The Method of Link-relatives

In the method of link-relatives, each successive variate is expressed as a percent of the preceding variate. The "typical link-relatives," representing the medians of the respective specific link-relatives, can be expressed as a chain, the first entry (for January) being set equal to 100. Thus, if the medians for January, February, March are 98.2, 99.0, 100.7, respectively, the corresponding numbers in the chain are 100 for January, 99.0% of 100 = 99.0 for February, and 100.7% of 99.0 = 99.7 for March. From the entry in the chain for December (say, 97.5), a number for January can be obtained: 98.2% of 97.5 = 95.7. But the entry for January was initially set equal to 100. Any difference between the two figures for January represents the presence within the time series of the effects of factors other than seasonal. In order to eliminate these influences we shall assume that variations other than seasonal have taken place uniformly over the whole period, i.e. that they can be approximated by a straight line, which implies that the change in any two consecutive months has been constant. For the above example, this constant change is

$$\frac{95.7 - 100}{12} = -0.36,$$

which should be applied to each of the entries of the chain in order to eliminate the other-than-seasonal influences. Thus we obtain the following adjusted chain: 100 for January, $99.0 - (-0.36) = 99.4$ for February, $99.7 - 2(-0.36) = 100.4$ for March. From this adjusted chain, the typical seasonals are obtained as index numbers having as a base the mean of the twelve entries of the adjusted chain.

Frequently, the typical seasonals obtained by the method of link-relatives are in fairly good agreement with those obtained by the ratio-to-moving-average method. Obviously, the latter method gives more accurate results, since it employs a more representative base for calculating the typical seasonals; on the other hand, the method of link-relatives requires considerably less computation, since the base for each link-relative is simply the previous observation.

15.3.4 Adjustment of a Time Series for Seasonal Variation

Typical seasonals give information about the usual effect of a season on a particular observation in that season. For the data of Table 15-1, the typical seasonal of 61 found for January in Example 15-8 tells us that in January the number of farm workers hired is 61% of that of the average month. In many situations, it is important to determine what the time series would have been if there had been no seasonal variation—we desire to eliminate the seasonal variation from the series. Knowing the typical seasonals, we can do this easily. If we denote the typical seasonal, also called seasonal index, corresponding to a given time by s and the corresponding observation by Y, we have

$$Y = \frac{s}{100}(T + C + E), \tag{15-4}$$

since the typical seasonal is the percent of the result in the average month
(when only components other than seasonal are present). To find the theoretical value which would have occurred in the absence of any seasonal variation, we solve Equation (15-4) for $(T + C + E)$. This value of $(T + C + E)$
represents the sum of all nonseasonal influences.

Example 15-9. Using the typical seasonals obtained by the ratio-to-moving-
average method in Table 15-7, adjust for seasonal variation the data for 1940
of the time series of Table 15-1.

Solution. To calculate the adjusted value for January 1940 we find, since
$Y = 146$ and $s = 61$,

$$T + C + E = \frac{146}{0.61} = 239.$$

This and the adjusted values for all months of 1940 are listed in the third
column of Table 15-8.

Table 15-8
Number of hired farm workers on Pacific Coast in 1940 adjusted for seasonal
variation and trend.

Month	Y	Adjusted for seasonal variation $(T + C + E)$	Trend value (T)	Sum of cyclical and random influences $(C + E)$	Seasonally adjusted number as a percent of the trend
Jan.	146	239	237	2	101
Feb.	157	253	238	15	106
Mar.	159	248	238	10	104
Apr.	202	301	239	62	126
May	227	289	239	50	121
June	232	227	240	−13	95
July	283	214	241	−27	89
Aug.	260	179	242	−63	74
Sept.	345	211	243	−32	87
Oct.	340	270	243	27	111
Nov.	195	172	244	−72	70
Dec.	151	178	245	−67	73

Many time series issued monthly by the United States Government are
published on a seasonally adjusted basis in order that any changes that might
occur between consecutive months can be ascribed to other-than-seasonal
factors.

Typical seasonals are rarely constant over long periods of time, though in
a short period their variation is ordinarily so small that they can be assumed

to be constant. However, a particular event may have sufficient impact to alter materially the typical seasonals. For example, in a time series giving the quantity of oranges produced each month in recent years in the United States, the Florida freeze of 1975 changes the typical seasonals considerably. Therefore a new set of typical seasonals should be calculated for the period following the freeze.

Now that we can eliminate both the trend and the seasonal variation from a time series, we can calculate for each observation the sum of the cyclical and random components.

Example 15-10. Calculate the sum of the cyclical and random components for the data of the time series of Table 15-1 for 1940.

Solution. The values of $(T + C + E)$ and the trend values obtained from the regression equation found in Example 15-1 are listed in the third and fourth columns of Table 15-8. Subtracting the T-values from the $(T + C + E)$-values gives the $(C + E)$-values, which are listed in the fifth column of Table 15-8.

Frequently, in order to indicate the relative importance of the cyclical and random influences, the $(T + C + E)$-values are expressed as a percent of the trend. These percents for the data of Example 15-10 are listed in the last column of Table 15-8.

15.4 THE CYCLICAL FLUCTUATION

The study of cyclical fluctuations is quite similar to the study of seasonal variations except that the latter refers to short cycles whereas the former refers to long cycles. In order to determine the length of the cycles, it is advisable to study the scatter diagram carefully and to try to find from it the period between successive tops or bottoms. Frequently, the study of cyclical fluctuations is complicated by the fact that cycles of different lengths may be superimposed upon one another, making the isolation of the individual cycles difficult. We shall consider only a single, discernible, long-term cycle in a time series. The procedures used in its isolation from the time series are analogous to the method of moving averages used in the study of the seasonal variation. The moving averages therefore represent the hypothetical values which would be observed in the absence of any cyclical fluctuation. Consequently, in a yearly time series (in which there is no seasonal variation), the random or irregular variation can be obtained by subtracting the trend values from the corresponding moving averages. Sometimes, particularly when the random variation is very small, the moving averages themselves can be employed as representing the trend of the original time series. These principles are illustrated with fictitious data in Example 15-11.

Example 15-11. The number of items of a certain product sold within a country is given in Table 15-9. Find the length of the cyclical fluctuation, the moving averages, the trend values, and the random variations, and graph the trend line and the moving averages.

Table 15-9
Number of items sold (in thousands).

Year	Sales	Year	Sales	Year	Sales
1930	458	1937	650	1944	308
1931	672	1938	329	1945	507
1932	611	1939	365	1946	693
1933	266	1940	574	1947	762
1934	207	1941	754	1948	415
1935	410	1942	601	1949	368
1936	720	1943	365	1950	575

Solution.   First we construct the scatter diagram for the data of Table 15-9. A curve drawn through the scatter diagram (Figure 15-4) reveals that the time between successive tops or bottoms of the graph is about 5 years.

Applying a moving average of 5 years to the observations of the time series of Table 15-9, and centering each average at the middle observation of the five used in the average, we obtain the moving averages listed in Table 15-10.

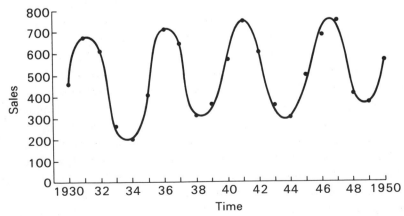

Figure 15-4
Scatter diagram for data of Table 15-9.

These represent the $(T + E)$-values for the time series. Calculating the trend line by use of Equations (12-42), we find

$$Y_e = 503 + 7.01 \ x,$$

from which we obtain the trend values given in Table 15-10. Finally, we calcu-

Table 15-10
Number of items sold (in thousands) adjusted for cyclical fluctuation and trend.

Year	Y	Moving average $(T + E)$	Trend value (T)	Random variation (E)	Year	Y	Moving average $(T + E)$	Trend value (T)	Random variation (E)
1930	458				1941	754	532	510	22
1931	672				1942	601	520	517	3
1932	611	443	447	−4	1943	365	507	524	−17
1933	266	433	454	−21	1944	308	495	531	−36
1934	207	443	461	−18	1945	507	527	538	−11
1935	410	451	468	−17	1946	693	537	545	−8
1936	720	463	475	−12	1947	762	549	552	−3
1937	650	495	482	13	1948	415	563	559	4
1938	329	528	489	39	1949	368			
1939	365	534	496	38	1950	575			
1940	574	525	503	22					

late the values of the random variations by subtracting the trend values, T, from the corresponding moving averages, $T + E$. These are also listed in Table 15-10. Graphing both the trend line and the moving averages, we see from Figure 15-5 that, owing to the small values of random variation, there is close agreement between the curve of moving averages and the trend line.

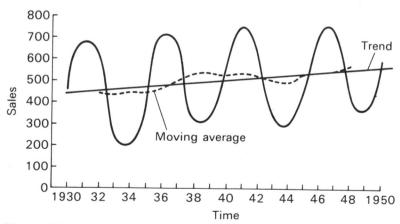

Figure 15-5
Graph of moving averages and trend line for data of Table 15-9.

Considerable attention has been devoted to business cycles that may exist in the over-all economy of a country. Various claims concerning the existence of such cycles have been made. Although some of these claims might have been valid for periods up to World War II, they have not proved valid for the period thereafter. The existence of such cycles in an entire economy is being regarded, therefore, with increasing skepticism. Some modern economists have completely abandoned the idea of business cycles in the economy of the United States.

EXERCISES

1. The following data gives the net income per share (in dollars) of Eastman Kodak Company for the years 1964–1974.

1964	1965	1966	1967	1968	1969	1970	1971	1972	1973	1974
1.28	1.71	2.15	2.19	2.33	2.49	2.50	2.60	3.39	4.05	3.90

(a) Find the equation of the trend line by the method of least squares and graph that line.
(b) Graph the trend line by the method of semiaverages.
(c) Find the trend value for each of the years listed.

2. The following data show average total hourly earnings, in dollars (including employee benefits), of workers employed by Ford Motor Company for the period 1965–1974.

1965	1966	1967	1968	1969	1970	1971	1972	1973	1974
4.65	4.81	5.28	5.46	5.78	6.40	7.21	7.83	8.43	9.49

(a) Find the equation of the trend line by the method of least squares and graph that line.
(b) Graph the trend line by the method of semiaverages.
(c) Find the trend value for each of the years listed.

3. Calculate the specific seasonals and typical seasonals for the following time series (showing the number of absences in a factory) whose variations are due entirely to seasonal influences.

Month	1973	1974	1975	1976
Jan.	337	321	339	319
Feb.	312	320	321	305
Mar.	309	301	315	305
Apr.	308	309	299	305
May	312	301	296	300
June	298	279	297	300
July	275	281	281	290
Aug.	278	299	275	279
Sept.	262	260	265	269
Oct.	295	301	292	299
Nov.	299	303	309	300
Dec.	321	331	317	335

4. The following data give the cold storage holdings of frozen eggs in millions of pounds.

Month	1941	1942	1943	1944
Jan.	73.3	95.5	82.9	102.3
Feb.	53.8	76.3	59.8	81.7
Mar.	45.2	73.8	56.5	98.6
Apr.	63.4	107.4	99.2	148.6
May	99.5	159.6	172.3	218.0
June	142.0	223.8	251.5	292.4
July	178.6	278.5	323.2	354.2
Aug.	195.2	290.5	351.2	388.5
Sept.	194.0	272.0	343.6	371.6
Oct.	178.4	234.9	306.2	332.6
Nov.	153.8	180.3	242.3	279.2
Dec.	129.5	126.3	172.4	219.8

(a) Calculate the specific seasonals by the ratio-to-moving-average method for all months of the table, where such calculation is possible.
(b) Use the specific seasonals obtained in part (a) to calculate the typical seasonal for each month.
(c) Adjust the time series for the year 1942 for seasonal variation.
(d) Find the trend line by the method of least squares.
(e) Determine the trend value for each month of 1942.
(f) Find the sum of the cyclical and random components for each month of 1942.

5. The following data give the fresh lemon shipments from California and Arizona in thousands of boxes.

Year	Nov.–Jan.	Feb.–Apr.	May–July	Aug.–Oct.
1930–1931	906	1249	2816	1133
31–32	869	1293	2336	869
32–33	892	941	2448	1125
33–34	952	1233	2957	1219
34–35	1139	1614	3036	1515
1935–1936	1124	1628	3116	1508
36–37	1203	1467	2784	1188
37–38	988	1437	3152	1767
38–39	1155	1595	3451	1798
39–40	1242	1705	3113	1739
1940–1941	1549	1789	3801	2081
41–42	1361	1684	3106	1836
42–43	1415	1945	3961	2293
43–44	1783	1928	4007	1767
44–45	1850	2173	3616	1955
1945–1946	1837	2029	3534	1755
46–47	1480	1818	3546	2330
47–48	1413	1702	3612	1763
48–49	1287	1619	3473	1572
49–50	1049	1563	3313	1459

(a) Calculate the specific seasonals by the ratio-to-moving-average method for the period 1931–1932 to 1948–1949.
(b) Use the specific seasonals obtained in part (a) to calculate the typical seasonals for the four quarters Nov.–Jan., Feb.–Apr., May–July, Aug.–Oct.
(c) Adjust the time series for the period 1931–1932 and 1932–1933 for seasonal variation.
(d) Make up a yearly time series from the given data; graph these data; and observe a change in trend in 1939–1940. (Therefore, two different trend lines should be used—one for the period 1931–1932 to 1939–1940, and another for the period 1940–1941 to 1948–1949.) Find the equations of these two trend lines by the method of least squares, and graph them.
(e) Determine the trend value for each period of 1931–1932 and 1932–1933.
(f) Find the sum of the cyclical and random components for each period of 1931–1932 and 1932–1933.

6. The number of items of a certain product imported into the United States is given on the next page in thousands of units.

Year	Number	Year	Number	Year	Number
1910	170	1917	182	1924	60
1911	210	1918	90	1925	107
1912	188	1919	92	1926	140
1913	98	1920	141	1927	154
1914	83	1921	183	1928	61
1915	131	1922	141	1929	44
1916	205	1923	80	1930	92

(a) Find the length of the cyclical fluctuation, and apply a moving average of that length to these data.

(b) Find the trend line by use of these moving averages.

(c) Calculate the trend values from the equation obtained in part (b), and find for each of the years of the table the random component.

16

The *F*-Distribution

16.1 DEFINITION

Another very important distribution frequently used in making tests of hypotheses is the *F*-distribution. In Section 10.4 it was pointed out in studying the distribution of the difference between means that the *F*-distribution is used to determine the validity of the assumption of identical standard deviations of two populations. It is also the distribution on which the whole analysis of variance procedure, discussed briefly in Chapter 17, is based.

Definition 16-1. If the distribution of a nonnegative variable *F* is given by

$$y = \frac{cF^{(v_1 - 2)/2}}{(v_2 + v_1 F)^{(v_1 + v_2)/2}},$$
(16-1)

the distribution is called an *F-distribution*, where both v_1 and v_2 are numbers of degrees of freedom, and *c* is a constant dependent on v_1 and v_2 and is chosen in such a way that the total area under the probability curve given by Equation (16-1) is equal to 1.

This distribution primarily distinguishes itself from distributions studied in previous chapters in that it involves two numbers of degrees of freedom, v_1 and v_2. Since v_2 appears only in the denominator of the fraction given above, it is frequently referred to as the denominator degrees of freedom, and v_1 is referred to as the numerator degrees of freedom.

The graphs of (16-1) vary considerably as v_1 and v_2 change. Figure 16-1

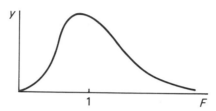

Figure 16-1
A typical F-distribution ($v_1 > 2$).

shows the graph of Equation (16-1) for a typical case. If v_1 is greater than 2, the curve reaches its highest point at

$$F = \frac{v_1 - 2}{v_1} \cdot \frac{v_2}{v_2 + 2}, \tag{16-2}$$

which is always less than 1, since each of the two fractions in (16-2) is smaller than 1. But as both v_1 and v_2 become larger and larger both fractions approach 1, so that for large v_1 and v_2 the highest point of (16-1) occurs very nearly at $F = 1$.

The F-distribution has the following important property: If a quantity F satisfies an F-distribution with v_1 and v_2 degrees of freedom for numerator and denominator, respectively, then the reciprocal of that quantity, $1/F$, also satisfies an F-distribution, but with the numerator degrees of freedom now being v_2 and the denominator degrees of freedom being v_1.

By introducing the notation $F(v_1, v_2)$ for a quantity that satisfies the F-distribution with v_1 and v_2 degrees of freedom, respectively, we have the following theorem.

Theorem 16-1. If $F(v_1, v_2)$ satisfies the F-distribution, then

$$F'(v_2, v_1) = \frac{1}{F(v_1, v_2)}$$

satisfies the F-distribution also.

The proof of this theorem requires calculus and will therefore not be given here.

Construction of tables for the F-distribution is complicated by the fact that two degrees of freedom, v_1 and v_2, are involved. Table VIIIa in the Appendix gives for various values of v_1 and v_2 (v_1 being listed in the first row and v_2 in the first column) the value of F required such that the total area under the curve to the left of that F-value is 0.95. This F-value will from now on be denoted by $F_{.95}$. Thus, from Table VIIIa, for $v_1 = 7$, $v_2 = 10$, we find $F_{.95} = 3.14$, which means that in this case the total area under the curve from $F = 0$ to $F = 3.14$ equals 0.95. Table VIIIb gives values of $F_{.975}$ and Table VIIIc values of $F_{.99}$.

Although Table VIII does not give values of $F_{.05}$, $F_{.025}$, and $F_{.01}$, these values of F can easily be obtained by means of the formula

$$F_\alpha(v_1, v_2) = \frac{1}{F_{1-\alpha}(v_2, v_1)}. \tag{16-3}$$

Proof of Formula (16-3). To fix our ideas let $\alpha = 0.025$, and let $P[F < F_{.025}]$ denote the probability that $F(v_1, v_2)$ is less than $F_{.025}$. Then, since this probability equals the area under the curve to the left of $F_{.025}$, we have

$$0.025 = P[F < F_{.025}] = P\left[\frac{1}{F} > \frac{1}{F_{.025}}\right] = P\left[F' > \frac{1}{F_{.025}}\right]. \tag{16-4}$$

But, since we know from Theorem 16-1 that F' satisfies the F-distribution with the number of degrees of freedom reversed and since from (16-4)

$$P\left[F' > \frac{1}{F_{.025}}\right] = 0.025,$$

$1/F_{.025}$ is the point in the distribution of $F(v_2, v_1)$ for which the total area to the right is 0.025, or the total area to the left is 0.975. Thus,

$$F_{.025}(v_1, v_2) = \frac{1}{F_{.975}(v_2, v_1)}.$$

Formula (16-3) can be proved in general by replacing in the statements above 0.025 by α, and 0.975 by $1 - \alpha$.

Example 16-1.   Find $F_{.025}(3, 5)$.

Solution.   From Formula (16-3) we find

$$F_{.025}(3, 5) = \frac{1}{F_{.975}(5, 3)} = \frac{1}{14.9} = 0.067.$$

An important relationship exists between Student's t-distribution and the F-distribution for the special case where $v_1 = 1$. It is expressed in Theorem 16-2.

Theorem 16-2.   If a quantity t satisfies Student's t-distribution with v_2 degrees of freedom, then its square, t^2, satisfies the F-distribution with $v_1 = 1$ and v_2 degrees of freedom for the denominator.

From Theorem 16-2, it can be proved that

$$F_{1-\alpha}(1, v_2) = t_\alpha^2(v_2), \qquad\qquad (16\text{-}5)$$

where $t_\alpha(v_2)$ denotes the t-value for v_2 degrees of freedom for which the two tail areas total α.

Proof of Formula (16-5).   Since $F_{1-\alpha}(1, v_2)$ is the point of the F-distribution for which the area under the curve to the right equals α, we have

$$\alpha = P[F > F_{1-\alpha}(1, v_2)]$$
$$= P(t^2 > F_{1-\alpha})$$
$$= P\left(t > \sqrt{F_{1-\alpha}} \text{ or } t < -\sqrt{F_{1-\alpha}}\right). \qquad (16\text{-}6)$$

From (16-6), we see that $\sqrt{F_{1-\alpha}}$ is that point in the t-distribution for which the two tail areas (that is, to the right of $\sqrt{F_{1-\alpha}}$ and to the left of $-\sqrt{F_{1-\alpha}}$) total α, which means that

$$\sqrt{F_{1-\alpha}(1, v_2)} = t_\alpha(v_2),$$

and, squaring both sides,

$$F_{1-\alpha}(1, v_2) = t_\alpha^2(v_2).$$

By means of this formula it is possible to find from Table III all values of F of the columns for $v_1 = 1$ in Table VIII.

Example 16-2.   Use the t-table (Table III) to find $F_{.95}(1, 10)$.

Solution.

$$F_{.95}(1, 10) = t_{.05}^2(10) = (2.228)^2 = 4.964,$$

which agrees with the value of $F_{.95}(1, 10)$ given in Table VIIIa.

16.2   TESTING THE HOMOGENEITY OF TWO VARIANCES

Definition 16-2.   The variances of two samples are said to be *homogeneous* if the populations from which the two samples are taken have identical variances (or identical standard deviations).

In many problems, as was already indicated in Chapter 10, it is important to have a test that helps to decide whether two variances are homogeneous. Such a test will now be developed.

Consider two populations with identical standard deviations. Let N denote the number of all possible pairs of samples which can be taken from these populations such that the first of each pair is from the first population and has n_1 variates and the other is from the second population and has n_2 variates. Let the means of the samples from the first population be denoted by $\overline{X}_1, \overline{X}_2, \ldots, \overline{X}_N$, and those from the second by $\overline{Y}_1, \overline{Y}_2, \ldots, \overline{Y}_N$. By use of Formula (10-2), calculate from each of these samples the estimate of the standard deviation of the population, and denote the estimates obtained from samples of the first population by $s_{x,1}, s_{x,2}, \ldots, s_{x,N}$ and those obtained from samples of the second population by $s_{y,1}, s_{y,2}, \ldots, s_{y,N}$, that is,

$$s_{x,1} = \sqrt{\frac{\sum(X - \overline{X}_1)^2}{n_1 - 1}}, \qquad s_{y,1} = \sqrt{\frac{\sum(Y - \overline{Y}_1)^2}{n_2 - 1}}.$$

Calculate for each pair the ratio of the squares of the estimates of the standard deviations of the two populations; thereby N such ratios are obtained, which we shall denote by $F_1, F_2, \ldots, F_N$, as shown in Figure 16-2.

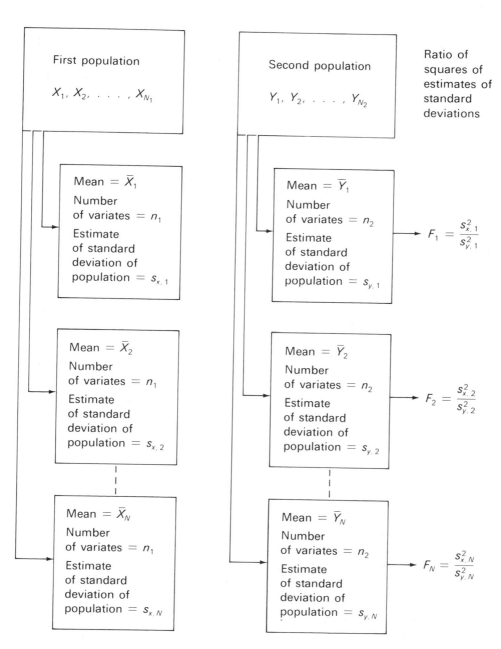

Figure 16-2

Distribution of variance ratios for two populations with identical standard deviations.

The F-values thus obtained satisfy an F-distribution with $v_1 = n_1 - 1$ and $v_2 = n_2 - 1$. This is restated in the following theorem.

Theorem 16-3. If X and Y are normally distributed with identical standard deviations, and if all possible pairs of samples are drawn from these populations (as shown in Figure 16-2), then the ratio of the squares of the estimates of the standard deviations, that is,

$$F = \frac{s_x^2}{s_y^2},\qquad(16\text{-}7)$$

where

$$s_x = \sqrt{\frac{\sum(X - \overline{X})^2}{n_1 - 1}},\qquad s_y = \sqrt{\frac{\sum(Y - \overline{Y})^2}{n_2 - 1}}\qquad(16\text{-}8)$$

satisfies an F-distribution with $v_1 = n_1 - 1$ and $v_2 = n_2 - 1$.

From Theorem 16-2 it follows immediately that, for the arrangement described in Figure 16-2, also the reciprocal of F, that is,

$$F' = \frac{s_y^2}{s_x^2}\qquad(16\text{-}9)$$

satisfies the F-distribution with $v_1 = n_2 - 1$ and $v_2 = n_1 - 1$.

To test for the homogeneity of variances, we set up the hypothesis that the standard deviations of the two populations under consideration are identical. Thus the value of F as obtained from (16-7) is expected to be near 1. If a very large or very small value of F is obtained, it would be an indication that the hypothesis of equal standard deviations should be rejected, or, more precisely, we reject the hypothesis if the calculated value of F is such that the two tail areas corresponding to it are less than 0.05 (assuming a 5% level of significance). This means that we reject the hypothesis if the calculated F-value is either larger than $F_{.975}$ or less than $F_{.025}$; otherwise we accept it.

A two-tailed test is always applied in this case, since the problem is to find whether s_x and s_y differ significantly. We must be prepared to reject the hypothesis of identical population standard deviations if F is either significantly larger or significantly smaller than 1. Here, the hypothesis $\sigma_x = \sigma_y$ is considered as having the alternative $\sigma_x \neq \sigma_y$. In many other problems, indeed in most problems where the F-test is applied, as in the analysis of variance (see Chapter 17), one variance is assumed given, and the question is raised whether another variance exceeds the given one by a significant amount. This situation calls for a one-tailed test, as the alternative to the hypothesis $\sigma_x = \sigma_y$ then is $\sigma_x > \sigma_y$, where σ_y is the given or fixed variance.

Since Appendix Table VIII gives only values of $F_{.95}$, $F_{.975}$, $F_{.99}$, which are critical points on the right side of the distributions, it is advisable in calculating F to use (16-7) or (16-9), whichever gives the larger value. Clearly, this occurs when the larger standard deviation is placed in the numerator. The procedure then in testing for the homogeneity of two variances is always to divide the larger variance by the smaller, which results in an F-value larger than 1, which can then be compared with the critical F-value given in Table VIII.

Example 16-3. Assuming the data of Example 10-4, determine whether the variances of the two samples can be considered homogeneous on the basis of the 5% level of significance.

Solution. From the solution of Example 10-4 we find

$$s_x^2 = \frac{\sum (X - \bar{X})^2}{n_1 - 1} = \frac{110}{3} = 36.67,$$

$$s_y^2 = \frac{\sum (Y - \bar{Y})^2}{n_2 - 1} = \frac{56}{5} = 11.20,$$

and

$$F = \frac{36.67}{11.20} = 3.27.$$

From Appendix Table VIIIb we find

$$F_{.975}(3, 5) = 7.76.$$

Since the calculated F-value does not exceed $F_{.975}(3, 5)$, we conclude that the two variances can be assumed to be homogeneous. These two samples can be considered as coming from two populations with identical standard deviations, and we see that the assumption made in Example 10-4 is valid.

It may appear surprising that in Example 16-3 where the variance of one sample is more than three times that of the other, the hypothesis of equal standard deviations of the two populations is still acceptable. Indeed, the value of $F_{.975}(3, 5) = 7.76$ indicates that one sample variance could have been more than seven times that of the other without violating the hypothesis of homogeneous variances. This shows that, for small samples taken from populations with identical standard deviations, the variances can differ very greatly. Consequently, the assumption of homogeneous variances can safely be made in most situations and, in particular, in applying Theorem 10-2.

If, however, the *F*-value indicates that the standard deviations of the two populations are different, then Theorem 10-2 should not be used to test for significant differences between means. A special theorem, not discussed here, is applicable in that case.

EXERCISES

1. Find the following:
   (a) $F_{.01}(5, 9)$;   (b) $F_{.025}(6, 8)$;   (c) $F_{.05}(20, 15)$.

2. Use the *t*-table to find, accurate to three decimal places:
   (a) $F_{.99}(1, 10)$;   (b) $F_{.95}(1, 15)$;   (c) $F_{.05}(8, 1)$.

3. Can the variances for the data of Exercise 14, Chapter 10, be considered as being homogeneous (5% level)?

4. From a sample of 13 variates the estimate of the standard deviation of the population was calculated as 8.32, and from another sample of 15 variates it was calculated to be 4.47. Can these samples be considered as having homogeneous variances?

5. In an experiment to test the effect of a large amount of chalk or lime on marigolds, William J. Cochran obtained the following numbers of plants per plot for control and treated plots.

   Control: 140, 142, 36, 129, 49, 37, 114, 125
   Treated: 117, 137, 137, 143, 130, 112, 130, 121.

   Test for homogeneity of variances.

6. To test the effect of phosphorus as a fertilizer, 10 plots were treated with phosphorus, and 10 other plots were not treated. The yields in bushels of corn for the treated plots and those used as control are as follows:

   Treated: 6.2, 5.7, 6.5, 6.0, 6.3, 5.8, 5.7, 6.0, 6.0, 5.8
   Control: 5.6, 5.9, 5.6, 5.7, 5.8, 5.7, 6.0, 5.5, 5.7, 5.5

   Does this indicate that the fertilizer increased the yield significantly on the basis of the 5% level of significance (test for homogeneity of variances first)?

7. In a sample of 11 variates the estimate of the standard deviation of the population was calculated to be $s_x = 9.76$. If another sample has 16 variates, find the limits within which s_y must lie in order that the two samples can be considered to have homogeneous variances on the basis of the 5% level of significance.

8. If the variates in a sample are 80, 70, 65, 69, 74, 79, 53 and those in another sample are 60, 45, 49, 70, X, find the limits within which X must be in order that the two samples can be considered as having homogeneous variances (5% level).

9. For a sample of 21 variates the sum of the squares of the deviations from the mean was found to be $\Sigma(X - \bar{X})^2 = 483$. From a sample taken from another population the estimate of the standard deviation of that population was found to be 2.59. Within what range must the size of the second sample be in order that the variances of the two populations can be considered different at the 5% level of significance?

10. Two samples of 7 and 14 variates, respectively, are checked for homogeneity of variances. If the estimate of the standard deviation of the population as obtained from the sample of 7 variates is exactly 3 units larger than the estimate of the standard deviation obtained from the other sample, what are the smallest values the estimates of the two standard deviations can have in order that the variances can be considered homogeneous at the 5% level of significance?

17

The Analysis
of Variance

17.1 THE ANALYSIS OF VARIANCE WITH ONE CRITERION OF CLASSIFICATION (ONE-WAY ANALYSIS OF VARIANCE). INTRODUCTION

In Chapter 10 we saw that by means of Student's t-test it is possible to determine whether the means of two given samples differ significantly. In this chapter we shall consider the problem of determining whether among a given set of more than two samples (say k samples) there are means that differ significantly.

At first glance this problem might appear to be easily solved by taking all possible pairs of samples and testing them individually for significant differences by means of Student's t-test. Closer inspection, however, will show that such a procedure must be rejected not only on the grounds that it is laborious, but much more importantly on the grounds that it has a good chance of leading to a false conclusion.

Let us consider seven samples taken from populations with identical means. If we were to test all possible pairs of samples for significant differences among their means, we would have to carry out the t-test on $C_{7,2} = 21$ pairs, for each of which the probability of arriving at the correct conclusion of no significant difference would be $0.95 = 19/20$ (assuming that we use the 5% level of significance). Consequently, the probability that we would arrive at the correct conclusion for all 21 pairs is $(19/20)^{21}$, which means that the probability of obtaining at least one incorrect conclusion is given by

$$1 - \left(\frac{19}{20}\right)^{21} = 1 - 0.340 = 0.660.$$

Thus, 66% of the time, we would be drawing a wrong conclusion, i.e. committing a type 1 error. Clearly a statistical test in which the probability of obtaining an incorrect conclusion is of such magnitude cannot be considered an acceptable test. For larger numbers of samples this probability would be still higher, and even for smaller numbers of samples it may be quite high. Thus, if only five samples were considered, it can easily be shown that we would be wrong 40.1% of the time by testing pairs of samples individually, assuming that they all came from populations with identical means.

It is therefore necessary to develop a different method, one that takes into consideration all the sample means together. This procedure is called the *analysis of variance* and is based on two principles:

1. The partitioning of the sum of squares.
2. The estimating of the standard deviation of the population by two methods and a comparison of these estimates.

These two principles will now be discussed separately.

17.2   PARTITIONING OF THE SUM OF SQUARES, EQUAL SAMPLE SIZES

Consider k samples, all of which have the same number of variates, say n. Denote the total of all variates in the first sample by T_1, the total of all variates in the second sample by T_2, and so on, and the total of all variates, or *grand total*, by T; similarly denote the mean of the first sample by $\overline{X}_1$, that of the second sample by $\overline{X}_2$, and so on, and the mean of all variates, or *grand mean*, by $\overline{X}$. Since in most practical applications with which we shall be concerned the various samples represent the results of different treatments that have been applied, we shall consider hereafter the term treatment as having the same meaning as sample. The notation applicable to this case is summarized in Table 17-1.

Table 17-1
The case of k samples of n variates each.

	1st sample or 1st treatment	2nd sample or 2nd treatment	$\cdots$	kth sample or kth treatment	
	X_{11}	X_{12}	$\cdots$	X_{1k}	
	X_{21}	X_{22}	$\cdots$	X_{2k}	
	$\vdots$	$\vdots$		$\vdots$	
	X_{n1}	X_{n2}	$\cdots$	X_{nk}	
Total	T_1	T_2	$\cdots$	T_k	Grand total $= T$
Mean	$\bar{X}_1$	$\bar{X}_2$	$\cdots$	$\bar{X}_k$	Grand mean $= \bar{X}$

In Table 17-1 we introduce double subscript notation. The first subscript identifies the particular variate within a sample, the second subscript the sample to which the variate belongs. Thus, X_{21} denotes the second variate in the first sample, in general X_{ij} denotes the ith variate in the jth sample.

In order to explain the principle of the partitioning of the sum of squares it is necessary first to introduce a set of definitions.

Definition 17-1. The *total sum of squares* is defined as

$$S.S. = \sum_{i=1}^{n} \sum_{j=1}^{k} (X_{ij} - \bar{X})^2. \tag{17-1}$$

The double sum appearing in Formula (17-1) means that the sum is to be taken for all integral values of the first subscript from 1 to n and all integral values of the second subscript from 1 to k. In other words, the total sum of squares is the sum of the squares of the deviations from the grand mean for all the variates; it is therefore a measure of the dispersion of all the variates about the grand mean.

Definition 17-2. The *between means* (or *among-means*) *sum of squares* (or *sum of squares for treatments*) is defined as

$$S.S.T. = n \sum_{j=1}^{k} (\bar{X}_j - \bar{X})^2. \tag{17-2}$$

Since the sum appearing in Formula (17-2) is the sum of the squares of the deviations of the sample (or treatment) means about the grand mean, $S.S.T.$ is a measure of the dispersion of the sample means about the grand mean. The less the sample means differ from each other, the smaller $S.S.T.$ will be. If, on the other hand, the treatments show widely differing effects, there will be a large difference among the sample means and consequently a large value of $S.S.T.$

Definition 17-3. The *within-samples sum of squares* (or *sum of squares for error*) is defined as

$$S.S.E. = \sum_{i=1}^{n}(X_{i1} - \bar{X}_1)^2 + \sum_{i=1}^{n}(X_{i2} - \bar{X}_2)^2 + \cdots + \sum_{i=1}^{n}(X_{ik} - \bar{X}_k)^2$$

$$= \sum_{i=1}^{n}\sum_{j=1}^{k}(X_{ij} - \bar{X}_j)^2. \tag{17-3}$$

In the first of the k sums appearing on the right-hand side of Formula (17-3) the deviations are taken about the mean of the first sample; this term therefore represents a measure of the dispersion within the first sample. Similarly, the second term represents a measure of the dispersion within the second sample, and so on. Consequently, $S.S.E.$ is a measure of the dispersion of the variates within the samples about their respective sample means and is independent of any differences between sample means; its value depends only on the random changes occurring in almost any experiment and thus represents the experimental error of the experiment.

Example 17-1. Consider the following five samples of three variates each:

	3	5	7	6	4	
	2	8	8	8	9	
	4	8	6	7	5	
Total	9	21	21	21	18	$T = 90$
Mean	3	7	7	7	6	$\bar{X} = 6$

Calculate for these data the total sum of squares, the between-means sum of squares, and the within-samples sum of squares by use of the definitions of these quantities.

Solution.

$$S.S. = (3-6)^2 + (2-6)^2 + (4-6)^2 + (5-6)^2 + (8-6)^2$$
$$+ \cdots + (5-6)^2 = 9 + 16 + 4 + 1 + 4 + \cdots + 1 = 62;$$

$$S.S.T. = 3[(3-6)^2 + (7-6)^2 + (7-6)^2 + (7-6)^2 + (6-6)^2]$$
$$= 3(9+1+1+1+0) = 36;$$

$$S.S.E. = (3-3)^2 + (2-3)^2 + (4-3)^2 + (5-7)^2 + (8-7)^2 + (8-7)^2$$
$$+ \cdots + (5-6)^2 = 26.$$

The procedure used to calculate $S.S.$, $S.S.T.$, and $S.S.E.$ in Example 17-1 is not the one commonly employed in practice since it is much easier to obtain these quantities, especially in machine calculations, by use of the following formulae, sometimes referred to as their computing forms:

$$S.S. = \sum_{i=1}^{n} \sum_{j=1}^{k} X_{ij}^2 - \frac{T^2}{kn}; \tag{17-4}$$

$$S.S.T. = \frac{\sum_{j=1}^{k} T_j^2}{n} - \frac{T^2}{kn}; \tag{17-5}$$

$$S.S.E. = \sum_{i=1}^{n} \sum_{j=1}^{k} X_{ij}^2 - \frac{\sum_{j=1}^{k} T_j^2}{n}. \tag{17-6}$$

Formulae (17-4), (17-5), and (17-6) will be proved following Example 17-2.

Example 17-2. Calculate $S.S.$, $S.S.T.$, and $S.S.E.$ for the data of Example 17-1 by use of their computing forms.

Solution.

$$S.S. = 3^2 + 2^2 + 4^2 + 5^2 + 8^2 + \cdots + 5^2 - \frac{8100}{15} = 602 - 540 = 62;$$

$$S.S.T. = \frac{9^2 + 21^2 + 21^2 + 21^2 + 18^2}{3} - \frac{8100}{15} = 576 - 540 = 36;$$

$$S.S.E. = 3^2 + 2^2 + 4^2 + 5^2 + 8^2 + \cdots + 5^2 - \frac{9^2 + 21^2 + 21^2 + 21^2 + 18^2}{3}$$

$$= 602 - 576 = 26.$$

Proof of Formulae (17-4), (17-5), and (17-6). All three formulae are consequences of the result of Example 3-4, namely that for a given set of N numbers, whose mean is m,

$$\sum_{i=1}^{N}(X_i - m)^2 = \sum_{i=1}^{N} X_i^2 - \frac{\left(\sum_{i=1}^{N} X_i\right)^2}{N}. \tag{17-7}$$

For purposes of application, it is more convenient to state (17-7) in words: For a set of N numbers $X_1, X_2, \ldots, X_N$, the sum of squares of the deviations from the mean equals the sum of the squares of these numbers minus $1/N$ times the square of the total of these numbers.

From Definition 17-1 we see that $S.S.$ represents the sum of the squares of the deviations of the kn numbers $X_{11}, X_{21}, \ldots, X_{nk}$ from their mean; in accordance with the above statement this sum can be rewritten as the sum of the squares of these numbers minus $1/kn$ times T^2, the square of the total of these numbers. Formula (17-4) therefore follows from (17-1).

To prove (17-5) from (17-2) we apply (17-7) to the k numbers $\bar{X}_1, \bar{X}_2, \ldots, \bar{X}_k$, whose mean is $\bar{X}$; (17-2) can then be written as

$$S.S.T. = n\left[\sum_{j=1}^{k}\bar{X}_j^2 - \frac{\left(\sum_{j=1}^{k}\bar{X}_j\right)^2}{k}\right] = n\left[\sum_{j=1}^{k}\frac{T_j^2}{n^2} - \frac{\left(\sum_{j=1}^{k}T_j\right)^2}{kn^2}\right] = \frac{\sum_{j=1}^{k}T_j^2}{n} - \frac{T^2}{kn},$$

which proves (17-5).

To prove (17-6) we apply (17-7) to each of the k sums appearing in (17-3) and obtain

$$S.S.E. = \sum_{i=1}^{n}X_{i1}^2 - \frac{\left(\sum_{i=1}^{n}X_{i1}\right)^2}{n} + \sum_{i=1}^{n}X_{i2}^2 - \frac{\left(\sum_{i=1}^{n}X_{i2}\right)^2}{n}$$

$$+ \cdots + \sum_{i=1}^{n}X_{ik}^2 - \frac{\left(\sum_{i=1}^{n}X_{ik}\right)^2}{n}$$

$$= \sum_{i=1}^{n}\sum_{j=1}^{k}X_{ij}^2 - \frac{T_1^2}{n} - \frac{T_2^2}{n} - \cdots - \frac{T_k^2}{n} = \sum_{i=1}^{n}\sum_{j=1}^{k}X_{ij}^2 - \frac{\sum_{j=1}^{k}T_j^2}{n}.$$

From the solution of Example 17-1 the reader may have observed that the sum of $S.S.T.$ and $S.S.E.$ equals $S.S.$, that is, $36 + 26 = 62$, and he may wonder whether this property is true for any set of data. This indeed is the case, and it is the partitioning principle stated in Theorem 17-1.

Theorem 17-1. The total sum of squares for a set of variates as shown in Table 17-1 is equal to the sum of the between-means sum of squares and the within-samples sum of squares, or

$$S.S. = S.S.T. + S.S.E. \tag{17-8}$$

Before proceeding to its proof, we note that Theorem 17-1 means that we can always partition the total dispersion in such a set of data into two components, one due to the dispersion existing between the sample means, the other due to the dispersion existing within the samples.

Proof of Theorem 17-1. Formula (17-8) is immediately seen to be true if we add the expressions for *S.S.T.* and *S.S.E.* as given by (17-5) and (17-6), respectively, and obtain

$$S.S.T. + S.S.E. = \frac{\sum_{j=1}^{k} T_j^2}{n} - \frac{T^2}{kn} + \sum_{i=1}^{n} \sum_{j=1}^{k} X_{ij}^2 - \frac{\sum_{j=1}^{k} T_j^2}{n}$$

$$= \sum_{i=1}^{n} \sum_{j=1}^{k} X_{ij}^2 - \frac{T^2}{kn} = S.S.$$

From (17-1) we see that if $\bar{X} = 0$, then *S.S.* reduces to $\sum_{i=1}^{n} \sum_{j=1}^{k} X_{ij}^2$, that is, the first term appearing on the right-hand side of (17-4). Thus, the term T^2/kn corrects for the fact that the sum of squares is required about the grand mean $\bar{X}$ and not about 0. This results in Definition 17-4.

Definition 17-4. The *correction term* for a set of variates as shown in Table 17-1 is defined as

$$C = \frac{T^2}{kn}. \tag{17-9}$$

17.3   PARTITIONING OF THE SUM OF SQUARES, UNEQUAL SAMPLE SIZES

The case where the various samples do not all have the same number of variates arises less frequently and will therefore be discussed only briefly. Again let us consider k samples, but now with n_1 variates in the first sample, n_2 variates in the second sample, and so on; the meaning of $T_1, T_2, \ldots, T_k$, $T, \bar{X}_1, \bar{X}_2, \ldots, \bar{X}_k, \bar{X}$ will be the same as before. The notation for this case is summarized in Table 17-2.

Table 17-2

The case of k samples with $n_1, n_2, \ldots, n_k$ variates, respectively.

	1st sample or 1st treatment	2nd sample or 2nd treatment		kth sample or kth treatment	
	X_{11}	X_{12}	$\cdots$	X_{1k}	
	X_{21}	X_{22}	$\cdots$	X_{2k}	
	$\vdots$	$\vdots$		$\vdots$	
	$X_{n_1,1}$	$X_{n_2,2}$	$\cdots$	$X_{n_k,k}$	
Total	T_1	T_2	$\cdots$	T_k	Grand total $= T$
Mean	$\bar{X}_1$	$\bar{X}_2$	$\cdots$	$\bar{X}_k$	Grand mean $= \bar{X}$

The reader should not be deceived by the appearance of equal length of all columns; this was done for conciseness in presenting the table. In any actual case of unequal number of variates the columns would be of unequal lengths; for a numerical case see Table 17-7, where $n_1 = 6$, $n_2 = 4$, $n_3 = 5$, $n_4 = 4$.

The expressions defined by (17-1), (17-2), and (17-3) must be modified very slightly to apply to an unequal number of variates. Here the definitions take the following forms:

$$S.S. = \sum_{i=1}^{n_1} (X_{i1} - \bar{X})^2 + \sum_{i=1}^{n_2} (X_{i2} - \bar{X})^2 + \cdots + \sum_{i=1}^{n_k} (X_{ik} - \bar{X})^2. \qquad (17\text{-}10)$$

$$S.S.T. = n_1(\bar{X}_1 - \bar{X})^2 + n_2(\bar{X}_2 - \bar{X})^2 + \cdots + n_k(\bar{X}_k - \bar{X})^2. \qquad (17\text{-}11)$$

$$S.S.E. = \sum_{i=1}^{n_1} (X_{i1} - \bar{X}_1)^2 + \sum_{i=1}^{n_2} (X_{i2} - \bar{X}_2)^2 + \cdots + \sum_{i=1}^{n_k} (X_{ik} - \bar{X}_k)^2. \qquad (17\text{-}12)$$

Clearly, the interpretations of (17-10), (17-11), and (17-12) are identical to those for (17-1), (17-2), and (17-3). Also, each of the above expressions again can be written in a form suitable for machine calculation. This results in the following formulae:

$$S.S. = \sum_{i=1}^{n_1} X_{i1}^2 + \sum_{i=1}^{n_2} X_{i2}^2 + \cdots + \sum_{i=1}^{n_k} X_{ik}^2 - \frac{T^2}{n_1 + n_2 + \cdots + n_k}. \qquad (17\text{-}13)$$

$$S.S.T. = \frac{T_1^2}{n_1} + \frac{T_2^2}{n_2} + \cdots + \frac{T_k^2}{n_k} - \frac{T^2}{n_1 + n_2 + \cdots + n_k}. \qquad (17\text{-}14)$$

$$S.S.E. = \sum_{i=1}^{n_1} X_{i1}^2 + \sum_{i=1}^{n_2} X_{i2}^2 + \cdots + \sum_{i=1}^{n_k} X_{ik}^2 - \left(\frac{T_1^2}{n_1} + \frac{T_2^2}{n_2} + \cdots + \frac{T_k^2}{n_k} \right). \qquad (17\text{-}15)$$

The proofs of the preceding three formulae are so similar to those of formulae (17-4), (17-5), and (17-6) that they will be left as an exercise for the student (see Exercise 6).

Similarly, it is obvious that the principle of the partitioning of the sum of squares holds also here, since the sum of (17-14) and (17-15) equals that of (17-13). We thus have the following theorem, which is completely analogous to Theorem 17-1, and which can be proved in a similar manner.

Theorem 17-2.   The total sum of squares for a set of variates as shown in Table 17-2 is equal to the sum of the between-means sum of squares and the within-samples sum of squares.

As before, the second term appearing on the right-hand sides of both (17-13) and (17-14) is referred to as the correction term.

17.4   COMPARISON OF VARIANCES IN ANALYSIS OF VARIANCE TABLE, EQUAL SAMPLE SIZES

Considering again k samples of n variates each, i.e. the situation of Table 17-1, we make the hypothesis that all of the k samples are taken from the same population. We then can estimate the variance of that population, σ^2, in two ways:

1. We can estimate σ^2 by pooling the variates of the k samples by a generalization of the procedures used for two samples where we applied Formula (10-7). The numerator then becomes the sum of the squares of the deviations of the variates of all k samples from their respective sample means and the denominator $n_1 + n_2 + \cdots + n_k - k$, or, since all samples are assumed to have the same number of variates, n, we have for the denominator $kn - k = k(n-1)$, so that we obtain the following estimate of σ^2 by pooling the sums of the squares of the deviations:

$$s_p^2 = \frac{\sum_{i=1}^{n}(X_{i1} - \bar{X}_1)^2 + \sum_{i=1}^{n}(X_{i2} - \bar{X}_2)^2 + \cdots + \sum_{i=1}^{n}(X_{ik} - \bar{X}_k)^2}{k(n-1)}.$$

$$(17\text{-}16)$$

We use the subscript p in s_p^2 to indicate that this estimate was obtained by pooling the sums of the squares of the deviations.

Since the numerator of the fraction appearing in (17-16) equals $S.S.E.$ we can rewrite (17-16) as

$$s_p^2 = \frac{S.S.E.}{k(n-1)} .\qquad (17\text{-}17)$$

2. Another estimate of σ^2 can be obtained by estimating first the variance of the sample means, $\sigma_{\bar{x}}^2$. An estimate of $\sigma_{\bar{x}}^2$ can be obtained by considering $\bar{X}_1, \bar{X}_2, \ldots, \bar{X}_k$ as a sample of all possible treatment means. From Formula (10-2) we have

$$s_{\bar{x}}^2 = \frac{\sum_{j=1}^{k} (\bar{X}_j - \bar{X})^2}{k-1} .\qquad (17\text{-}18)$$

But, since $\sigma_{\bar{x}}^2 = \sigma^2/n$, or $\sigma^2 = n\sigma_{\bar{x}}^2$, we have a second estimate of σ^2 from the sample (or treatment) means, which we shall denote by s_t^2 (since it was obtained by use of the treatment means), which is given by

$$s_t^2 = ns_{\bar{x}}^2 = \frac{n\sum_{j=1}^{k} (\bar{X}_j - \bar{X})^2}{k-1} .\qquad (17\text{-}19)$$

Since the numerator of the fraction appearing in (17-19) equals $S.S.T.$, we can rewrite (17-19) as

$$s_t^2 = \frac{S.S.T.}{k-1} .\qquad (17\text{-}20)$$

The analysis of variance procedure consists in comparing these two estimates of σ^2. Since both quantities represent estimates of the same variance σ^2, their ratio should not differ significantly from 1. More precisely, the following can be proved: If repeatedly k samples of n variates each are taken from a population and for each of these sets of k samples the ratio s_t^2/s_p^2 is calculated, these ratios satisfy the F-distribution which results in the following theorem.

Theorem 17-3. If all possible sets of k samples of n variates each are taken from a normally distributed population, then the quantity

$$F = \frac{s_t^2}{s_p^2} ,\qquad (17\text{-}21)$$

where s_p^2 and s_t^2 are given by (17-17) and (17-20), respectively, satisfies the F-distribution with $v_1 = k - 1$ and $v_2 = k(n-1)$.

It is important to note that the sum of the numbers of degrees of freedom associated with s_t^2, that is, $k-1$, and with s_p^2, that is, $k(n-1)$, add up to $k-1+k(n-1) = kn-1$, which is the number of degrees of freedom associated with the total number of variates. Consequently, there is not only a partitioning principle for the sum of squares, but also for the degrees of freedom.

If now our hypothesis fails to be true, i.e. if the k samples are in fact taken from populations with different means, then s_t^2 will be considerably greater than s_p^2, owing to the wider dispersion of the sample means about the grand mean. If s_t^2 is so large that in comparison with s_p^2 it yields an F-value exceeding $F_{.95}$, we consider the sample means significantly different and conclude that there must be at least one pair of samples whose means differ significantly. It should be carefully observed that a significant result can occur only if s_t^2 exceeds s_p^2, i.e., if F, as defined by (17-21), exceeds 1; thus, a one-tailed test must always be used in the analysis of variance. Whenever s_t^2 is less than or equal to s_p^2, the result is never considered significant; the dispersion of the sample means about the grand mean would be no more than that expected ordinarily from a set of k samples, all of which were taken from the same population. Thus, whenever $s_t^2 \leq s_p^2$, it is unnecessary to calculate an F-value, since it would obviously be insignificant.

The use of the analysis of variance in testing for significant differences between sample means requires, therefore, the calculation of s_p^2 and s_t^2 and the determination, on the basis of a one-tailed test, whether their ratio is significantly different from 1. This calculation is most conveniently performed in the following sequence of steps.

Step 1.   Find $C = \dfrac{T^2}{kn}$;

Step 2.   Find $S.S. = \displaystyle\sum_{i=1}^{n} \sum_{j=1}^{k} X_{ij}^2 - C$;

Step 3.   Find $S.S.T. = \dfrac{\displaystyle\sum_{j=1}^{k} T_j^2}{n} - C$;

Step 4.   Find $S.S.E. = S.S. - S.S.T.$;

Step 5.   Find $d.f.$ associated with each of the above expressions;

Step 6.   Find $s_t^2 = \dfrac{S.S.T.}{k-1}$;

Step 7. Find $s_p^2 = \dfrac{S.S.E.}{k(n-1)}$;

Step 8. Find $F = \dfrac{s_t^2}{s_p^2}$;

Step 9. Find $F_{.95}$, the tabular F-value for $v_1 = k - 1$ and $v_2 = k(n-1)$, and compare with the F-value obtained in Step 8.

It is customary to summarize these calculations in a so-called analysis of variance table as shown in Table 17-3.

Table 17-3

Analysis of variance table for equal sample sizes.

Source of variation	Sum of squares	d.f.	Mean square	F	$F_{.95}$
Total	$S.S. = \sum\limits_{i=1}^{n} \sum\limits_{j=1}^{k} X_{ij}^2 - C$	$kn - 1$			
Between-means (or treatments)	$S.S.T. = \dfrac{\sum\limits_{j=1}^{k} T_j^2}{n} - C$	$k - 1$	$s_t^2 = \dfrac{S.S.T}{k-1}$.	$\dfrac{s_t^2}{s_p^2}$	Tabular value
Within-samples (or error)	$S.S.E. = S.S. - S.S.T.$	$k(n-1)$	$s_p^2 = \dfrac{S.S.E.}{k(n-1)}$		

Example 17-3. On the basis of the 5% level of significance, are there significant differences between the means of the five samples given in Table 17-4?

Table 17-4

An experiment with five samples of four variates each.

			Sample (or Treatment)			
	A	B	C	D	E	
	10	12	9	14	10	
	4	10	4	9	6	
	6	13	4	10	8	
	4	7	5	11	4	
Total	24	42	22	44	28	Grand total $= 160$
Mean	6.0	10.5	5.5	11.0	7.0	Grand mean $= 8.0$

Solution.   Following the suggested procedure, we find

1. $C = \dfrac{160^2}{20} = 1280$;

2. $S.S. = 10^2 + 4^2 + \cdots + 4^2 - C = 1482 - 1280 = 202$;

3. $S.S.T. = \dfrac{24^2 + 42^2 + \cdots + 28^2}{4} - C = 1386 - 1280 = 106$;

4. $S.S.E. = 202 - 106 = 96$.

The results can now be entered into an analysis of variance table (see Table 17-5) and the remaining calculations completed in the table.

Table 17-5
Analysis of variance table for the data of Table 17-4.

Source of variation	Sum of squares	d.f.	Mean square	F	$F_{.95}$
Total	202	19			
Between-means (or treatments)	106	4	26.5	4.14	3.06
Within-samples (or error)	96	15	6.4		

Since the calculated F-value of 4.14 exceeds the tabular value of $F_{.95}$ for $v_1 = 4$, $v_2 = 15$, we conclude that the hypothesis that all of the given samples were taken from the same population is not justified at the 5% level of significance; thus, there is at least one pair of samples whose means differ significantly. Since, however, $F_{.99}(4, 15) = 4.89$, the differences in means are not highly significant. A further question that arises is the problem of deciding which pairs of samples of Table 17-4 have significantly different means. This can be answered quite easily by a method to be shown in Section 17-5.

17.5   THE LEAST SIGNIFICANT DIFFERENCE

If the F-ratio indicates significant differences between sample means, it becomes important to determine which means cause the significance. A simple procedure to answer this question is available for the case of equal sample sizes. It is based on the following definition.

Definition 17-5.　For a set of k samples of n variates each, showing significant differences between sample means as indicated by the F-value in the analysis of variance, the *least significant difference*, denoted by *L.S.D.*, is defined as the smallest difference which could exist between two significantly different sample means.

A formula for the *L.S.D.* is easily derived. Let $\overline{X}_1$ and $\overline{X}_2$ be any two sample means (not necessarily the first two sample means), then by Chapter 10, $\overline{X}_1$ and $\overline{X}_2$ differ significantly—see Formula (10-8)—if

$$t = \frac{\overline{X}_1 - \overline{X}_2}{s_{\bar{x}_1 - \bar{x}_2}}$$

exceeds $t_{.05}$ (assuming the 5% level of significance). Thus, the *L.S.D.* is the difference between two sample means, $\overline{X}_1 - \overline{X}_2$, for which $t = t_{.05}$, that is

$$L.S.D. = t_{.05}\, s_{\bar{x}_1 - \bar{x}_2}.$$

But, from Chapter 10, for $n_1 = n_2 = n$,

$$s_{\bar{x}_1 - \bar{x}_2} = \sqrt{s_{\bar{x}_1}^2 + s_{\bar{x}_2}^2} = \sqrt{\frac{s^2}{n} + \frac{s^2}{n}} = \sqrt{\frac{2s^2}{n}},$$

where s equals s_p as defined by (17-16). From here on we shall omit the subscript p and understand s to equal s_p. The formula for the *L.S.D.*, the least significant difference, then becomes

$$L.S.D. = t_{.05}\sqrt{\frac{2s^2}{n}}, \tag{17-22}$$

where $t_{.05}$ is the tabular value of t at the 5% level for the number of degrees of freedom, on which s is based, which according to (17-16) equals $k(n-1)$.

Example 17-4.　On the basis of the 5% level of significance, which of the pairs of samples in Example 17-3 differ significantly in their means?

Solution.　From (17-22) we find the least significant difference to be

$$L.S.D. = 2.131\sqrt{\frac{2(6.4)}{4}}$$

$$= 3.812,$$

since $s^2 = 6.4$ from Table 17-5 and $t_{.05} = 2.131$ for $\nu = 15$. Thus, a look at Table 17-4 shows that the first two sample means differ significantly, since $\bar{X}_1 - \bar{X}_2 = 4.5$ exceeds 3.812. A complete investigation of all possible pairs reveals that for the following pairs the sample means differ significantly: *AB, AD, BC, CD, DE*.

It is important to realize that the formula for the *L.S.D.* should never be used when the *F*-value obtained in the analysis of variance is not significant. It is entirely possible—and even very likely—that if the formula for the *L.S.D.* is mistakenly used in a situation where the *F*-value is insignificant, some differences between sample means will exceed the *L.S.D.* In such situations we could not call these differences significant, but would have to regard them as the kind of large differences expected to occur in samples from populations with identical means.

Use of the *L.S.D.* has the advantage that for the special case of two samples its use will always result in precisely the same conclusion as obtained in a Student *t*-test (see Section 17.8). If, however, the analysis of variance involves more than two samples, it is clear that the two extreme means differ more widely just by chance alone than do other means. Consequently, the value given by the usual formula for an *L.S.D.* is too small to compare the two extreme means.

Two possible solutions suggest themselves to resolve this difficulty:

1. Increase the size of the *L.S.D.* so that all possible pairs of treatment means are compared with a fixed value which exceeds that given by the *L.S.D.*
2. Use different significant values (called *multiple ranges*) depending on the number of means which are compared.

The type of test resulting from the first solution is usually referred to as a *fixed range test* since only one range is used for all possible comparisons. The disadvantage of such a test is evident: Two adjacent means that by the ordinary *L.S.D.* test would be called significantly different might not appear significantly different when compared to a somewhat larger range. Thus, the chance of accepting an incorrect hypothesis, or comitting a type 2 error, has been increased for two adjacent means while, on the other hand, the chance of rejecting a correct hypothesis, or committing a type 1 error, has been decreased for nonadjacent means. Tests of this type have been given by R. A. Fisher in 1935 and J. W. Tukey in 1952.

Tests of the type suggested by the second solution are referred to as *multiple range tests.* The first such test was proposed by D. Newman in 1939 and was modified slightly by M. Keuls in 1952. Tukey proposed a test in 1953 that can be considered a compromise between the Newman-Keuls test and the earlier test proposed by Tukey. Finally in 1953 and 1955 D. B. Duncan proposed a test making use of so-called special *protection levels* based upon degrees of freedom.

17.6   DUNCAN'S NEW MULTIPLE RANGE TEST

This test provides a series of shortest significant ranges with which to compare differences between means. The shortest significant range R_p for comparing the largest and smallest of p means, arranged in order of magnitude, is given by $Q_p s_{\bar{x}}$, where the value of Q_p can be found in Appendix Table IX with the number of degrees of freedom for the error variance. The standard error of any mean determined from n individual variates is

$$s_{\bar{x}} = \sqrt{\frac{\text{error variance}}{n}}.$$

If the means being compared are arranged in order of magnitude, adjacent means having a difference greater than R_2 are considered significantly different. The difference between the largest and smallest of any three consecutive means is considered significant if it exceeds R_3; or, in general, the difference between the largest and smallest of any p consecutive means is considered significant if it exceeds R_p. The only exception to this rule is that no difference between two means is considered significant if the means are both contained in a subset of means with a nonsignificant range. The term "subset" is used here to designate any group of means (from 2 to k) of the given set of k means.

As indicated in Section 17-5, the *L.S.D.* should never be applied unless the *F*-value obtained in the analysis of variance is significant. In the case of Duncan's new multiple range test, most of its users do not calculate an *F*-value first, but proceed immediately to determine which means, if any, are significantly different. In most cases, significant differences between means will be found only if the *F*-value is also significant, but it can happen that, by use of Duncan's new multiple range test, one or more pairs of means will be found to be significantly different whereas the *F*-value is not significant (such a situation, for the case of two criteria of classification, is shown in Example 17-7). To avoid such contradictions, it might, therefore, be well to proceed in accordance with the statement that follows, which appears in an article

(ARS-20-3, May 1957; reprinted 1964) entitled "Mean Separation by the Functional Analysis of Variance and Multiple Comparisons," by E. L. LeClerg, Chief, Biometrical Services, Agricultural Research Service, U.S. Department of Agriculture: "The necessity for limiting the use of this multiple range test to cases where the F-test is significant has not been definitely established. However, in order to prevent contradictions between the range test and the F-test, it is recommended that this range test be used only when the F-test is significant."

Accordingly, we shall follow this recommendation. For a further discussion of this point, see the discussion following Example 17-7.

Example 17-5.   Use Duncan's new multiple range test to solve Example 17-4.

Solution.   From Table 17-5 we know that the F-value is significant so that there must be some pairs of samples with significantly different means. To determine which means are significantly different, the tabular values of R_p for 15 degrees of freedom are each multiplied by the standard error, $\sqrt{6.4/4} = 1.265$. The results are summarized in Table 17-6.

Table 17-6
New multiple range test for data of Table 17-4.

Shorest significant ranges					*Comparisons*				
p:	2	3	4	5	Sample :	D	B	E	A	C
Q_p:	3.01	3.16	3.25	3.31	Mean:	11.0	10.5	7.0	6.0	5.5
R_p:	3.81	4.00	4.11	4.19						

For the difference between the means of two adjacent samples—that is, D and B, B and E, E and A, and A and C—to be significant, it must exceed 3.81. For the difference between the largest and smallest of three consecutive sample means—that is, D and E, B and A, and E and C—to be significant, it must exceed 4.00; differences between D and A and B and C are significant if they exceed 4.11; and the difference between D and C is significant if it exceeds 4.19. These results are summarized in the right-hand part of Table 17-6, where any two means not underscored by the same line are significantly different, and any two means underscored by the same line are not significantly different. This means that the following pairs of sample means are significantly different: DA, DC, BA, BC, and this agrees with the result obtained in Example

17-4, except that the means of samples D and E are not significantly different when Duncan's new multiple range test is used. Note that for $p = 2$, the value of R_p is the same as that of the L.S.D., which is always the case.

17.7 COMPARISON OF VARIANCES IN ANALYSIS OF VARIANCE TABLE, UNEQUAL SAMPLE SIZES

Modifications necessary in the formulae developed for equal numbers of variates are very minor if unequal numbers of variates are involved. Consider again the case where the first sample has n_1 variates, the second n_2 variates, and so on, i.e. the situation of Table 17-2. The two estimates of the variance of the population, σ^2, then take the following forms:

$$s_p^2 = \frac{\sum_{i=1}^{n_1}(X_{i1} - \overline{X}_1)^2 + \sum_{i=1}^{n_2}(X_{i2} - \overline{X}_2)^2 + \cdots + \sum_{i=1}^{n_k}(X_{ik} - \overline{X}_k)^2}{n_1 + n_2 + \cdots + n_k - k}$$

$$= \frac{S.S.E.}{n_1 + n_2 + \cdots + n_k - k}, \tag{17-23}$$

$$s_t^2 = \frac{n_1(\overline{X}_1 - \overline{X})^2 + n_2(\overline{X}_2 - \overline{X})^2 + \cdots + n_k(\overline{X}_k - \overline{X})^2}{k - 1}$$

$$= \frac{S.S.T.}{k - 1} \tag{17-24}$$

Theorem 17-3 can then be restated for the case of unequal numbers of variates.

Theorem 17-4. If all possible sets of k samples of n_1, n_2, ..., n_k variates, respectively, are taken from a normally distributed population, then the quantity

$$F = \frac{s_t^2}{s_p^2},$$

where s_p^2 and s_t^2 are given by (17-23) and (17-24), respectively, satisfies the F-distribution with $v_1 = k - 1$ and $v_2 = n_1 + n_2 + \cdots + n_k - k$.

It should be noted that again there is a partitioning principle for the numbers of degrees of freedom in that the sum of the numbers of degrees of freedom $k - 1$ and $n_1 + n_2 + \cdots + n_k - k$, associated with s_t^2 and s_p^2, repectively, add up to $k - 1 + n_1 + n_2 + \cdots + n_k - k = n_1 + n_2 + \cdots + n_k - 1$, the number of degrees of freedom associated with the total number of variates.

The calculations necessary to obtain the F-value are again best arranged in a sequence of steps that are analogous to those listed for equal sample sizes and that can be summarized in an analysis of variance table similar to Table 17-3, but modified in accordance with the slightly different formulae for the sums of squares and degrees of freedom.

Example 17-6. Are there significant differences between the means of the four samples given in Table 17-7 on the basis of the 5% level of significance?

Table 17-7
An experiment with four samples of unequal sizes.

	\multicolumn{4}{c}{Sample (or treatment)}	
	A	B	C	D	
	12	12	9	12	
	10	16	7	8	
	7	15	6	8	
	8	9	11	10	
	9		7		
	14				
Total	60	52	40	38	Grand total = 190
Mean	10.0	13.0	8.0	9.5	Grand mean = 10.0

Solution. Following the suggested procedure outlined for equal numbers of variates with the necessary modifications, we find

1. $C = \dfrac{190^2}{19} = 1900;$

2. $S.S. = 12^2 + 10^2 + \cdots + 10^2 - C = 2048 - 1900 = 148;$

3. $S.S.T. = \dfrac{60^2}{6} + \dfrac{52^2}{4} + \dfrac{40^2}{5} + \dfrac{38^2}{4} - C = 1957 - 1900 = 57;$

4. $S.S.E. = 148 - 57 = 91.$

These results and the remaining calculations necessary to arrive at the required F-value are shown in Table 17-8.

Table 17-8
Analysis of variance table for the data of Table 17-7.

Source of variation	Sum of squares	d.f.	Mean square	F	$F_{.95}$
Total	148	18			
Between-means (or treatments)	57	3	19.0	3.13	3.29
Within-samples (or error)	91	15	6.07		

Since the F-value of 3.13 is less than the tabular value of $F_{.95}$ for $v_1 = 3$ and $v_2 = 15$, assuming the 5% level of significance, we accept the hypothesis that all of the given samples are taken from the same population.

Whenever unequal numbers of variates occur in the samples and the F-value indicates significance, there is no simple way to decide which pairs of sample means differ significantly. We must test all possible pairs individually by means of the t-test, as shown in Chapter 10, each time using the appropriate number of variates for the samples being compared.

17.8   THE SPECIAL CASE OF TWO SAMPLES

Clearly, the analysis of variance for two samples, whose variates are not paired, must lead to conclusions identical with those of the t-test of Chapter 10. This will be shown for equal numbers of variates. The case where the two samples have unequal numbers of variates is left as an exercise (see Exercise 12).

For $k = 2$ we must show that the F-ratio as given by (17-21) reduces to the square of the t-value as given by Formula (10-8). This follows from Theorem 16-2, since, if we let $F = t^2$, where t satisfies the t-distribution, then F must satisfy the F-distribution with $v_1 = 1$. Of course, for this case in Formula (10-8) we have $n_1 = n_2 = n$ and $m_{\bar{x}-\bar{y}} = 0$, and we must replace $\bar{X}$ by $\bar{X}_1$ and $\bar{Y}$ by $\bar{X}_2$. Formula (10-8) then takes the form

$$t = \frac{\bar{X}_1 - \bar{X}_2}{\sqrt{s_{\bar{x}_1}^2 + s_{\bar{x}_2}^2}} = \frac{\bar{X}_1 - \bar{X}_2}{\sqrt{\dfrac{s^2}{n} + \dfrac{s^2}{n}}} = \frac{\bar{X}_1 - \bar{X}_2}{\sqrt{\dfrac{2s^2}{n}}}, \qquad (17\text{-}25)$$

where

$$s^2 = \frac{\sum\limits_{i=1}^{n} (X_{i1} - \bar{X}_1)^2 + \sum\limits_{i=1}^{n} (X_{i2} - \bar{X}_2)^2}{2(n-1)}. \qquad (17\text{-}26)$$

Now, for $k = 2$, Formula (17-16) reduces also to (17-26), while (17-19) becomes

$$s^2 = n[(\bar{X}_1 - \bar{X})^2 + (\bar{X}_2 - \bar{X})^2]$$

$$= n\left[\left(\bar{X}_1 - \frac{\bar{X}_1 + \bar{X}_2}{2}\right)^2 + \left(\bar{X}_2 - \frac{\bar{X}_1 + \bar{X}_2}{2}\right)^2\right]$$

$$= \frac{n}{2}(\bar{X}_1 - \bar{X}_2)^2,$$

and the F-ratio of (17-21) takes the form

$$F = \frac{\frac{n}{2}(\bar{X}_1 - \bar{X}_2)^2}{s^2} = \frac{(\bar{X}_1 - \bar{X}_2)^2}{\dfrac{2s^2}{n}}, \qquad (17\text{-}27)$$

which is the square of the t-value given by (17-25).

No numerical example of the analysis of variance for two samples need be carried out here since it is just a special case of the procedure already discussed. However, the reader is asked, in Exercise 11, to set up an analysis of variance table for Example 10-4.

The analysis of variance as discussed in this chapter can therefore be considered as a generalization of the procedure for testing two samples for significant differences between their means when the variates are not paired. We have consistently assumed here—although we did not say so explicitly each time—that the variates were not paired or that there was no other common criterion of classification between them. We assumed that the variates were classified only according to the samples or treatments to which they belonged or that there was only one criterion of classification, namely the treatments.

Definition 17-6.   A design in which there is only one criterion of classification for the variates—namely the treatments—is called a *completely randomized design.*

17.9   ASSUMPTIONS MADE IN THE ANALYSIS OF VARIANCE WITH ONE CRITERION OF CLASSIFICATION

In discussing the procedures of this chapter various assumptions were stated that have to be satisfied to allow use of the analysis of variance. It may be well, however, to write these assumptions in a more concise way by means of a so-called *model*, which is a formula expressing the hypotheses in terms of an equation.

For the analysis of variance with one criterion of classification the model can be stated by the formula

$$X_{ij} = m + \tau_j + \varepsilon_{ij}, \tag{17-28}$$

where

X_{ij} = the ith variate in the jth sample,
m = the population grand mean,
τ_j = the differential effect of the jth treatment over the mean m,
ε_{ij} = the random sampling effect, also called *residual*.

It is further assumed that

$$\sum_{j=1}^{k} \tau_j = 0, \tag{17-29}$$

i.e. that the sum of all treatment effects equals zero, and that the ε_{ij} satisfy a normal distribution with mean 0 and standard deviation σ.

If we consider a common application of the analysis of variance, where the X_{ij} represent crop yields of plots under various treatments, each plot yield, by hypothesis, would be the sum of the overall mean yield, the effect due to the particular treatment, and the unexplainable effects caused by random factors, or

Plot yield = Overall mean + Treatment effect + Error.

The model, as given by (17-28), automatically implies that we assume the effect, if any, of each of the various treatments to be such that a certain fixed amount is added to the overall mean, or that the treatments have additive effects. If, on the other hand, the data clearly indicate that the effects are in fact not additive, the analysis of variance procedure, as outlined in this chapter, must be modified. When the effects are multiplicative the difficulty can easily be overcome by considering the logarithms of the X_{ij} instead of the X_{ij} themselves; clearly, under the hypothesis of multiplicative effects in the X_{ij} there will be additive effects in the log X_{ij}, and the analysis of variance can be applied to the latter. In most nonadditive cases, however, other than the multiplicative one, the assumption of additive effects serves as an excellent approximation to the actual case. Indeed, whenever the largest sample mean is less than 50% greater than the smallest, the assumption of additivity can safely be made.

The reader will recall that the use of the F-test in the analysis of variance also requires that all samples be taken from the same normal population or, which is equivalent, that all samples belong to normal populations, all of which have identical standard deviations. A procedure for testing the homogeneity of two variances was given in Chapter 16; procedures for testing the homogeneity of more than two variances are also available, but are not discussed in this text. One of the more frequently used tests of this type is *Bartlett's test of homogeneity of variances*. If a large number of variates is available, the normality of the samples can be tested by the procedure shown in Section 13.5. Study has shown, however, that for cases showing only minor departures from the assumptions of equal variances and normal distributions the analysis of variance procedure still gives quite adequate results.

17.10 TWO CRITERIA OF CLASSIFICATION (TWO-WAY ANALYSIS OF VARIANCE). THE RANDOMIZED COMPLETE-BLOCK DESIGN

It should be noted carefully that the model, as given by (17-28), does not allow for any effects such as those present in the case of pairing of two samples or, more generally, effects that are common to specific variates of all samples. In such cases a second criterion of classification is present. This occurs, for example, in field trials where the yields of a crop under different conditions are to be tested. In such trials, land is usually laid out in blocks, each block being subdivided into as many plots as there are treatments to be studied. In such an arrangement, called a *randomized complete-block experiment*, since each treatment has to appear in every block, all samples must be of equal size. Clearly, in the randomized complete-block experiment, the chance that certain treatments are given only to low-yielding soils while others are given only to high-yielding soils is considerably smaller than for a completely randomized arrangement as discussed in this chapter.

Though the term randomized complete-block is obviously named after its most frequent application, the field trial, it is always applied whenever a set of variates is classified according to two criteria of classification, even though one criterion is not a "block" in the literal sense of the word. Thus, a feeding experiment to determine the effects of various diets is frequently performed by giving to a certain number of groups of animals each of the diets to be studied since the animals in each group can be kept under more uniform conditions than could the large number of animals under study in all the groups. Here each group is considered a "block."

In a study of comparison of prices in various types of stores (supermarkets, chain stores, small independent stores, and so on), an economist undoubtedly would select, in various towns, one of each of these types of stores. The "treatments" would be the types of store, the "blocks" the various towns.

The analysis of a randomized complete-block experiment is quite similar to the procedure discussed in this chapter for a design with one criterion of classification. The only modification necessary is to take account of the variation between block means. This is necessary both in the partitioning of the sum of squares and in the analysis of variance table. The procedure is illustrated in Example 17-7.

Example 17-7.   Five judges score 4 products on a 10-point scale. The results are shown in Table 17-9. Analyze the results using the 5% level of significance.

Table 17-9
Five judges score 4 products.

Judge (or block)	\multicolumn{4}{c}{Product (or treatment)}	Total
	A	B	C	D	
I	7	10	7	8	32
II	9	10	5	6	30
III	8	8	5	7	28
IV	7	8	4	4	23
V	8	9	6	4	27
Total	39	45	27	29	140
Mean	7.8	9.0	5.4	5.8	

Solution.   In this case, the products play the role of the treatments and the judges the role of the blocks. We calculate the sum of squares for blocks, $S.S.B.$, in a way analogous to that used to calculate $S.S.T.$ and find

1. $C = \dfrac{140^2}{20} = 980$;

2. $S.S. = 7^2 + 9^2 + \cdots + 4^2 - C = 1048 - 980 = 68.0$;

3. $S.S.T. = \dfrac{39^2 + 45^2 + 27^2 + 29^2}{5} - C = 1023.2 - 980 = 43.2;$

4. $S.S.B. = \dfrac{32^2 + 30^2 + 28^2 + 23^2 + 27^2}{4} - C = 991.5 - 980 = 11.5;$

5. $S.S.E. = 68.0 - 43.2 - 11.5 = 13.3.$

These results and the remaining calculations are shown in Table 17-10.

Table 17-10
Analysis of variance table for the data of Table 17-9.

Source of variation	Sum of squares	d.f.	Mean square	F	$F_{.95}$
Total	68.0	19			
Between-means of products	43.2	3	14.4	12.97	3.49
Between-means of judges	11.5	4	2.88	2.59	3.26
Error	13.3	12	1.11		

Since the F-value of 2.59 is less than the tabular value of $F_{.95}$ for $v_1 = 4$ and $v_2 = 12$, no significant differences between the mean scores of judges are indicated, but since the F-value of 12.97 is greater than the tabular value of $F_{.95}$ for $v_1 = 3$ and $v_2 = 12$, there are significant differences between the mean scores of the products. Since the F-value for products is significant, we now proceed to test which of these means are significantly different. Duncan's new multiple range test will be used to find which pairs of products have significant differences between their means. The standard error for the means of products is $s_{\bar{x}} = \sqrt{1.11/5} = \sqrt{0.222} = 0.471$, and the results are summarized in Table 17-11.

Table 17-11
New multiple range test for means of products of data of Table 17-9.

Shortest significant ranges				Comparisons				
p:	2	3	4	Product:	B	A	D	C
Q_p:	3.08	3.22	3.31	Mean Score:	9.0	7.8	5.8	5.4
R_p:	1.45	1.52	1.56					

Therefore, at the 5% level, there is no significant difference between the mean scores of products A and B, and none between D and C. The mean scores for both A and B, however, are significantly better than those for C and D.

It might be of interest to determine what conclusion concerning possible differences between mean scores of judges would have been obtained had we proceeded immediately to use Duncan's new multiple range test without first determining whether the F-value for differences between mean scores of judges is significant. In that case, the appropriate value of R_p for determining the pairs of mean scores of judges that are significantly different is $Q_p s_{\bar{x}}$, where $s_{\bar{x}} = \sqrt{1.11/4} = \sqrt{0.278} = 0.527$. The results are summarized in Table 17-12.

Table 17-12
New multiple range test for mean scores of judges of Table 17-9.

Shortest significant ranges					*Comparisons*					
p:	2	3	4	5	Judges:	I	II	III	V	IV
Q_p:	3.08	3.22	3.31	3.37	Mean Score:	8.00	7.50	7.00	6.75	5.75
R_p:	1.62	1.70	1.74	1.78						

Therefore, at the 5% level, the mean score of Judge IV is significantly lower than those of Judges I and II, in spite of the fact that the F-value for differences between scores of judges is not significant. This contradiction can be explained as follows: The small F-value of 2.59 is due to the fact that, except for Judge IV, the mean scores of judges were very close together, that is, there was very little variation in the mean scores of these four judges. The question then arises: Does this experiment indicate that the mean scores of judges will generally be very close together so that the mean score of Judge IV should be considered as different from the others (as the Duncan new multiple range test tells us) or would there normally be a wider dispersion in the mean scores of judges than occurred here among Judges I, II, III, and V, in which case the mean score of Judge IV should not be considered significantly different (as the F-test tells us). The answer to this question cannot be decided from the data of this single experiment. The only way to decide it would be to repeat the experiment and analyze its results.

In the experiment above the nonsignificant F-value of 2.59 was not much below the tabular value of $F_{.95} = 3.26$, whereas, on the other hand, the difference between the mean scores of Judges II and IV, namely 1.75, was only slightly above the applicable shortest significance range of $R_4 = 1.74$. This means that both results were close to the 5% level, in one case on the side of nonsignificance, in the other on the side of significance. It is clear that, whenever a result is close to the level of significance, extreme caution should be exercised in drawing any conclusions; almost always a repetition of the experiment would be well advised.

The analysis of variance has been applied here to the simplest cases. It is also applicable to much more complicated designs, but these procedures will not be discussed here.

17.11   TRANSFORMATIONS

Assumptions made in the use of the analysis of variance have already been discussed. Valid tests of significance require that the treatment and environmental effects be additive and that the experimental errors be independent, have equal variance, and be normally distributed. Departure from any of these assumptions results in actual probabilities different from those found by means of the F-distribution. Fortunately, the measurements in much experimental work conform sufficiently well to these assumptions that the resulting analysis of variance leads to valid conclusions. It is not unusual, however, to encounter samples from populations in which one or more of the assumptions is violated. In such cases, the experimenter can resort to nonparametric methods, or he may change the scale of measurement by means of a transformation. Transformations are also used to convert to a more workable scale data which are not recorded in a form very suitable for statistical analysis. An example of this was encountered in Chapter 12 where, in order to apply a test of significance involving a well-known distribution, the coefficient of correlation, r, was transformed into

$$Z = \frac{1}{2} \ln \frac{1+r}{1-r},$$

which had an approximately normal distribution and a variance dependent only on the sample size.

Since nonnormality, nonadditivity, and heterogeneity of variance frequently occur together, often a single transformation that corrects one of these will, at the same time, result in improvements in the others. Sometimes it is found that a definite relationship exists between the treatment means and the corresponding treatment variances. When that is so, it is often possible to make a simple transformation of the data that will make the variance homogeneous and independent of the means and will result in more normally distributed data for each of the treatments. It should be pointed out, however, that, after a test or comparison has been made on the basis of transformed data, retransformation to the original units is often necessary for interpretable information.

The more common transformations are the square root, logarithmic, reciprocal, and angular or inverse sine transformations.

17.11.1   The Square Root Transformation, $\sqrt{X}$

If data consist of small whole numbers, for which the treatment mean and variance are proportional (or possibly even equal, as in a Poisson distribution), each of the original observations should be replaced by its square root before proceeding with the analysis of variance. Bacterial colonies in a plate count or the number of plants or insects in a given area lead to data of this kind. If some of the counts are small (under 10) and especially when zeros are present, $\sqrt{X + 0.5}$ is the appropriate transformation. The square root transformation may also be used for data expressed as proportions or percentages where the range is from 0 to 20 or 80 to 100 but not both. Percentages between 80 and 100 should be subtracted from 100 before applying the transformation.

Example 17-8.   Table 17-13 shows weed counts for five treatments on five plots each to control weed infestation. Compare the means and variances for each treatment with counts in the original form and after applying the square root transformation.

Table 17-13
Comparison of weed counts as observed and after applying the square root transformation.

| | Original data | | | | | Transformed data | | | | |
| | Treatment | | | | | Treatment | | | | |
	A	B	C	D	E	A	B	C	D	E
	28	7	6	177	184	5.3	2.6	2.4	13.3	13.6
	22	11	9	151	146	4.7	3.3	3.0	12.3	12.1
	54	30	26	110	131	7.3	5.5	5.1	10.5	11.4
	19	6	7	117	110	4.4	2.4	2.6	10.8	10.5
	32	11	7	135	134	5.7	3.3	2.6	11.6	11.6
Mean	31	13	11	138	141	5.48	3.42	3.14	11.70	11.84
Variance	152.8	76.4	57.2	584.8	596.8	1.03	1.21	1.00	1.04	1.04

Solution. From the data in their original form it is apparent that the rank order of the means and variances is identical, indicating that the two are correlated. Consequently, we apply the square root transformation and note that there no longer is a correlation between the means and variances of the treatments.

If the error mean square were calculated for data in their original form, it would be, on the average, too small for comparing treatments D and E, and too large for comparing A, B, and C. This difficulty is eliminated by the transformation. The analysis of variance can now be applied to the transformed data. By stabilizing the variance, we have eliminated the correlation between the mean and the variances, and hence, the mean squares between treatments and within treatments will be independent.

17.11.2 The Logarithmic Transformation, Log X

If variances are proportional to the squares of the treatment means, the logarithmic transformation equalizes the variances. It is also used if effects are known to be proportional instead of additive. Proportional effects are common in economic data and in certain biological populations. When zero values are present, the transformation $\log(X + 1)$ is appropriate.

Example 17-9. Calculate the means and variances of each of the four treatments of five variates each shown in Table 17-14. Apply the logarithmic transformation and compare the variances.

Table 17-14
Original data and logarithmic transformed data for four groups of five values each.

| | Original data | | | | Transformed data | | | |
| | Treatment | | | | Treatment | | | |
	A	B	C	D	A	B	C	D
	5	3	15	10	0.6990	0.4771	1.1761	1.0000
	11	6	8	6	1.0414	0.7782	0.9031	0.7782
	7	3	10	6	0.8451	0.4771	1.0000	0.7782
	7	5	6	12	0.8451	0.6990	0.7782	1.0792
	5	3	13	6	0.6990	0.4771	1.1139	0.7782
Mean	7.0	4.0	10.4	8.0	0.82592	0.58170	0.99426	0.88276
Variance	4.8	1.6	10.6	6.4	0.01588	0.01704	0.02048	0.01703

Solution.   The variances of the original data are considerably different in value and are proportional to the squares of the means. The transformation has resulted in nearly identical variances for all groups. The analysis of variance can now be applied to the transformed data.

17.11.3   The Reciprocal Transformation, 1/X

If the standard deviations are proportional to the square of the treatment means, the reciprocal transformation may be used. It is sometimes useful in psychological experiments involving word association or reaction time or in studies where one of the variables recorded is time of reaction or time to completion.

17.11.4   The Angular or Inverse Sine Transformation, Arcsin $\sqrt{X}$ or Sin^{-1} $\sqrt{X}$

Frequently in experimental work and research studies, as well as in economic and social data, the variable being studied is expressed as a probability, a proportion, or a percentage. In such cases the distribution tends to be binomial in form and the inverse sine (or arc sine) transformation is appropriate. This is an important transformation and is especially recommended when the percentages cover a wide range of values. It weights more heavily the small percentages which have small variance. Appendix Table X gives, for various percentages, the corresponding angles expressed in degrees corresponding to arcsin $\sqrt{\text{percentage}}$. If all percentages are less than 20 or more than 80, the square root transformation may be used: For percentages that are all in the range 30 to 70, it is doubtful if any transformation is needed.

Example 17-10.   The percents of bunt infection determined from five sample collections of each of four varieties of wheat, together with the data transformed by use of the inverse sine transformation, are shown in Table 17-15. Apply the analysis of variance to the original and transformed data and compare the results.

Table 17-15
Percentage bunt infection in four varieties of wheat sampled five times.

| | Original data | | | | | Transformed data | | |
| | Variety | | | | | Variety | | |
	A	B	C	D	A	B	C	D
	0.8	4.0	9.8	6.0	5.13	11.54	18.24	14.18
	3.8	1.9	56.2	79.8	11.24	7.92	48.56	63.29
	0.0	0.7	66.0	7.0	0.0	4.80	54.33	15.34
	6.0	3.5	10.3	84.6	14.18	10.78	18.72	66.89
	1.7	3.2	9.2	2.8	7.49	10.31	17.66	9.63
Total	12.3	13.3	151.5	180.2	38.04	45.35	157.51	169.33
Mean	2.46	2.66	30.30	36.04	7.61	9.07	31.50	33.87

Solution.   The results of the calculations are shown in Table 17-16.

Table 17-16
Analysis of variance tables for both sets of data of Table 17-15.

| | Original data | | | | | Transformed data | | | | |
Source of variation	Sum of squares	d.f.	Mean square	F	Source of variation	Sum of squares	d.f.	Mean square	F	$F_{.95}$
Total	15,133	19			Total	7751	19			
Between means (or treatments)	4767	3	1589	2.45	Between means (or treatments)	2983	3	994	3.34	3.24
Within samples (or error)	10,366	16	648		Within samples (or error)	4768	16	298		

The F-value 2.45 for the original data is less than $F_{.95} = 3.24$, and, therefore, no significant differences between the mean bunt infections for the varieties are indicated. The F-value for the transformed data exceeds the tabular value and indicates significant differences. The transformation has, therefore, increased the sensitivity of the analysis.

For the transformed data,

$$L.S.D. = t_{.05}\sqrt{\frac{2s^2}{n}} = 2.12\sqrt{\frac{2 \cdot 298}{5}} = 23.15.$$

The mean infection in variety D is, therefore, significantly greater than those in varieties A and B and that in variety C is significantly greater than that in A.

The means of the bunt infections in terms of the transformed data are expressed in degrees and, to have meaning, must be transformed back to percentages.
Using Appendix Table X, these percentages are 1.8, 2.5, 27.3, and 31.1, which
are different from the means of the original percentages.

EXERCISES

1. A test is to be made on four given samples to determine whether they can all be
   assumed to have been taken from populations with identical means. An
   experimenter, unfamiliar with the analysis of variance, tests all possible pairs
   of these sample means separately by means of Student's t-test. If he uses the
   10% level of significance, what is the probability that
   (a) he will draw all wrong conclusions;
   (b) he will draw exactly one wrong conclusion;
   (c) he will draw at least one wrong conclusion;
   (d) he will draw at least two correct conclusions?

2. Three samples are to be tested for significant differences between their means.
   An experimenter, unfamiliar with the fact that he should use the F-distribution
   to make this test, tests all possible pairs of these samples separately for significant differences between means by use of the Student t-test at the 5% level of
   significance. If the populations from which the samples are taken actually all
   have the same means, what is the probability that
   (a) he will draw at least one wrong conclusion;
   (b) he will draw at least two wrong conclusions;
   (c) he will draw all wrong conclusions?

3. Consider the following four samples:

A	B	C	D
11	18	10	13
9	12	9	8
7	15	11	9
12	13	10	12
14	16	7	11
13	16	10	10

   Calculate by use of their definitions
   (a) the total sum of squares;
   (b) the between-means sum of squares;
   (c) the within-samples sum of squares.
   (d) Calculate the above three quantities by use of their computing forms.

4. Carry out the calculations required in Exercise 3 for the following set of five samples.

A	B	C	D	E
12	14	16	14	11
10	13	14	15	9
15	12	11	11	8
13	9	14	12	12
12	11	15	9	9

5. Carry out the calculations required in Exercise 3 for the following set of three samples.

A	B	C
16	10	14
10	11	17
12	9	13
13	6	15
11	7	12
13		14
		13

6. Prove Formulae (17-13), (17-14), and (17-15).

7. (a) Are there significant differences between the means of the four samples in Exercise 3 on the basis of the 5% level of significance?
   (b) If there are significant differences, use the L.S.D. to find which pairs of samples have significantly different means.
   (c) Do part (b) by use of Duncan's new multiple range test.

8. Answer the questions of Exercise 7 for the five samples of Exercise 4.

9. (a) Are there significantly different means among the three samples of Exercise 5 at the 5% level of significance?
   (b) If so, find the pairs of samples having significantly different means.

10. Can the five samples below be considered as having been taken from populations with identical means on the basis of the 5% level of significance?

A	B	C	D	E
10	12	9	10	10
4	10	4	6	6
6	13	4	9	8
4	7	5	11	4

11. Solve Example 10-4 by means of the analysis of variance, and check that the F-value obtained equals the square of the t-value obtained in Example 10-4.

12. Prove that for two samples of unequal size the value of F obtained in the analysis of variance equals the square of the t-value used in Student's t-test to determine whether the two sample means differ significantly.

13. The following experiment was designed to determine the relative merits of four different feeds in regard to the gain in weight of pigs. Twenty pigs are divided at random into four lots with five pigs in each. Each lot is given a different feed. The gain in weight (in pounds) of each of the pigs for a fixed length of time is given below. Make a complete analysis of the data, and state your conclusions.

A	B	C	D
133	163	210	195
144	148	233	184
135	152	220	199
149	146	226	187
143	157	229	193

14. An experimenter wishes to combine the results of similar experiments performed by different people. He needs from each of these experiments the sum of squares for error. In one of these experiments the error variance was not reported, and only the least significant difference was given; it was equal to 4.12 at the 5% level. If in this experiment there were four treatments each of which was replicated seven times, what was the sum of squares for error for this particular experiment?

15. From a certain population four samples of six variates are taken and the F-value is calculated. This process (i.e. taking four samples of six variates each from that population) is repeated many times and each time the F-value is calculated. Answer the following questions (give approximate answers where accurate answers are not possible).
    (a) What percent of the time would you expect to obtain an F-value less than 1?
    (b) What percent of the time would you expect to obtain an F-value less than 0.10?

16. Four methods of performing a certain operation have been tried, and we have 10 observations for each method. The mean productivities under each method are 60, 66, 70, and 72, respectively. Not having the original data from which to calculate the sum of squares we assume that the coefficient of variation (the pooled estimate of σ divided by the average of all observations) is 0.10.
    (a) On this assumption, are there significant differences between these methods (5% level)?
    (b) If so, find which methods give significantly different mean productivities by use of Duncan's new multiple range test.

17. Complete the analysis and prepare an analysis of variance table for both the original and transformed data of Example 17-8. If significant differences between means are found, compare the means by use of Duncan's new multiple range test.

18. Complete the analysis and prepare an analysis of variance table for both the original and transformed data of Example 17-9. If significant differences are found, compare the means by use of Duncan's new multiple range test.

19. The severity of dry rot in potatoes after inoculation into five different types of wounds are shown below.

A	B	C	D	E
73	46	36	40	14
61	50	32	22	23
69	44	46	44	31
60	40	47	34	28

(a) Perform an analysis of variance on the original data and on the data transformed by use of the inverse sine transformation to determine whether the inoculation has significantly different effects on the five types of wounds (5% level).

(b) If there are significantly different effects, determine for which types of wounds they are significantly different by use of Duncan's new multiple range test.

20. In an agricultural experiment to test yields of 4 varieties of wheat, 3 similar blocks of ground were each divided into 4 equal plots, each of which was planted to one of the varieties of wheat. The yields in pounds are shown below.

	Variety			
Block	A	B	C	D
1	28	30	22	20
2	25	24	21	20
3	20	21	21	18

(a) Are there significant differences between the means of varieties (5% level)?

(b) If so, which variety means are significantly different? Use Duncan's new multiple range test.

(c) Are there significant differences between the means of blocks?

(d) If so, which block means are significantly different? Use Duncan's new multiple range test.

21. In an educational experiment 5 students were rated as to ability in 4 subject-matter areas. The results are shown below.

	Subject			
Student	A	B	C	D
1	8	10	10	11
2	7	9	9	10
3	6	7	8	9
4	6	6	7	8
5	6	6	6	5	7

(a) Are there significant differences between the mean ratings of students (5% level)?
(b) Are there significant differences between the means of subject-matter areas?
(c) If significant differences between the mean ratings of students are indicated, use Duncan's new multiple range test to find which means differ significantly.
(d) If significant differences between the means of subject-matter areas are indicated, use Duncan's new multiple range test to find which means differ significantly.

22. As part of an experiment conducted by the Department of Viticulture and Enology at the University of Calfornia, Davis, 9 students were asked to indicate pleasantness or unpleasantness for the odor of wine to which 4 varying amounts of acetaldehyde and diacetyl had been added. A seven-point scale was used to indicate the degree of liking or disliking and the score card provided for the following answers: (1) like very much, (2) like much, (3) like slightly, (4) neither like nor dislike, (5) dislike slightly, (6) dislike much, and (7) dislike very much. The results are shown below.

	Concentration			
Student	A	B	C	D
1	5	2	6	6
2	6	1	3	5
3	6	4	4	3
4	3	3	6	5
5	3	3	4	5
6	2	3	4	4
7	5	6	5	5
8	2	3	2	3
9	3	4	4	5

(a) Are there significant differences between the means of concentrations (5% level)?

(b) If differences are indicated in part (a), use Duncan's new multiple range test to find which means differ significantly.

(c) Are there significant differences between mean scores of students?

(d) If differences are indicated in part (c), use Duncan's new multiple range test to find which means differ significantly.

23. In an experiment conducted by W. E. Howard of the University of California, Davis, the effectiveness of various pit traps to capture rodents was studied for a period of eleven years. The following table shows the number of rodents captured in six of these traps placed randomly at Hopland, California.

Trap Number

Year	1	2	3	4	5	6
1959	1	2	3	1	1	1
1960	6	15	13	9	31	15
1961	4	26	9	15	21	24
1962	10	20	34	20	6	18
1963	22	27	35	29	36	37
1964	39	42	87	29	71	58
1965	16	22	16	24	13	7
1966	12	11	25	12	20	12
1967	13	13	14	17	13	9
1968	30	24	33	16	17	13
1969	26	21	33	14	27	9

(a) Are there significant differences in the mean number of rodents captured in the six traps (5% level)?

(b) If there are significant differences among the traps, find by use of a least significant difference which pairs of traps have significantly different means.

(c) Do part (b) by use of Duncan's new multiple range test.

(d) Are there significant differences in the mean number of rodents captured in the eleven years of the study?

(e) If there are significant differences among the years, find by use of a least significant difference which pairs of years have significantly different means.

(f) Do part (e) by use of Duncan's new multiple range test.

Appendix

Selected Readings

Books on probability and statistics fall into two basic categories: those that use the notation and methods of the calculus, and those that do not. In each category we have listed, for further study, typical reference works and text-books. The list is not intended to include all the statistics works available; instead, for several major topics we have listed some titles we think will be useful.

A.1 BOOKS NOT USING THE CALCULUS

A.1.1 Analysis of Variance

After having completed a study of the material in this book, most readers will find analysis of variance the topic ranking next in order of importance and logical development. In Chapter 10, Student's t-distribution was used to determine whether the means of two samples were significantly different. If it is desired to determine whether the means of a set of more than two samples are significantly different, the appropriate procedure to employ is the analysis of variance, which makes use of the F-distribution. The following two books give brief introductions to the analysis of variance and the analysis of covariance.

Dixon, Wilfred J., and Massey, Frank J., Jr., *Introduction to Statistical Analysis*, 3rd ed., McGraw-Hill Book Co., New York, 1969.

Guenther, William C., *Analysis of Variance*, Prentice-Hall, Inc., Englewood Cliffs, New Jersey, 1964.

A considerably more extensive treatment of the analysis of variance is presented in

Ostle, Bernard, *Statistics in Research*, 2nd ed., The Iowa State University Press, Ames, Iowa, 1963.

A.1.2   Experimental Designs

Since many of the experimental designs in common use rely on an analysis of variance, most books on experimental design also include an introduction to the analysis of variance.

> Cochran, William G., Cox., Gertrude M., *Experimental Designs*, 2nd ed., John Wiley and Sons, Inc., New York, 1957.
>
> Finney, David J., *Experimental Design and Its Statistical Basis*, The University of Chicago Press, Chicago, 1955.
>
> Mendenhall, William, *Introduction to Linear Models and the Design and Analysis of Experiments*, Wadsworth Publishing Co., Inc. (Duxbury Press), Belmont, California, 1968.
>
> Snedecor, George W., and Cochran, William G., *Statistical Methods*, 6th ed., The Iowa State University Press, Ames, Iowa, 1967.
>
> Steel, Robert G., and Torrie, James H., *Principles and Procedures of Statistics*, McGraw-Hill Book Co., Inc., New York, 1960.
>
> Winer, B. J., *Statistical Principles in Experimental Design*, 2nd ed., McGraw-Hill Book Co., Inc., New York, 1971.

A.1.3   Regression, Correlation (Including Rank Correlation)

Chapter 12 presents an introduction to regression and correlation analysis with the main emphasis on linear regression. The reader desiring a more detailed treatment of linear regression as well as in introduction to nonlinear regression and multiple and partial correlation is referred to

> Draper, N. R., and Smith, H., *Applied Regression Analysis*, John Wiley and Sons, Inc., New York, 1966.
>
> Guilford, Joy P., and Fruchter, Benjamin, *Fundamental Statistics in Psychology and Education*, 5th ed., McGraw-Hill Book Co., Inc., New York, 1973.

In many problems, it is very difficult, or impossible, to obtain measurements of a variable; but it may be possible to order a set of such measurements according to magnitude (i.e. obtain their ranks). Rank correlation is the name given to correlation between the ranks of two variables, a topic which is discussed extensively in

> Kendall, Maurice G., *Rank Correlation Methods*, 4th ed., Hafner Press, New York, 1962.

A.1.4 Index Numbers and Time Series

A very detailed study of index numbers is contained in

Mudgett, Bruce D., *Index Numbers*, John Wiley and Sons, Inc., New York, 1952.

Time series are treated thoroughly in

Davis, Harold T., *The Analysis of Economic Time Series*, Trinity University Press, San Antonio, 1941.
Fisher, Irving, *The Making of Index Numbers*, 3rd ed. (reprint of 1927 ed.), Augustus M. Kelley Press, Clifton, N.J.

A.2 BOOKS USING THE CALCULUS

A.2.1 Fundamental Concepts

The reader—in particular, the prospective statistician or mathematician—who wishes to gain a thorough understanding of the foundations of probability and statistics without pursuing a great many topics is referred to

Lindgren, B. W., and McElrath, G. W., *Introduction to Probability and Statistics*, 3rd ed., The Macmillan Company, New York, 1969.
Neyman, Jerzy, *First Course in Probability and Statistics*, Henry Holt and Company, Inc., New York, 1950.

A.2.2 General Statistics

The following five books use the calculus in treating most of the topics covered in this text.

Dwass, Meyer, *Probability and Statistics: An Undergraduate Course*, W. A. Benjamin, Inc., New York, 1970.
Freund, John E., *Mathematical Statistics*, 2nd ed., Prentice-Hall, Inc., Englewood Cliffs, New Jersey, 1971.
Hoel, Paul G., *Introduction to Mathematical Statistics*, 4th ed., John Wiley and Sons, Inc., New York, 1971.
Kendall, Maurice G., and Stuart, A., *The Advanced Theory of Statistics:* Vol. I, *Distribution Theory*, 3rd ed., 1969; Vol. II, *Statistical Inference and Statistical Relationship*, revised 3rd ed., 1974; Vol. III, *Design and Analysis, and Time Series*, 2nd ed., 1974. Hafner Press, New York.
Lindgren, B. W., *Statistical Theory*, 2nd ed., The Macmillan Company, New York, 1968.

A.2.3   Analysis of Variance and Experimental Designs

The analysis of variance is introduced in some of the books listed in the previous list. A very careful and rigorous treatment of this topic, in addition to many experimental designs and a number of other topics frequently encountered in statistics, is found in the following five books.

Anderson, Richard L., and Bancroft, Theodore A., *Statistical Theory in Research*, McGraw-Hill Book Co., Inc., New York, 1952.

Fisher, Ronald A., and Prance, Ghiuean T., *The Design of Experiments*, revised 8th ed. (reprint of 6th ed.), Hafner Press, New York.

Graybill, Franklin A., *An Introduction to Linear Statistical Models*, Vol. I, McGraw-Hill Book Co., Inc., New York, 1961.

Kempthorne, Oscar, *The Design and Analysis of Experiments*, John Wiley and Sons, Inc., New York, 1952.

Scheffé, Henry, *The Analysis of Variance*, John Wiley and Sons, Inc., New York, 1959.

A.2.4   Industrial Statistics

Statistical methods are employed extensively in engineering and industry. The following five works treat this topic.

Brownlee, K. A., *Statistical Theory and Methodology in Science and Engineering*, 2nd ed., John Wiley and Sons, Inc., New York, 1965.

Burr, Irving W., *Engineering Statistics and Quality Control*, McGraw-Hill Book Co., Inc., New York, 1953.

Cowden, Dudley J., *Statistical Methods in Quality Control*, Prentice-Hall, Inc., New Jersey, 1957.

Davies, O. L., *Design and Analysis of Industrial Experiments*, 2nd revised ed. (reprint of 1956 ed.), Hafner Press, New York, 1967.

Graybill, Franklin A., *An Introduction to Linear Statistical Models*, Vol. I, McGraw-Hill Book Co., Inc., New York, 1961.

Table I
Areas Under the Normal
Probability Curve

The entries under A denote the area between the line of symmetry (that is, $z = 0$) and the given z-value.

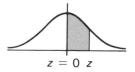

$z = 0$ z

z	A	z	A	z	A	z	A
0.00	0.0000	0.30	0.1179	0.60	0.2258	0.90	0.3159
.01	.0040	.31	.1217	.61	.2291	.91	.3186
.02	.0080	.32	.1255	.62	.2324	.92	.3212
.03	.0120	.33	.1293	.63	.2357	.93	.3238
.04	.0160	.34	.1331	.64	.2389	.94	.3264
.05	.0199	.35	.1368	.65	.2422	.95	.3289
.06	.0239	.36	.1406	.66	.2454	.96	.3315
.07	.0279	.37	.1443	.67	.2486	.97	.3340
.08	.0319	.38	.1480	.68	.2518	.98	.3365
.09	.0359	.39	.1517	.69	.2549	.99	.3389
.10	.0398	.40	.1554	.70	.2580	1.00	.3413
.11	.0438	.41	.1591	.71	.2612	1.01	.3438
.12	.0478	.42	.1628	.72	.2642	1.02	.3461
.13	.0517	.43	.1664	.73	.2673	1.03	.3485
.14	.0557	.44	.1700	.74	.2704	1.04	.3508
.15	.0596	.45	.1736	.75	.2734	1.05	.3531
.16	.0636	.46	.1772	.76	.2764	1.06	.3554
.17	.0675	.47	.1808	.77	.2794	1.07	.3577
.18	.0714	.48	.1844	.78	.2823	1.08	.3599
.19	.0754	.49	.1879	.79	.2852	1.09	.3621
.20	.0793	.50	.1915	.80	.2881	1.10	.3643
.21	.0832	.51	.1950	.81	.2910	1.11	.3665
.22	.0871	.52	.1985	.82	.2939	1.12	.3686
.23	.0910	.53	.2019	.83	.2967	1.13	.3708
.24	.0948	.54	.2054	.84	.2996	1.14	.3729
.25	.0987	.55	.2088	.85	.3023	1.15	.3749
.26	.1026	.56	.2123	.86	.3051	1.16	.3770
.27	.1064	.57	.2157	.87	.3079	1.17	.3790
.28	.1103	.58	.2190	.88	.3106	1.18	.3810
.29	.1141	.59	.2224	.89	.3133	1.19	.3830

Table I.   Areas Under the Normal Probability Curve

  (*continued*)

z	A	z	A	z	A	z	A
1.20	0.3849	1.55	0.4394	1.90	0.4713	2.25	0.4878
1.21	.3869	1.56	.4406	1.91	.4719	2.26	.4881
1.22	.3888	1.57	.4418	1.92	.4726	2.27	.4884
1.23	.3907	1.58	.4430	1.93	.4732	2.28	.4887
1.24	.3925	1.59	.4441	1.94	.4738	2.29	.4890
1.25	.3944	1.60	.4452	1.95	.4744	2.30	.4893
1.26	.3962	1.61	.4463	1.96	.4750	2.31	.4896
1.27	.3980	1.62	.4474	1.97	.4756	2.32	.4898
1.28	.3997	1.63	.4485	1.98	.4762	2.33	.4901
1.29	.4015	1.64	.4495	1.99	.4767	2.34	.4904
1.30	.4032	1.65	.4505	2.00	.4773	2.35	.4906
1.31	.4049	1.66	.4515	2.01	.4778	2.36	.4909
1.32	.4066	1.67	.4525	2.02	.4783	2.37	.4911
1.33	.4082	1.68	.4535	2.03	.4788	2.38	.4913
1.34	.4099	1.69	.4545	2.04	.4793	2.39	.4916
1.35	.4155	1.70	.4554	2.05	.4798	2.40	.4918
1.36	.4131	1.71	.4564	2.06	.4803	2.41	.4920
1.37	.4147	1.72	.4573	2.07	.4808	2.42	.4922
1.38	.4162	1.73	.4582	2.08	.4812	2.43	.4925
1.39	.4177	1.74	.4591	2.09	.4817	2.44	.4927
1.40	.4192	1.75	.4599	2.10	.4821	2.45	.4929
1.41	.4207	1.76	.4608	2.11	.4826	2.46	.4931
1.42	.4222	1.77	.4616	2.12	.4830	2.47	.4932
1.43	.4236	1.78	.4625	2.13	.4834	2.48	.4934
1.44	.4251	1.79	.4633	2.14	.4838	2.49	.4936
1.45	.4265	1.80	.4641	2.15	.4842	2.50	.4938
1.46	.4279	1.81	.4649	2.16	.4846	2.51	.4940
1.47	.4292	1.82	.4656	2.17	.4850	2.52	.4941
1.48	.4306	1.83	.4664	2.18	.4854	2.53	.4943
1.49	.4319	1.84	.4671	2.19	.4857	2.54	.4945
1.50	.4332	1.85	.4678	2.20	.4861	2.55	.4946
1.51	.4345	1.86	.4686	2.21	.4865	2.56	.4948
1.52	.4357	1.87	.4693	2.22	.4868	2.57	.4949
1.53	.4370	1.88	.4700	2.23	.4871	2.58	.4951
1.54	.4382	1.89	.4706	2.24	.4875	2.59	.4952

Table I. Areas Under the Normal Probability Curve

(*concluded*)

z	A	z	A	z	A	z	A
2.60	0.4953	2.95	0.4984	3.30	0.4995	3.65	0.4999
2.61	.4955	2.96	.4985	3.31	.4995	3.66	.4999
2.62	.4956	2.97	.4985	3.32	.4996	3.67	.4999
2.63	.4957	2.98	.4986	3.33	.4996	3.68	.4999
2.64	.4959	2.99	.4986	3.34	.4996	3.69	.4999
2.65	.4960	3.00	.4987	3.35	.4996	3.70	.4999
2.66	.4961	3.01	.4987	3.36	.4996	3.71	.4999
2.67	.4962	3.02	.4987	3.37	.4996	3.72	.4999
2.68	.4963	3.03	.4988	3.38	.4996	3.73	.4999
2.69	.4964	3.04	.4988	3.39	.4997	3.74	.4999
2.70	.4965	3.05	.4989	3.40	.4997	3.75	.4999
2.71	.4966	3.06	.4989	3.41	.4997	3.76	.4999
2.72	.4967	3.07	.4989	3.42	.4997	3.77	.4999
2.73	.4968	3.08	.4990	3.43	.4997	3.78	.4999
2.74	.4969	3.09	.4990	3.44	.4997	3.79	.4999
2.75	.4970	3.10	.4990	3.45	.4997	3.80	.4999
2.76	.4971	3.11	.4991	3.46	.4997	3.81	.4999
2.77	.4972	3.12	.4991	3.47	.4997	3.82	.4999
2.78	.4973	3.13	.4991	3.48	.4998	3.83	.4999
2.79	.4974	3.14	.4992	3.49	.4998	3.84	.4999
2.80	.4974	3.15	.4992	3.50	.4998	3.85	.4999
2.81	.4975	3.16	.4992	3.51	.4998	3.86	.4999
2.82	.4976	3.17	.4992	3.52	.4998	3.87	.5000
2.83	.4977	3.18	.4993	3.53	.4998	3.88	.5000
2.84	.4977	3.19	.4993	3.54	.4998	3.89	.5000
2.85	.4978	3.20	.4993	3.55	.4998		
2.86	.4979	3.21	.4993	3.56	.4998		
2.87	.4980	3.22	.4994	3.57	.4998		
2.88	.4980	3.23	.4994	3.58	.4998		
2.89	.4981	3.24	.4994	3.59	.4998		
2.90	.4981	3.25	.4994	3.60	.4998		
2.91	.4982	3.26	.4994	3.61	.4999		
2.92	.4983	3.27	.4995	3.62	.4999		
2.93	.4983	3.28	.4995	3.63	.4999		
2.94	.4984	3.29	.4995	3.64	.4999		

Table la
Ordinates (y-Values) for the
Normal Probability Curve

The entries under y denote the ordinates for the given z-values.

$z = 0$ z

z	y	z	y	z	y	z	y
0.00	0.3989	0.30	0.3814	0.60	0.3332	0.90	0.2661
.01	.3989	.31	.3802	.61	.3312	.91	.2637
.02	.3989	.32	.3790	.62	.3292	.92	.2613
.03	.3988	.33	.3778	.63	.3271	.93	.2589
.04	.3986	.34	.3765	.64	.3251	.94	.2565
.05	.3984	.35	.3752	.65	.3230	.95	.2541
.06	.3982	.36	.3739	.66	.3209	.96	.2516
.07	.3980	.37	.3725	.67	.3187	.97	.2492
.08	.3977	.38	.3711	.68	.3166	.98	.2468
.09	.3973	.39	.3697	.69	.3144	.99	.2444
.10	.3969	.40	.3683	.70	.3122	1.00	.2420
.11	.3965	.41	.3668	.71	.3101	1.01	.2395
.12	.3961	.42	.3653	.72	.3078	1.02	.2371
.13	.3956	.43	.3637	.73	.3056	1.03	.2347
.14	.3950	.44	.3621	.74	.3034	1.04	.2323
.15	.3945	.45	.3605	.75	.3011	1.05	.2299
.16	.3939	.46	.3589	.76	.2989	1.06	.2275
.17	.3932	.47	.3572	.77	.2966	1.07	.2251
.18	.3925	.48	.3555	.78	.2943	1.08	.2226
.19	.3918	.49	.3538	.79	.2920	1.09	.2202
.20	.3910	.50	.3521	.80	.2897	1.10	.2178
.21	.3902	.51	.3503	.81	.2874	1.11	.2155
.22	.3894	.52	.3485	.82	.2850	1.12	.2131
.23	.3885	.53	.3467	.83	.2827	1.13	.2107
.24	.3876	.54	.3448	.84	.2803	1.14	.2083
.25	.3867	.55	.3429	.85	.2780	1.15	.2059
.26	.3857	.56	.3410	.86	.2756	1.16	.2036
.27	.3847	.57	.3391	.87	.2732	1.17	.2012
.28	.3836	.58	.3372	.88	.2709	1.18	.1989
.29	.3825	.59	.3352	.89	.2685	1.19	.1965

Table Ia.   Ordinates (y-Values) for the Normal Probability Curve

(*continued*)

z	y	z	y	z	y	z	y
1.20	0.1942	1.55	0.1200	1.90	0.0656	2.25	0.0317
1.21	.1919	1.56	.1182	1.91	.0644	2.26	.0310
1.22	.1895	1.57	.1163	1.92	.0632	2.27	.0303
1.23	.1872	1.58	.1145	1.93	.0619	2.28	.0296
1.24	.1849	1.59	.1127	1.94	.0608	2.29	.0290
1.25	.1826	1.60	.1109	1.95	.0596	2.30	.0283
1.26	.1804	1.61	.1091	1.96	.0584	2.31	.0277
1.27	.1781	1.62	.1074	1.97	.0573	2.32	.0270
1.28	.1758	1.63	.1057	1.98	.0562	2.33	.0264
1.29	.1736	1.64	.1040	1.99	.0551	2.34	.0258
1.30	.1714	1.65	.1023	2.00	0.540	2.35	.0252
1.31	.1691	1.66	.1006	2.01	.0529	2.36	.0246
1.32	.1669	1.67	.0989	2.02	.0519	2.37	.0241
1.33	.1647	1.68	.0973	2.03	.0508	2.38	.0235
1.34	.1626	1.69	.0957	2.04	.0498	2.39	.0229
1.35	.1604	1.70	.0940	2.05	.0488	2.40	.0224
1.36	.1582	1.71	.0925	2.06	.0478	2.41	.0219
1.37	.1561	1.72	.0909	2.07	.0468	2.42	.0213
1.38	.1539	1.73	.0893	2.08	.0459	2.43	.0208
1.39	.1518	1.74	.0878	2.09	.0449	2.44	.0203
1.40	.1497	1.75	.0863	2.10	.0440	2.45	.0198
1.41	.1476	1.76	.0848	2.11	.0431	2.46	.0194
1.42	.1456	1.77	.0833	2.12	.0422	2.47	.0189
1.43	.1435	1.78	.0818	2.13	.0413	2.48	.0184
1.44	.1415	1.79	.0804	2.14	.0404	2.49	.0180
1.45	.1394	1.80	.0789	2.15	.0395	2.50	.0175
1.46	.1374	1.81	.0775	2.16	.0387	2.51	.0171
1.47	.1354	1.82	.0761	2.17	.0379	2.52	.0167
1.48	.1334	1.83	.0748	2.18	.0371	2.53	.0162
1.49	.1315	1.84	.0734	2.19	.0363	2.54	.0158
1.50	.1295	1.85	.0721	2.20	.0355	2.55	.0154
1.51	.1276	1.86	.0707	2.21	.0347	2.56	.0151
1.52	.1257	1.87	.0694	2.22	.0339	2.57	.0147
1.53	.1238	1.88	.0681	2.23	.0332	2.58	.0143
1.54	.1219	1.89	.0669	2.24	.0325	2.59	.0139

Table Ia.   Ordinates (y-Values) for the Normal Probability Curve

(concluded)

z	y	z	y	z	y	z	y
2.60	0.0136	2.95	0.0051	3.30	0.0017	3.65	0.0005
2.61	.0132	2.96	.0050	3.31	.0017	3.66	.0005
2.62	.0129	2.97	.0048	3.32	.0016	3.67	.0005
2.63	.0126	2.98	.0047	3.33	.0016	3.68	.0005
2.64	.0122	2.99	.0046	3.34	.0015	3.69	.0004
2.65	.0119	3.00	.0044	3.35	.0015	3.70	.0004
2.66	.0116	3.01	.0043	3.36	.0014	3.71	.0004
2.67	.0113	3.02	.0042	3.37	.0014	3.72	.0004
2.68	.0110	3.03	.0040	3.38	.0013	3.73	.0004
2.69	.0107	3.04	.0039	3.39	.0013	3.74	.0004
2.70	.0104	3.05	.0038	3.40	.0012	3.75	.0003
2.71	.0101	3.06	.0037	3.41	.0012	3.76	.0003
2.72	.0099	3.07	.0036	3.42	.0011	3.77	.0003
2.73	.0096	3.08	.0035	3.43	.0011	3.78	.0003
2.74	.0093	3.09	.0034	3.44	.0011	3.79	.0003
2.75	.0091	3.10	.0033	3.45	.0010	3.80	.0003
2.76	.0088	3.11	.0032	3.46	.0010	3.81	.0003
2.77	.0086	3.12	.0031	3.47	.0010	3.82	.0003
2.78	.0084	3.13	.0030	3.48	.0009	3.83	.0003
2.79	.0081	3.14	.0029	3.49	.0009	3.84	.0002
2.80	.0079	3.15	.0028	3.50	.0009	3.85	.0002
2.81	.0077	3.16	.0027	3.51	.0008	3.86	.0002
2.82	.0075	3.17	.0026	3.52	.0008	3.87	.0002
2.83	.0073	3.18	.0025	3.53	.0008	3.88	.0002
2.84	.0071	3.19	.0025	3.54	.0008	3.89	.0002
2.85	.0069	3.20	.0024	3.55	.0007	3.90	.0002
2.86	.0067	3.21	.0023	3.56	.0007	3.91	.0002
2.87	.0065	3.22	.0022	3.57	.0007	3.92	.0002
2.88	.0063	3.23	.0022	3.58	.0007	3.93	.0002
2.89	.0061	3.24	.0021	3.59	.0006	3.94	.0002
2.90	.0059	3.25	.0020	3.60	.0006	3.95	.0002
2.91	.0058	3.26	.0020	3.61	.0006	3.96	.0002
2.92	.0056	3.27	.0019	3.62	.0006	3.97	.0001
2.93	.0054	3.28	.0018	3.63	.0005	3.98	.0001
2.94	.0053	3.29	.0018	3.64	.0005	3.99	.0001

Table II
Poisson Distribution

Each number in this table represents the probability of obtaining at least X successes, or the area under the histogram to the right of and including the rectangle whose center is at X.

m	X=0	X=1	X=2	X=3	X=4	X=5	X=6	X=7	X=8	X=9	X=10	X=11	X=12	X=13	X=14
.10	1.000	.095	.005												
.20	1.000	.181	.018	.001											
.30	1.000	.259	.037	.004											
.40	1.000	.330	.062	.008	.001										
.50	1.000	.393	.090	.014	.002										
.60	1.000	.451	.122	.023	.003										
.70	1.000	.503	.156	.034	.006	.001									
.80	1.000	.551	.191	.047	.009	.001									
.90	1.000	.593	.228	.063	.013	.002									
1.00	1.000	.632	.264	.080	.019	.004	.001								
1.1	1.000	.667	.301	.100	.026	.005	.001								
1.2	1.000	.699	.337	.120	.034	.008	.002								
1.3	1.000	.727	.373	.143	.043	.011	.002								
1.4	1.000	.753	.408	.167	.054	.014	.003	.001							
1.5	1.000	.777	.442	.191	.066	.019	.004	.001							
1.6	1.000	.798	.475	.217	.079	.024	.006	.001							
1.7	1.000	.817	.507	.243	.093	.030	.008	.002							
1.8	1.000	.835	.537	.269	.109	.036	.010	.003	.001						
1.9	1.000	.850	.566	.296	.125	.044	.013	.003	.001						
2.0	1.000	.865	.594	.323	.143	.053	.017	.005	.001						
2.2	1.000	.889	.645	.377	.181	.072	.025	.007	.002						
2.4	1.000	.909	.692	.430	.221	.096	.036	.012	.003	.001					
2.6	1.000	.926	.733	.482	.264	.123	.049	.017	.005	.001					
2.8	1.000	.939	.769	.531	.308	.152	.065	.024	.008	.002	.001				
3.0	1.000	.950	.801	.577	.353	.185	.084	.034	.012	.004	.001				
3.2	1.000	.959	.829	.620	.397	.219	.105	.045	.017	.006	.002				
3.4	1.000	.967	.853	.660	.442	.256	.129	.058	.023	.008	.003	.001			
3.6	1.000	.973	.874	.697	.485	.294	.156	.073	.031	.012	.004	.001			
3.8	1.000	.978	.893	.731	.527	.332	.184	.091	.040	.016	.006	.002	.001		
4.0	1.000	.982	.908	.762	.567	.371	.215	.111	.051	.021	.008	.003	.001		
4.2	1.000	.985	.922	.790	.605	.410	.247	.133	.064	.028	.011	.004	.001		
4.4	1.000	.988	.934	.815	.641	.449	.280	.156	.079	.036	.015	.006	.002	.001	
4.6	1.000	.990	.944	.837	.674	.487	.314	.182	.095	.045	.020	.008	.003	.001	
4.8	1.000	.992	.952	.857	.706	.524	.349	.209	.113	.056	.025	.010	.004	.001	
5.0	1.000	.993	.960	.875	.735	.560	.384	.238	.133	.068	.032	.014	.005	.002	.001

Table III
Student's t-Distribution

The entries under A denote the sum of the two tail areas for the values of t given below. The values of ν denote the number of degrees of freedom (df).

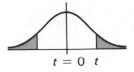

$t = 0 \quad t$

ν or $d.f.$	$A = 0.1$	$A = 0.05$	$A = 0.02$	$A = 0.01$	$A = 0.001$
1	6.314	12.706	31.821	63.657	636.619
2	2.920	4.303	6.965	9.925	31.598
3	2.353	3.182	4.541	5.841	12.941
4	2.132	2.776	3.747	4.604	8.610
5	2.015	2.571	3.365	4.032	6.859
6	1.943	2.447	3.143	3.707	5.959
7	1.895	2.365	2.998	3.499	5.405
8	1.860	2.306	2.896	3.355	5.041
9	1.833	2.262	2.821	3.250	4.781
10	1.812	2.228	2.764	3.169	4.587
11	1.796	2.201	2.718	3.106	4.437
12	1.782	2.179	2.681	3.055	4.318
13	1.771	2.160	2.650	3.012	4.221
14	1.761	2.145	2.624	2.977	4.140
15	1.753	2.131	2.602	2.947	4.073
16	1.746	2.120	2.583	2.921	4.015
17	1.740	2.110	2.567	2.898	3.965
18	1.734	2.101	2.552	2.878	3.922
19	1.729	2.093	2.539	2.861	3.883
20	1.725	2.086	2.528	2.845	3.850
21	1.721	2.080	2.518	2.831	3.819
22	1.717	2.074	2.508	2.819	3.792
23	1.714	2.069	2.500	2.807	3.767
24	1.711	2.064	2.492	2.797	3.745
25	1.708	2.060	2.485	2.787	3.725
26	1.706	2.056	2.479	2.779	3.707
27	1.703	2.052	2.473	2.771	3.690
28	1.701	2.048	2.467	2.763	3.674
29	1.699	2.045	2.462	2.756	3.659
30	1.697	2.042	2.457	2.750	3.646
40	1.684	2.021	2.423	2.704	3.551
60	1.671	2.000	2.390	2.660	3.460
120	1.658	1.980	2.358	2.617	3.373
∞	1.645	1.960	2.326	2.576	3.291

Table III is abridged from Table III of Fisher and Yates: *Statistical Tables for Biological, Agricultural, and Medical Research,* published by Longman Group Ltd., London (previously published by Oliver and Boyd Limited, Edinburgh), by permission of the authors and publishers.

Table IV—Wilcoxon Distribution (With No Pairing)

The numbers given in this table are the number of cases for which the sum of the ranks of the sample of size n_1 is less than or equal to W_1.

Values of U, where $U = W_1 - \frac{1}{2}n_1(n_1 + 1)$

n_1	n_2	C_{n,n_1}	0	1	2	3	4	5	6	7	8	9	10	11	12	13	14	15	16	17	18	19	20
3	3	20	1	2	4	7	10	13	16	18	19	20											
3	4	35	1	2	4	7	11	15	20	24	28	31	33	34	35								
4	4	70	1	2	4	7	12	17	24	31	39	46	53	58	63	66	68	69	70				
3	5	56	1	2	4	7	11	16	22	28	34	40	45	49	52	54	55	56					
4	5	126	1	2	4	7	12	18	26	35	46	57	69	80	91	100	108	114	119	122	124	125	126
5	5	252	1	2	4	7	12	19	28	39	53	69	87	106	126	146	165	183	199	213	224	233	240
3	6	84	1	2	4	7	11	16	23	30	38	46	54	61	68	73	77	80	82	83	84		
4	6	210	1	2	4	7	12	18	27	37	50	64	80	96	114	130	146	160	173	183	192	198	203
5	6	462	1	2	4	7	12	19	29	41	57	76	99	124	153	183	215	247	279	309	338	363	386
6	6	924	1	2	4	7	12	19	30	43	61	83	111	143	182	224	272	323	378	433	491	546	601
3	7	120	1	2	4	7	11	16	23	31	40	50	60	70	80	89	97	104	109	113	116	118	119
4	7	330	1	2	4	7	12	18	27	38	52	68	87	107	130	153	177	200	223	243	262	278	292
5	7	792	1	2	4	7	12	19	29	42	59	80	106	136	171	210	253	299	347	396	445	493	539
6	7	1,716	1	2	4	7	12	19	30	44	63	87	118	155	201	253	314	382	458	539	627	717	811
7	7	3,432	1	2	4	7	12	19	30	45	65	91	125	167	220	283	358	445	545	657	782	918	1,064
3	8	165	1	2	4	7	11	16	23	31	41	52	64	76	89	101	113	124	134	142	149	154	158
4	8	495	1	2	4	7	12	18	27	38	53	70	91	114	141	169	200	231	264	295	326	354	381
5	8	1,287	1	2	4	7	12	19	29	42	60	82	110	143	183	228	280	337	400	466	536	607	680
6	8	3,003	1	2	4	7	12	19	30	44	64	89	122	162	213	272	343	424	518	621	737	860	994
7	8	6,435	1	2	4	7	12	19	30	45	66	93	129	174	232	302	388	489	609	746	904	1,080	1,277
8	8	12,870	1	2	4	7	12	19	30	45	67	95	133	181	244	321	418	534	675	839	1,033	1,254	1,509

Table IV is reproduced from Table H of Hodges and Lehmann: *Basic Concepts of Probability and Statistics*, published by Holden-Day, San Francisco, by permission of the authors and publisher.

Table V
Wilcoxon Distribution (With Pairing)

The numbers given in these tables are the number of cases for which the sum of the ranks is less than or equal to W_1.

Table Va.   The case where $W_1 \leq n$.

W_1		W_1		W_1		W_1	
0	1	6	14	11	55	16	169
1	2	7	19	12	70	17	207
2	3	8	25	13	88	18	253
3	5	9	33	14	110	19	307
4	7	10	43	15	137	20	371
5	10						

Table Vb. The case where $W_1 > n$

																		n
$W_1 - n$	3	4	5	6	7	8	9	10	11	12	13	14	15	16	17	18	19	20
1	6	9	13	18	24	32	42	54	69	87	109	136	168	206	252	306	370	446
2	7	11	16	22	30	40	52	67	85	107	134	166	204	250	304	368	444	533
3	8	13	19	27	37	49	64	82	104	131	163	201	247	301	365	441	530	634
4		14	22	32	44	59	77	99	126	158	196	242	296	360	436	525	629	751
5		15	25	37	52	70	92	119	151	189	235	289	353	429	518	622	744	886
6		16	27	42	60	82	109	141	179	225	279	343	419	508	612	734	876	1,041
7			29	46	68	95	127	165	211	265	329	405	494	598	720	862	1,027	1,219
8			30	50	76	108	146	192	246	310	386	475	579	701	843	1,008	1,200	1,422
9			31	54	84	121	167	221	285	361	450	554	676	818	983	1,175	1,397	1,653
10			32	57	91	135	188	252	328	417	521	643	785	950	1,142	1,364	1,620	1,916
11				59	98	148	210	285	374	478	600	742	907	1,099	1,321	1,577	1,873	2,213
12				61	104	161	233	320	423	545	687	852	1,044	1,266	1,522	1,818	2,158	2,548
13				62	109	174	256	356	476	617	782	974	1,196	1,452	1,748	2,088	2,478	2,926
14				63	114	186	279	394	532	695	886	1,108	1,364	1,660	2,000	2,390	2,838	3,350
15				64	118	197	302	433	591	779	999	1,254	1,550	1,890	2,280	2,728	3,240	3,825
16					121	207	324	472	653	868	1,120	1,414	1,753	2,143	2,591	3,103	3,688	4,356
17					123	216	345	512	717	962	1,251	1,587	1,975	2,422	2,934	3,519	4,187	4,947
18					125	224	366	552	783	1,062	1,391	1,774	2,218	2,728	3,312	3,980	4,740	5,604
19					126	231	385	591	851	1,166	1,539	1,976	2,481	3,062	3,728	4,487	5,351	6,333
20					127	237	403	630	920	1,274	1,697	2,192	2,766	3,427	4,183	5,045	6,026	7,139
21					128	242	420	668	989	1,387	1,863	2,423	3,074	3,823	4,680	5,658	6,769	8,028
22						246	435	704	1,059	1,502	2,037	2,669	3,404	4,251	5,222	6,328	7,584	9,008
23						249	448	749	1,128	1,620	2,219	2,929	3,757	4,714	5,810	7,059	8,478	10,084
24						251	460	772	1,197	1,741	2,408	3,203	4,135	5,212	6,447	7,856	9,455	11,264
25						253	470	803	1,265	1,863	2,603	3,492	4,536	5,746	7,136	8,721	10,520	12,557
26						254	479	832	1,331	1,986	2,805	3,794	4,961	6,318	7,878	9,658	11,681	13,968
27						255	487	859	1,395	2,110	3,012	4,109	5,411	6,928	8,675	10,673	12,941	15,506
28						256	493	883	1,457	2,233	3,223	4,437	5,884	7,576	9,531	11,766	14,306	17,180
29							498	905	1,516	2,355	3,438	4,776	6,380	8,265	10,445	12,942	15,783	18,997
30							502	925	1,572	2,476	3,656	5,126	6,901	8,993	11,420	14,206	17,377	20,966

Table V is reproduced from Tables J and K of Hodges and Lehmann : *Basic Concepts of Probability and Statistics*, published by Holden-Day, Inc., San Francisco, by permission of the authors and publisher.

Table VI

Transformation of r to Z $\left(\text{that is, } Z = \tfrac{1}{2} \ln \dfrac{1+r}{1-r}\right)$

r	Z	r	Z	r	Z
.00	.000				
.01	.010	.36	.377	.71	.887
.02	.020	.37	.388	.72	.908
.03	.030	.38	.400	.73	.929
.04	.040	.39	.412	.74	.950
.05	.050	.40	.424	.75	.973
.06	.060	.41	.436	.76	.996
.07	.070	.42	.448	.77	1.020
.08	.080	.43	.460	.78	1.045
.09	.090	.44	.472	.79	1.071
.10	.100	.45	.485	.80	1.099
.11	.110	.46	.497	.81	1.127
.12	.121	.47	.510	.82	1.157
.13	.131	.48	.523	.83	1.188
.14	.141	.49	.536	.84	1.221
.15	.151	.50	.549	.85	1.256
.16	.161	.51	.563	.86	1.293
.17	.172	.52	.576	.87	1.333
.18	.182	.53	.590	.88	1.376
.19	.192	.54	.604	.89	1.422
.20	.203	.55	.618	.90	1.472
.21	.213	.56	.633	.91	1.528
.22	.224	.57	.648	.92	1.589
.23	.234	.58	.662	.93	1.658
.24	.245	.59	.678	.94	1.738
.25	.255	.60	.693	.95	1.832
.26	.266	.61	.709	.96	1.946
.27	.277	.62	.725	.97	2.092
.28	.288	.63	.741	.98	2.298
.29	.299	.64	.758	.99	2.647
.30	.310	.65	.775		
.31	.321	.66	.793		
.32	.332	.67	.811		
.33	.343	.68	.829		
.34	.354	.69	.848		
.35	.365	.70	.867		

Table VI is abridged from Table VII of Fisher and Yates: *Statistical Tables for Biological, Agricultural, and Medical Research*, published by Longman Group Ltd., London (previously published by Oliver and Boyd Limited, Edinburgh), by permission of the authors and publishers.

Table VII
Chi-Square Distribution

The entries under A denote the right tail area for the values of χ^2 given below. The values of v denote the number of degrees of freedom.

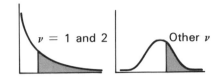

$v = 1$ and 2   Other v

v or d.f.	$A = 0.99$	$A = 0.98$	$A = 0.95$	$A = 0.90$	$A = 0.80$	$A = 0.70$	$A = 0.50$
1	.00016	.00063	.0039	.016	.064	.15	.46
2	.02	.04	.10	.21	.45	.71	1.39
3	.12	.18	.35	.58	1.00	1.42	2.37
4	.30	.43	.71	1.06	1.65	2.20	3.36
5	.55	.75	1.14	1.61	2.34	3.00	4.35
6	.87	1.13	1.64	2.20	3.07	3.83	5.35
7	1.24	1.56	2.17	2.83	3.82	4.67	6.35
8	1.65	2.03	2.73	3.49	4.59	5.53	7.34
9	2.09	2.53	3.32	4.17	5.38	6.39	8.34
10	2.56	3.06	3.94	4.86	6.18	7.27	9.34
11	3.05	3.61	4.58	5.58	6.99	8.15	10.34
12	3.57	4.18	5.23	6.30	7.81	9.03	11.34
13	4.11	4.76	5.89	7.04	8.63	9.93	12.34
14	4.66	5.37	6.57	7.79	9.47	10.82	13.34
15	5.23	5.98	7.26	8.55	10.31	11.72	14.34
16	5.81	6.61	7.96	9.31	11.15	12.62	15.34
17	6.41	7.26	8.67	10.08	12.00	13.53	16.34
18	7.02	7.91	9.39	10.86	12.86	14.44	17.34
19	7.63	8.57	10.12	11.65	13.72	15.35	18.34
20	8.26	9.24	10.85	12.44	14.58	16.27	19.34
21	8.90	9.92	11.59	13.24	15.44	17.18	20.34
22	9.54	10.60	12.34	14.04	16.31	18.10	21.34
23	10.20	11.29	13.09	14.85	17.19	19.02	22.34
24	10.86	11.99	13.85	15.66	18.06	19.94	23.34
25	11.52	12.70	14.61	16.47	18.94	20.87	24.34
26	12.20	13.41	15.38	17.29	19.82	21.79	25.34
27	12.88	14.12	16.15	18.11	20.70	22.72	26.34
28	13.56	14.85	16.93	18.94	21.59	23.65	27.34
29	14.26	15.57	17.71	19.77	22.48	24.58	28.34
30	14.95	16.31	18.49	20.60	23.36	25.51	29.34

Table VII.   Chi-Square Distribution

(*concluded*)

ν or *d.f.*	$A = 0.30$	$A = 0.20$	$A = 0.10$	$A = 0.05$	$A = 0.02$	$A = 0.01$	$A = 0.001$
1	1.07	1.64	2.71	3.84	5.41	6.64	10.83
2	2.41	3.22	4.60	5.99	7.82	9.21	13.82
3	3.66	4.64	6.25	7.82	9.84	11.34	16.27
4	4.88	5.99	7.78	9.49	11.67	13.28	18.46
5	6.06	7.29	9.24	11.07	13.39	15.09	20.52
6	7.23	8.56	10.64	12.59	15.03	16.81	22.46
7	8.38	9.80	12.02	14.07	16.62	18.48	24.32
8	9.52	11.03	13.36	15.51	18.17	20.09	26.12
9	10.66	12.24	14.68	16.92	19.68	21.67	27.88
10	11.78	13.44	15.99	18.31	21.16	23.21	29.59
11	12.90	14.63	17.28	19.68	22.62	24.72	31.26
12	14.01	15.81	18.55	21.03	24.05	26.22	32.91
13	15.12	16.98	19.81	22.36	25.47	27.69	34.53
14	16.22	18.15	21.06	23.68	26.87	29.14	36.12
15	17.32	19.31	22.31	25.00	28.26	30.58	37.70
16	18.42	20.46	23.54	26.30	29.63	32.00	39.25
17	19.51	21.62	24.77	27.59	31.00	33.41	40.79
18	20.60	22.76	25.99	28.87	32.35	34.80	42.31
19	21.69	23.90	27.20	30.14	33.69	36.19	43.82
20	22.78	25.04	28.41	31.41	35.02	37.57	45.32
21	23.86	26.17	29.62	32.67	36.34	38.93	46.80
22	24.94	27.30	30.81	33.92	37.66	40.29	48.27
23	26.02	28.43	32.01	35.17	38.97	41.64	49.73
24	27.10	29.55	33.20	36.42	40.27	42.98	51.18
25	28.17	30.68	34.38	37.65	41.57	44.31	52.62
26	29.25	31.80	35.56	38.88	42.86	45.64	54.05
27	30.32	32.91	36.74	40.11	44.14	46.96	55.48
28	31.39	34.03	37.92	41.34	45.42	48.28	56.89
29	32.46	35.14	39.09	42.56	46.69	49.59	58.30
30	33.53	36.25	40.26	43.77	47.96	50.89	59.70

Table VII is abridged from Table IV of Fisher and Yates: *Statistical Tables for Biological, Agricultural, and Medical Research*, published by Longman Group Ltd., London (previously published by Oliver and Boyd Limited, Edinburgh), by permission of the authors and publishers.

Table VIIIa
F-Distribution ($F_{.95}$)

The numbers given in this table are the values of F for which the area to the left equals 0.95 for Table VIIIa, 0.975 for Table VIIIb, and 0.99 for Table VIIIc for the indicated numerator and denominator degrees of freedom.

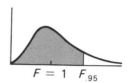

$F = 1$ $F_{.95}$

		Degrees of freedom for numerator								
	1	2	3	4	5	6	7	8	9	10
1	161	200	216	225	230	234	237	239	241	242
2	18.5	19.0	19.2	19.2	19.3	19.3	19.4	19.4	19.4	19.4
3	10.1	9.55	9.28	9.12	9.01	8.94	8.89	8.85	8.81	8.79
4	7.71	6.94	6.59	6.39	6.26	6.16	6.09	6.04	6.00	5.96
5	6.61	5.79	5.41	5.19	5.05	4.95	4.88	4.82	4.77	4.74
6	5.99	5.14.	4.76	4.53	4.39	4.28	4.21	4.15	4.10	4.06
7	5.59	4.74	4.35	4.12	3.97	3.87	3.79	3.73	3.68	3.64
8	5.32	4.46	4.07	3.84	3.69	3.58	3.50	3.44	3.39	3.35
9	5.12	4.26	3.86	3.63	3.48	3.37	3.29	3.23	3.18	3.14
10	4.96	4.10	3.71	3.48	3.33	3.22	3.14	3.07	3.02	2.98
11	4.84	3.98	3.59	3.36	3.20	3.09	3.01	2.95	2.90	2.85
12	4.75	3.89	3.49	3.26	3.11	3.00	2.91	2.85	2.80	2.75
13	4.67	3.81	3.41	3.18	3.03	2.92	2.83	2.77	2.71	2.67
14	4.60	3.74	3.34	3.11	2.96	2.85	2.76	2.70	2.65	2.60
15	4.54	3.68	3.29	3.06	2.90	2.79	2.71	2.64	2.59	2.54
16	4.49	3.63	3.24	3.01	2.85	2.74	2.66	2.59	2.54	2.49
17	4.45	3.59	3.20	2.96	2.81	2.70	2.61	2.55	2.49	2.45
18	4.41	3.55	3.16	2.93	2.77	2.66	2.58	2.51	2.46	2.41
19	4.38	3.52	3.13	2.90	2.74	2.63	2.54	2.48	2.42	2.38
20	4.35	3.49	3.10	2.87	2.71	2.60	2.51	2.45	2.39	2.35
21	4.32	3.47	3.07	2.84	2.68	2.57	2.49	2.42	2.37	2.32
22	4.30	3.44	3.05	2.82	2.66	2.55	2.46	2.40	2.34	2.30
23	4.28	3.42	3.03	2.80	2.64	2.53	2.44	2.37	2.32	2.27
24	4.26	3.40	3.01	2.78	2.62	2.51	2.42	2.36	2.30	2.25
25	4.24	3.39	2.99	2.76	2.60	2.49	2.40	2.34	2.28	2.24
30	4.17	3.32	2.92	2.69	2.53	2.42	2.33	2.27	2.21	2.16
40	4.08	3.23	2.84	2.61	2.45	2.34	2.25	2.18	2.12	2.08
60	4.00	3.15	2.76	2.53	2.37	2.25	2.17	2.10	2.04	1.99
120	3.92	3.07	2.68	2.45	2.29	2.18	2.09	2.02	1.96	1.91
∞	3.84	3.00	2.60	2.37	2.21	2.10	2.01	1.94	1.88	1.83

Degrees of freedom for denominator

TABLE VIIIa. *F*-Distribution ($F_{.95}$)

(*concluded*)

		12	15	20	24	30	40	60	120	∞
		\multicolumn{9}{c}{Degrees of freedom for numerator}								
	1	244	246	248	249	250	251	252	253	254
	2	19.4	19.4	19.4	19.5	19.5	19.5	19.5	19.5	19.5
	3	8.74	8.70	8.66	8.64	8.62	8.59	8.57	8.55	8.53
	4	5.91	5.86	5.80	5.77	5.75	5.72	5.69	5.66	5.63
	5	4.68	4.62	4.56	4.53	4.50	4.46	4.43	4.40	4.37
	6	4.00	3.94	3.87	3.84	3.81	3.77	3.74	3.70	3.67
	7	3.57	3.51	3.44	3.41	3.38	3.34	3.30	3.27	3.23
	8	3.28	3.22	3.15	3.12	3.08	3.04	3.01	2.97	2.93
	9	3.07	3.01	2.94	2.90	2.86	2.83	2.79	2.75	2.71
	10	2.91	2.85	2.77	2.74	2.70	2.66	2.62	2.58	2.54
	11	2.79	2.72	2.65	2.61	2.57	2.53	2.49	2.45	2.40
	12	2.69	2.62	2.54	2.51	2.47	2.43	2.38	2.34	2.30
	13	2.60	2.53	2.46	2.42	2.38	2.34	2.30	2.25	2.21
	14	2.53	2.46	2.39	2.35	2.31	2.27	2.22	2.18	2.13
	15	2.48	2.40	2.33	2.29	2.25	2.20	2.16	2.11	2.07
	16	2.42	2.35	2.28	2.24	2.19	2.15	2.11	2.06	2.01
	17	2.38	2.31	2.23	2.19	2.15	2.10	2.06	2.01	1.96
	18	2.34	2.27	2.19	2.15	2.11	2.06	2.02	1.97	1.92
	19	2.31	2.23	2.16	2.11	2.07	2.03	1.98	1.93	1.88
	20	2.28	2.20	2.12	2.08	2.04	1.99	1.95	1.90	1.84
	21	2.25	2.18	2.10	2.05	2.01	1.96	1.92	1.87	1.81
	22	2.23	2.15	2.07	2.03	1.98	1.94	1.89	1.84	1.78
	23	2.20	2.13	2.05	2.01	1.96	1.91	1.86	1.81	1.76
	24	2.18	2.11	2.03	1.98	1.94	1.89	1.84	1.79	1.73
	25	2.16	2.09	2.01	1.96	1.92	1.87	1.82	1.77	1.71
	30	2.09	2.01	1.93	1.89	1.84	1.79	1.74	1.68	1.62
	40	2.00	1.92	1.84	1.79	1.74	1.69	1.64	1.58	1.51
	60	1.92	1.84	1.75	1.70	1.65	1.59	1.53	1.47	1.39
	120	1.83	1.75	1.66	1.61	1.55	1.50	1.43	1.35	1.25
	∞	1.75	1.67	1.57	1.52	1.46	1.39	1.32	1.22	1.00

Degrees of freedom for denominator

Table VIIIb
F-Distribution ($F_{.975}$)

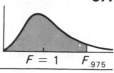

$F = 1 \qquad F_{.975}$

Degrees of freedom for numerator

Degrees of freedom for denominator

	1	2	3	4	5	6	7	8	9	10
1	648	800	864	900	922	937	948	957	963	969
2	38.5	39.0	39.2	39.2	39.3	39.3	39.4	39.4	39.4	39.4
3	17.4	16.0	15.4	15.1	14.9	14.7	14.6	14.5	14.5	14.4
4	12.2	10.6	9.98	9.60	9.36	9.20	9.07	8.98	8.90	8.84
5	10.0	8.43	7.76	7.39	7.15	6.98	6.85	6.76	6.68	6.62
6	8.81	7.26	6.60	6.23	5.99	5.82	5.70	5.60	5.52	5.46
7	8.07	6.54	5.89	5.52	5.29	5.12	4.99	4.90	4.82	4.76
8	7.57	6.06	5.42	5.05	4.82	4.65	4.53	4.43	4.36	4.30
9	7.21	5.71	5.08	4.72	4.48	4.32	4.20	4.10	4.03	3.96
10	6.94	5.46	4.83	4.47	4.24	4.07	3.95	3.85	3.78	3.72
11	6.72	5.26	4.63	4.28	4.04	3.88	3.76	3.66	3.59	3.53
12	6.55	5.10	4.47	4.12	3.89	3.73	3.61	3.51	3.44	3.37
13	6.41	4.97	4.35	4.00	3.77	3.60	3.48	3.39	3.31	3.25
14	6.30	4.86	4.24	3.89	3.66	3.50	3.38	3.28	3.21	3.15
15	6.20	4.77	4.15	3.80	3.58	3.41	3.29	3.20	3.12	3.06
16	6.12	4.69	4.08	3.73	3.50	3.34	3.22	3.12	3.05	2.99
17	6.04	4.62	4.01	3.66	3.44	3.28	3.16	3.06	2.98	2.92
18	5.98	4.56	3.95	3.61	3.38	3.22	3.10	3.01	2.93	2.87
19	5.92	4.51	3.90	3.56	3.33	3.17	3.05	2.96	2.88	2.82
20	5.87	4.46	3.86	3.51	3.29	3.13	3.01	2.91	2.84	2.77
21	5.83	4.42	3.82	3.48	3.25	3.09	2.97	2.87	2.80	2.73
22	5.79	4.38	3.78	3.44	3.22	3.05	2.93	2.84	2.76	2.70
23	5.75	4.35	3.75	3.41	3.18	3.02	2.90	2.81	2.73	2.67
24	5.72	4.32	3.72	3.38	3.15	2.99	2.87	2.78	2.70	2.64
25	5.69	4.29	3.69	3.35	3.13	2.97	2.85	2.75	2.68	2.61
30	5.57	4.18	3.59	3.25	3.03	2.87	2.75	2.65	2.57	2.51
40	5.42	4.05	3.46	3.13	2.90	2.74	2.62	2.53	2.45	2.39
60	5.29	3.93	3.34	3.01	2.79	2.63	2.51	2.41	2.33	2.27
120	5.15	3.80	3.23	2.89	2.67	2.52	2.39	2.30	2.22	2.16
∞	5.02	3.69	3.12	2.79	2.57	2.41	2.29	2.19	2.11	2.05

TABLE VIIIb. *F*-Distribution ($F_{.975}$)
(*concluded*)

		Degrees of freedom for numerator								
		12	15	20	24	30	40	60	120	∞
Degrees of freedom for denominator	1	977	985	993	997	1,001	1,006	1,010	1,014	1,018
	2	39.4	39.4	39.4	39.5	39.5	39.5	39.5	39.5	39.5
	3	14.3	14.3	14.2	14.1	14.1	14.0	14.0	13.9	13.9
	4	8.75	8.66	8.56	8.51	8.46	8.41	8.36	8.31	8.26
	5	6.52	6.43	6.33	6.28	6.23	6.18	6.12	6.07	6.02
	6	5.37	5.27	5.17	5.12	5.07	5.01	4.96	4.90	4.85
	7	4.67	4.57	4.47	4.42	4.36	4.31	4.25	4.20	4.14
	8	4.20	4.10	4.00	3.95	3.89	3.84	3.78	3.73	3.67
	9	3.87	3.77	3.67	3.61	3.56	3.51	3.45	3.39	3.33
	10	3.62	3.52	3.42	3.37	3.31	3.26	3.20	3.14	3.08
	11	3.43	3.33	3.23	3.17	3.12	3.06	3.00	2.94	2.88
	12	3.28	3.18	3.07	3.02	2.96	2.91	2.85	2.79	2.72
	13	3.15	3.05	2.95	2.89	2.84	2.78	2.72	2.66	2.60
	14	3.05	2.95	2.84	2.79	2.73	2.67	2.61	2.55	2.49
	15	2.96	2.86	2.76	2.70	2.64	2.59	2.52	2.46	2.40
	16	2.89	2.79	2.68	2.63	2.57	2.51	2.45	2.38	2.32
	17	2.82	2.72	2.62	2.56	2.50	2.44	2.38	2.32	2.25
	18	2.77	2.67	2.56	2.50	2.44	2.38	2.32	2.26	2.19
	19	2.72	2.62	2.51	2.45	2.39	2.33	2.27	2.20	2.13
	20	2.68	2.57	2.46	2.41	2.35	2.29	2.22	2.16	2.09
	21	2.64	2.53	2.42	2.37	2.31	2.25	2.18	2.11	2.04
	22	2.60	2.50	2.39	2.33	2.27	2.21	2.14	2.08	2.00
	23	2.57	2.47	2.36	2.30	2.24	2.18	2.11	2.04	1.97
	24	2.54	2.44	2.33	2.27	2.21	2.15	2.08	2.01	1.94
	25	2.51	2.41	2.30	2.24	2.18	2.12	2.05	1.98	1.91
	30	2.41	2.31	2.20	2.14	2.07	2.01	1.94	1.87	1.79
	40	2.29	2.18	2.07	2.01	1.94	1.88	1.80	1.72	1.64
	60	2.17	2.06	1.94	1.88	1.82	1.74	1.67	1.58	1.48
	120	2.05	1.95	1.82	1.76	1.69	1.61	1.53	1.43	1.31
	∞	1.94	1.83	1.71	1.64	1.57	1.48	1.39	1.27	1.00

Table VIIIc
F-Distribution ($F_{.99}$)

$F = 1$ $F_{.99}$

Degrees of freedom for numerator

		1	2	3	4	5	6	7	8	9	10
Degrees of freedom for denominator	1	4,052	5,000	5,403	5,625	5,764	5,859	5,928	5,982	6,023	6,056
	2	98.5	99.0	99.2	99.2	99.3	99.3	99.4	99.4	99.4	99.4
	3	34.1	30.8	29.5	28.7	28.2	27.9	27.7	27.5	27.3	27.2
	4	21.2	18.0	16.7	16.0	15.5	15.2	15.0	14.8	14.7	14.5
	5	16.3	13.3	12.1	11.4	11.0	10.7	10.5	10.3	10.2	10.1
	6	13.7	10.9	9.78	9.15	8.75	8.47	8.26	8.10	7.98	7.87
	7	12.2	9.55	8.45	7.85	7.46	7.19	6.99	6.84	6.72	6.62
	8	11.3	8.65	7.59	7.01	6.63	6.37	6.18	6.03	5.91	5.81
	9	10.6	8.02	6.99	6.42	6.06	5.80	5.61	5.47	5.35	5.26
	10	10.0	7.56	6.55	5.99	5.64	5.39	5.20	5.06	4.94	4.85
	11	9.65	7.21	6.22	5.67	5.32	5.07	4.89	4.74	4.63	4.54
	12	9.33	6.93	5.95	5.41	5.06	4.82	4.64	4.50	4.39	4.30
	13	9.07	6.70	5.74	5.21	4.86	4.62	4.44	4.30	4.19	4.10
	14	8.86	6.51	5.56	5.04	4.70	4.46	4.28	4.14	4.03	3.94
	15	8.68	6.36	5.42	4.89	4.56	4.32	4.14	4.00	3.89	3.80
	16	8.53	6.23	5.29	4.77	4.44	4.20	4.03	3.89	3.78	3.69
	17	8.40	6.11	5.19	4.67	4.34	4.10	3.93	3.79	3.68	3.59
	18	8.29	6.01	5.09	4.58	4.25	4.01	3.84	3.71	3.60	3.51
	19	8.19	5.93	5.01	4.50	4.17	3.94	3.77	3.63	3.52	3.43
	20	8.10	5.85	4.94	4.43	4.10	3.87	3.70	3.56	3.46	3.37
	21	8.02	5.78	4.87	4.37	4.04	3.81	3.64	3.51	3.40	3.31
	22	7.95	5.72	4.82	4.31	3.99	3.76	3.59	3.45	3.35	3.26
	23	7.88	5.66	4.76	4.26	3.94	3.71	3.54	3.41	3.30	3.21
	24	7.82	5.61	4.72	4.22	3.90	3.67	3.50	3.36	3.26	3.17
	25	7.77	5.57	4.68	4.18	3.86	3.63	3.46	3.32	3.22	3.13
	30	7.56	5.39	4.51	4.02	3.70	3.47	3.30	3.17	3.07	2.98
	40	7.31	5.18	4.31	3.83	3.51	3.29	3.12	2.99	2.89	2.80
	60	7.08	4.98	4.13	3.65	3.34	3.12	2.95	2.82	2.72	2.63
	120	6.85	4.79	3.95	3.48	3.17	2.96	2.79	2.66	2.56	2.47
	∞	6.63	4.61	3.78	3.32	3.02	2.80	2.64	2.51	2.41	2.32

TABLE VIIIc. ***F*-Distribution ($F_{.99}$)**
(*concluded*)

		12	15	20	24	30	40	60	120	∞
		\multicolumn{9} Degrees of freedom for numerator								
	1	6,106	6,157	6,209	6,235	6,261	6,287	6,313	6,339	6,366
	2	99.4	99.4	99.4	99.5	99.5	99.5	99.5	99.5	99.5
	3	27.1	26.9	26.7	26.6	26.5	26.4	26.3	26.2	26.1
	4	14.4	14.2	14.0	13.9	13.8	13.7	13.7	13.6	13.5
	5	9.89	9.72	9.55	9.47	9.38	9.29	9.20	9.11	9.02
	6	7.72	7.56	7.40	7.31	7.23	7.14	7.06	6.97	6.88
	7	6.47	6.31	6.16	6.07	5.99	5.91	5.82	5.74	5.65
	8	5.67	5.52	5.36	5.28	5.20	5.12	5.03	4.95	4.86
	9	5.11	4.96	4.81	4.73	4.65	4.57	4.48	4.40	4.31
	10	4.71	4.56	4.41	4.33	4.25	4.17	4.08	4.00	3.91
	11	4.40	4.25	4.10	4.02	3.94	3.86	3.78	3.69	3.60
	12	4.16	4.01	3.86	3.78	3.70	3.62	3.54	3.45	3.36
	13	3.96	3.82	3.66	3.59	3.51	3.43	3.34	3.25	3.17
	14	3.80	3.66	3.51	3.43	3.35	3.27	3.18	3.09	3.00
	15	3.67	3.52	3.37	3.29	3.21	3.13	3.05	2.96	2.87
	16	3.55	3.41	3.26	3.18	3.10	3.02	2.93	2.84	2.75
	17	3.46	3.31	3.16	3.08	3.00	2.92	2.83	2.75	2.65
	18	3.37	3.23	3.08	3.00	2.92	2.84	2.75	2.66	2.57
	19	3.30	3.15	3.00	2.92	2.84	2.76	2.67	2.58	2.49
	20	3.23	3.09	2.94	2.86	2.78	2.69	2.61	2.52	2.42
	21	3.17	3.03	2.88	2.80	2.72	2.64	2.55	2.46	2.36
	22	3.12	2.98	2.83	2.75	2.67	2.58	2.50	2.40	2.31
	23	3.07	2.93	2.78	2.70	2.62	2.54	2.45	2.35	2.26
	24	3.03	2.89	2.74	2.66	2.58	2.49	2.40	2.31	2.21
	25	2.99	2.85	2.70	2.62	2.53	2.45	2.36	2.27	2.17
	30	2.84	2.70	2.55	2.47	2.39	2.30	2.21	2.11	2.01
	40	2.66	2.52	2.37	2.29	2.20	2.11	2.02	1.92	1.80
	60	2.50	2.35	2.20	2.12	2.03	1.94	1.84	1.73	1.60
	120	2.34	2.19	2.03	1.95	1.86	1.76	1.66	1.53	1.38
	∞	2.18	2.04	1.88	1.79	1.70	1.59	1.47	1.32	1.00

Degrees of freedom for denominator (row label at left)

Table VIII is reproduced with the permission of Professor E. S. Pearson from M. Merrington, and C. M. Thompson, "Tables of percentage points of the inverted beta (F) distribution," *Biometrika*, Vol. 33 (1943), p. 73.

Table IXa
Duncan's New Multiple Ranges (5% Level).

The numbers given in this table are the values of Q_p used to find $R_p = Q_p s_{\bar{x}}$. The value of R_p, then, is the shortest significant range for comparing the largest and smallest of p means arranged in order of magnitude.

Degrees of Freedom	p = number of means within range being tested											
	2	3	4	5	6	7	8	9	10	20	50	100
1	18.00	18.00	18.00	18.00	18.00	18.00	18.00	18.00	18.00	18.00	18.00	18.00
2	6.08	6.08	6.08	6.08	6.08	6.08	6.08	6.08	6.08	6.08	6.08	6.08
3	4.50	4.52	4.52	4.52	4.52	4.52	4.52	4.52	4.52	4.52	4.52	4.52
4	3.93	4.01	4.03	4.03	4.03	4.03	4.03	4.03	4.03	4.03	4.03	4.03
5	3.64	3.75	3.80	3.81	3.81	3.81	3.81	3.81	3.81	3.81	3.81	3.81
6	3.46	3.59	3.65	3.68	3.69	3.70	3.70	3.70	3.70	3.70	3.70	3.70
7	3.34	3.48	3.55	3.59	3.61	3.62	3.63	3.63	3.63	3.63	3.63	3.63
8	3.26	3.40	3.48	3.52	3.55	3.57	3.58	3.58	3.58	3.58	3.58	3.58
9	3.20	3.34	3.42	3.47	3.50	3.52	3.54	3.54	3.55	3.55	3.55	3.55
10	3.15	3.29	3.38	3.43	3.46	3.49	3.50	3.52	3.52	3.53	3.53	3.53
11	3.11	3.26	3.34	3.40	3.44	3.46	3.48	3.49	3.50	3.51	3.51	3.51
12	3.08	3.22	3.31	3.37	3.41	3.44	3.46	3.47	3.48	3.50	3.50	3.50
13	3.06	3.20	3.29	3.35	3.39	3.42	3.44	3.46	3.47	3.49	3.49	3.49
14	3.03	3.18	3.27	3.33	3.37	3.40	3.43	3.44	3.46	3.48	3.48	3.48
15	3.01	3.16	3.25	3.31	3.36	3.39	3.41	3.43	3.45	3.48	3.48	3.48
16	3.00	3.14	3.24	3.30	3.34	3.38	3.40	3.42	3.44	3.48	3.48	3.48
17	2.98	3.13	3.22	3.28	3.33	3.37	3.39	3.41	3.43	3.48	3.48	3.48
18	2.97	3.12	3.21	3.27	3.32	3.36	3.38	3.40	3.42	3.47	3.47	3.47
19	2.96	3.11	3.20	3.26	3.31	3.35	3.38	3.40	3.42	3.47	3.47	3.47
20	2.95	3.10	3.19	3.26	3.30	3.34	3.37	3.39	3.41	3.47	3.47	3.47
30	2.89	3.04	3.13	3.20	3.25	3.29	3.32	3.35	3.37	3.47	3.49	3.49
40	2.86	3.01	3.10	3.17	3.22	3.27	3.30	3.33	3.35	3.47	3.50	3.50
60	2.83	2.98	3.07	3.14	3.20	3.24	3.28	3.31	3.33	3.47	3.54	3.54
120	2.80	2.95	3.04	3.12	3.17	3.22	3.25	3.29	3.31	3.47	3.58	3.60
∞	2.77	2.92	3.02	3.09	3.15	3.19	3.23	3.26	3.29	3.47	3.64	3.74

Table IXb
Duncan's New Multiple Ranges (1% Level)

Degrees of Freedom	p = number of means within range being tested											
	2	3	4	5	6	7	8	9	10	20	50	100
1	90.00	90.00	90.00	90.00	90.00	90.00	90.00	90.00	90.00	90.00	90.00	90.00
2	14.00	14.00	14.00	14.00	14.00	14.00	14.00	14.00	14.00	14.00	14.00	14.00
3	8.26	8.32	8.32	8.32	8.32	8.32	8.32	8.32	8.32	8.32	8.32	8.32
4	6.51	6.68	6.74	6.76	6.76	6.76	6.76	6.76	6.76	6.76	6.76	6.76
5	5.70	5.89	6.00	6.04	6.06	6.07	6.07	6.07	6.07	6.07	6.07	6.07
6	5.25	5.44	5.55	5.61	5.66	5.68	5.69	5.70	5.70	5.70	5.70	5.70
7	4.95	5.14	5.26	5.33	5.38	5.42	5.44	5.45	5.46	5.47	5.47	5.47
8	4.75	4.94	5.06	5.14	5.19	5.23	5.26	5.28	5.29	5.32	5.32	5.32
9	4.60	4.79	4.91	4.99	5.04	5.09	5.12	5.14	5.16	5.21	5.21	5.21
10	4.48	4.67	4.79	4.87	4.93	4.98	5.01	5.04	5.06	5.12	5.12	5.12
11	4.39	4.58	4.70	4.78	4.84	4.89	4.92	4.95	4.98	5.06	5.06	5.06
12	4.32	4.50	4.62	4.71	4.77	4.82	4.85	4.88	4.91	5.01	5.01	5.01
13	4.26	4.44	4.56	4.64	4.71	4.76	4.79	4.82	4.85	4.96	4.97	4.97
14	4.21	4.39	4.51	4.59	4.65	4.70	4.74	4.78	4.80	4.92	4.94	4.94
15	4.17	4.35	4.46	4.55	4.61	4.66	4.70	4.73	4.76	4.89	4.91	4.91
16	4.13	4.31	4.42	4.51	4.57	4.62	4.66	4.70	4.72	4.86	4.89	4.89
17	4.10	4.28	4.39	4.48	4.54	4.59	4.63	4.66	4.69	4.83	4.87	4.87
18	4.07	4.25	4.36	4.44	4.51	4.56	4.60	4.64	4.66	4.81	4.86	4.86
19	4.05	4.22	4.34	4.42	4.48	4.53	4.58	4.61	4.64	4.79	4.84	4.84
20	4.02	4.20	4.31	4.40	4.46	4.51	4.55	4.59	4.62	4.77	4.83	4.83
30	3.89	4.06	4.17	4.25	4.31	4.37	4.41	4.44	4.48	4.65	4.77	4.78
40	3.82	3.99	4.10	4.18	4.24	4.30	4.34	4.38	4.41	4.59	4.74	4.76
60	3.76	3.92	4.03	4.11	4.17	4.23	4.27	4.31	4.34	4.53	4.71	4.76
120	3.70	3.86	3.96	4.04	4.11	4.16	4.20	4.24	4.27	4.47	4.67	4.77
∞	3.64	3.80	3.90	3.98	4.04	4.09	4.14	4.17	4.20	4.41	4.64	4.78

Tables IXa,b are abridged from those compiled by D. B. Duncan, *Biometrics* 11, 1–42 (1955) and modified by H. L. Harter, *Biometrics* 16, 671–685 (1960) and *Biometrics* 17, 321–324 (1961) and are used by permission of the authors and editors.

Table X
Transformation of Percentage to Arcsin $\sqrt{\text{Percentage}}$

The numbers in this table are the angles (in degrees) corresponding to given percentages
under the transformation arcsin $\sqrt{\text{percentage}}$.

%	0	1	2	3	4	5	6	7	8	9
0.0	0	0.57	0.81	0.99	1.15	1.28	1.40	1.52	1.62	1.72
0.1	1.81	1.90	1.99	2.07	2.14	2.22	2.29	2.36	2.43	2.50
0.2	2.56	2.63	2.69	2.75	2.81	2.87	2.92	2.98	3.03	3.09
0.3	3.14	3.19	3.24	3.29	3.34	3.39	3.44	3.49	3.53	3.58
0.4	3.63	3.67	3.72	3.76	3.80	3.85	3.89	3.93	3.97	4.01
0.5	4.05	4.09	4.13	4.17	4.21	4.25	4.29	4.33	4.37	4.40
0.6	4.44	4.48	4.52	4.55	4.59	4.62	4.66	4.69	4.73	4.76
0.7	4.80	4.83	4.87	4.90	4.93	4.97	5.00	5.03	5.07	5.10
0.8	5.13	5.16	5.20	5.23	5.26	5.29	5.32	5.35	5.38	5.41
0.9	5.44	5.47	5.50	5.53	5.56	5.59	5.62	5.65	5.68	5.71
1	5.74	6.02	6.29	6.55	6.80	7.04	7.27	7.49	7.71	7.92
2	8.13	8.33	8.53	8.72	8.91	9.10	9.28	9.46	9.63	9.81
3	9.98	10.14	10.31	10.47	10.63	10.78	10.94	11.09	11.24	11.39
4	11.54	11.68	11.83	11.97	12.11	12.25	12.39	12.52	12.66	12.79
5	12.92	13.05	13.18	13.31	13.44	13.56	13.69	13.81	13.94	14.06
6	14.18	14.30	14.42	14.54	14.65	14.77	14.89	15.00	15.12	15.23
7	15.34	15.45	15.56	15.68	15.79	15.89	16.00	16.11	16.22	16.32
8	16.43	16.54	16.64	16.74	16.85	16.95	17.05	17.16	17.26	17.36
9	17.46	17.56	17.66	17.76	17.85	17.95	18.05	18.15	18.24	18.34
10	18.44	18.53	18.63	18.72	18.81	18.91	19.00	19.09	19.19	19.28
11	19.37	19.46	19.55	19.64	19.73	19.82	19.91	20.00	20.09	20.18
12	20.27	20.36	20.44	20.53	20.62	20.70	20.79	20.88	20.96	21.05
13	21.13	21.22	21.30	21.39	21.47	21.56	21.64	21.72	21.81	21.89
14	21.97	22.06	22.14	22.22	22.30	22.38	22.46	22.55	22.63	22.71
15	22.79	22.87	22.95	22.03	23.11	23.19	23.26	23.34	23.42	23.50
16	23.58	23.66	23.73	23.81	23.89	23.97	24.04	24.12	24.20	24.27
17	24.35	24.43	24.50	24.58	24.65	24.73	24.80	24.88	24.95	25.03
18	25.10	25.18	25.25	25.33	25.40	25.48	25.55	25.62	25.70	25.77
19	25.84	25.92	25.99	26.06	26.13	26.21	26.28	26.35	26.42	26.49
20	26.56	26.64	26.71	26.78	26.85	26.92	26.99	27.06	27.13	27.20
21	27.28	27.35	27.42	27.49	27.56	27.63	27.69	27.76	27.83	27.90
22	27.97	28.04	28.11	28.18	28.25	28.32	28.38	28.45	28.52	28.59
23	28.66	28.73	28.79	28.86	28.93	29.00	29.06	29.13	29.20	29.27
24	29.33	29.40	29.47	29.53	29.60	29.67	29.73	29.80	29.87	29.93
25	30.00	30.07	30.13	30.20	30.26	30.33	30.40	30.46	30.53	30.59
26	30.66	30.72	30.79	30.85	30.92	30.98	31.05	31.11	31.18	31.24
27	31.31	31.37	31.44	31.50	31.56	31.63	31.69	31.76	31.82	31.88
28	31.95	32.01	32.08	32.14	32.20	32.27	32.33	32.39	32.46	32.52
29	32.58	32.65	32.71	32.77	32.83	32.90	32.96	33.02	33.09	33.15
30	33.21	33.27	33.34	33.40	33.46	33.52	33.58	33.65	33.71	33.77

Table X. Transformation of Percentage to Arcsin √Percentage
(*continued*)

%	0	1	2	3	4	5	6	7	8	9
31	33.83	33.89	33.96	34.02	34.08	34.14	34.20	34.27	34.33	34.39
32	34.45	34.51	34.57	34.63	34.70	34.76	34.82	34.88	34.94	35.00
33	35.06	35.12	35.18	35.24	35.30	35.37	35.43	35.49	35.55	35.61
34	35.67	35.73	35.79	35.85	35.91	35.97	36.03	36.09	36.15	36.21
35	36.27	36.33	36.39	36.45	36.51	36.57	36.63	36.69	36.75	36.81
36	36.87	36.93	36.99	37.05	37.11	37.17	37.23	37.29	37.35	37.41
37	37.47	37.52	37.58	37.64	37.70	37.76	37.82	37.88	37.94	38.00
38	38.06	38.12	38.17	38.23	38.29	38.35	38.41	38.47	38.53	38.59
39	38.65	38.70	38.76	38.82	38.88	38.94	39.00	39.06	39.11	39.17
40	39.23	39.29	39.35	39.41	39.47	39.52	39.58	39.64	39.70	39.76
41	39.82	39.87	39.93	39.99	40.05	40.11	40.16	40.22	40.28	40.34
42	40.40	40.46	40.51	40.57	40.63	40.69	40.74	40.80	40.86	40.92
43	40.98	41.03	41.09	41.15	41.21	41.27	41.32	41.38	41.44	41.50
44	41.55	41.61	41.67	41.73	41.78	41.84	41.90	41.96	42.02	42.07
45	42.13	42.19	42.25	42.30	42.36	42.42	42.48	42.53	42.59	42.65
46	42.71	42.76	42.82	42.88	42.94	42.99	43.05	43.11	43.17	43.22
47	43.28	43.34	43.39	43.45	43.51	43.57	43.62	43.68	43.74	43.80
48	43.85	43.91	43.97	44.03	44.08	44.14	44.20	44.25	44.31	44.37
49	44.43	44.48	44.54	44.60	44.66	44.71	44.77	44.83	44.89	44.94
50	45.00	45.06	45.11	45.17	45.23	45.29	45.34	45.40	45.46	45.52
51	45.57	45.63	45.69	45.75	45.80	45.86	45.92	45.97	46.03	46.09
52	46.15	46.20	46.26	46.32	46.38	46.43	46.49	46.55	46.61	46.66
53	46.72	46.78	46.83	46.89	46.95	47.01	47.06	47.12	47.18	47.24
54	47.29	47.35	47.41	47.47	47.52	47.58	47.64	47.70	47.75	47.81
55	47.87	47.93	47.98	48.04	48.10	48.16	48.22	48.27	48.33	48.39
56	48.45	48.50	48.56	48.62	48.68	48.73	48.79	48.85	48.91	48.97
57	49.02	49.08	49.14	49.20	49.26	49.31	49.37	49.43	49.49	49.54
58	49.60	49.66	49.72	49.78	49.84	49.89	49.95	50.01	50.07	50.13
59	50.18	50.24	50.30	50.36	50.42	50.48	50.53	50.59	50.65	50.71
60	50.77	50.83	50.89	50.94	51.00	51.06	51.12	51.18	51.24	51.30
61	51.35	51.41	51.47	51.53	51.59	51.65	51.71	51.77	51.83	51.88
62	51.94	52.00	52.06	52.12	52.18	52.24	52.30	52.36	52.42	52.48
63	52.53	52.59	52.65	52.71	52.77	52.83	52.89	52.95	53.01	53.07
64	53.13	53.19	53.25	53.31	53.37	53.43	53.49	53.55	53.61	53.67
65	53.73	53.79	53.85	53.91	53.97	54.03	54.09	54.15	54.21	54.27
66	54.33	54.39	54.45	54.51	54.57	54.63	54.70	54.76	54.82	54.88
67	54.94	55.00	55.06	55.12	55.18	55.24	55.30	55.37	55.43	55.49
68	55.55	55.61	55.67	55.73	55.80	55.86	55.92	55.98	56.04	56.11
69	56.17	56.23	56.29	56.35	56.42	56.48	56.54	56.60	56.66	56.73
70	56.79	56.85	56.91	56.98	57.04	57.10	57.17	57.23	57.29	57.35

Table X.   Transformation of Percentage to Arcsin $\sqrt{\text{Percentage}}$
(*concluded*)

%	0	1	2	3	4	5	6	7	8	9
71	57.42	57.48	57.54	57.61	57.67	57.73	57.80	57.86	57.92	57.99
72	58.05	58.12	58.18	58.24	58.31	58.37	58.44	58.50	58.56	58.63
73	58.69	58.76	58.82	58.89	58.95	59.02	59.08	59.15	59.21	59.28
74	59.34	59.41	59.47	59.54	59.60	59.67	59.74	59.80	59.87	59.93
75	60.00	60.07	60.13	60.20	60.27	60.33	60.40	60.47	60.53	60.60
76	60.67	60.73	60.80	60.87	60.94	61.00	61.07	61.14	61.21	61.27
77	61.34	61.41	61.48	61.55	61.62	61.68	61.75	61.82	61.89	61.96
78	62.03	62.10	62.17	62.24	62.31	62.37	62.44	62.51	62.58	62.65
79	62.72	62.80	62.87	62.94	63.01	63.08	63.15	63.22	63.29	63.36
80	63.44	63.51	63.58	63.65	63.72	63.79	63.87	63.94	64.01	64.08
81	64.16	64.23	64.30	64.38	64.45	64.52	64.60	64.67	64.75	64.82
82	64.90	64.97	65.05	65.12	65.20	65.27	65.35	65.42	65.50	65.57
83	65.65	65.73	65.80	65.88	65.96	66.03	66.11	66.19	66.27	66.34
84	66.42	66.50	66.58	66.66	66.74	66.81	66.89	66.97	67.05	67.13
85	67.21	67.29	67.37	67.45	67.54	67.62	67.70	67.78	67.86	67.94
86	68.03	68.11	68.19	68.28	68.36	68.44	68.53	68.61	68.70	68.78
87	68.87	68.95	69.04	69.12	69.21	69.30	69.38	69.47	69.56	69.64
88	69.73	69.82	69.91	70.00	70.09	70.18	70.27	70.36	70.45	70.54
89	70.63	70.72	70.81	70.91	71.00	71.09	71.19	71.28	71.37	71.47
90	71.56	71.66	71.76	71.85	71.95	72.05	72.15	72.24	72.34	72.44
91	72.54	72.64	72.74	72.84	72.95	73.05	73.15	73.26	73.36	73.46
92	73.57	73.68	73.78	73.89	74.00	74.11	74.21	74.32	74.44	74.55
93	74.66	74.77	74.88	75.00	75.11	75.23	75.35	75.46	75.58	75.70
94	75.82	75.94	76.06	76.19	76.31	76.44	76.56	76.69	76.82	76.95
95	77.08	77.21	77.34	77.48	77.61	77.75	77.89	78.03	78.17	78.32
96	78.46	78.61	78.76	78.91	79.06	79.22	79.37	79.53	79.69	79.86
97	80.02	80.19	80.37	80.54	80.72	80.90	81.09	81.28	81.47	81.67
98	81.87	82.08	82.29	82.51	82.73	82.96	83.20	83.45	83.71	83.98
99.0	84.26	84.29	84.32	84.35	84.38	84.41	84.44	84.47	84.50	84.53
99.1	84.56	84.59	84.62	84.65	84.68	84.71	84.74	84.77	84.80	84.84
99.2	84.87	84.90	84.93	84.97	85.00	85.03	85.07	85.10	85.13	85.17
99.3	85.20	85.24	85.27	85.31	85.34	85.38	85.41	85.45	85.48	85.52
99.4	85.56	85.60	85.63	85.67	85.71	85.75	85.79	85.83	85.87	85.91
99.5	85.95	85.99	86.03	86.07	86.11	86.15	86.20	86.24	86.28	86.33
99.6	86.37	86.42	86.47	86.51	86.56	86.61	86.66	86.71	86.76	86.81
99.7	86.86	86.91	86.97	87.02	87.08	87.13	87.19	87.25	87.31	87.37
99.8	87.44	87.50	87.57	87.64	87.71	87.78	87.86	87.93	88.01	88.10
99.9	88.19	88.28	88.38	88.48	88.60	88.72	88.85	89.01	89.19	89.43
100.0	90.00	—	—	—	—	—	—	—	—	—

Table X appeared in *Plant Protection* (Leningrad), Vol. 12 (1937), p. 67, and is reproduced with permission of the author, C. I. Bliss.

Table XI
Squares and Square Roots.

N	N^2	$\sqrt{N}$	$\sqrt{10N}$	N	N^2	$\sqrt{N}$	$\sqrt{10N}$
1.00	1.0000	1.00000	3.16228	1.45	2.1025	1.20416	3.80789
1.01	1.0201	1.00499	3.17805	1.46	2.1316	1.20830	3.82099
1.02	1.0404	1.00995	3.19374	1.47	2.1609	1.21244	3.83406
1.03	1.0609	1.01489	3.20936	1.48	2.1904	1.21655	3.84708
1.04	1.0816	1.01980	3.22490	1.49	2.2201	1.22066	3.86005
1.05	1.1025	1.02470	3.24037	1.50	2.2500	1.22474	3.87298
1.06	1.1236	1.02956	3.25576	1.51	2.2801	1.22882	3.88587
1.07	1.1449	1.03441	3.27109	1.52	2.3104	1.23288	3.89872
1.08	1.1664	1.03923	3.28634	1.53	2.3409	1.23693	3.91152
1.09	1.1881	1.04403	3.30151	1.54	2.3716	1.24097	3.92428
1.10	1.2100	1.04881	3.31662	1.55	2.4025	1.24499	3.93700
1.11	1.2321	1.05357	3.33167	1.56	2.4336	1.24900	3.94968
1.12	1.2544	1.05830	3.34664	1.57	2.4649	1.25300	3.96232
1.13	1.2769	1.06301	3.36155	1.58	2.4964	1.25698	3.97492
1.14	1.2996	1.06771	3.37639	1.59	2.5281	1.26095	3.98748
1.15	1.3225	1.07238	3.39116	1.60	2.5600	1.26491	4.00000
1.16	1.3456	1.07703	3.40588	1.61	2.5921	1.26886	4.01248
1.17	1.3689	1.08167	3.42053	1.62	2.6244	1.27279	4.02492
1.18	1.3924	1.08628	3.43511	1.63	2.6569	1.27671	4.03733
1.19	1.4161	1.09087	3.44964	1.64	2.6896	1.28062	4.04969
1.20	1.4400	1.09545	3.46410	1.65	2.7225	1.28452	4.06202
1.21	1.4641	1.10000	3.47851	1.66	2.7556	1.28841	4.07431
1.22	1.4884	1.10454	3.49285	1.67	2.7889	1.29228	4.08656
1.23	1.5129	1.10905	3.50714	1.68	2.8224	1.29615	4.09878
1.24	1.5376	1.11355	3.52136	1.69	2.8561	1.30000	4.11096
1.25	1.5625	1.11803	3.53553	1.70	2.8900	1.30384	4.12311
1.26	1.5876	1.12250	3.54965	1.71	2.9241	1.30767	4.13521
1.27	1.6129	1.12694	3.56371	1.72	2.9584	1.31149	4.14729
1.28	1.6384	1.13137	3.57771	1.73	2.9929	1.31529	4.15933
1.29	1.6641	1.13578	3.59166	1.74	3.0276	1.31909	4.17133
1.30	1.6900	1.14018	3.60555	1.75	3.0625	1.32288	4.18330
1.31	1.7161	1.14455	3.61939	1.76	3.0976	1.32665	4.19524
1.32	1.7424	1.14891	3.63318	1.77	3.1329	1.33041	4.20714
1.33	1.7689	1.15326	3.64692	1.78	3.1684	1.33417	4.21900
1.34	1.7956	1.15758	3.66060	1.79	3.2041	1.33791	4.23084
1.35	1.8225	1.16190	3.67423	1.80	3.2400	1.34164	4.24264
1.36	1.8496	1.16619	3.68782	1.81	3.2761	1.34536	4.25441
1.37	1.8769	1.17047	3.70135	1.82	3.3124	1.34907	4.26615
1.38	1.9044	1.17473	3.71484	1.83	3.3489	1.35277	4.27785
1.39	1.9321	1.17898	3.72827	1.84	3.3856	1.35647	4.28952
1.40	1.9600	1.18322	3.74166	1.85	3.4225	1.36015	4.30116
1.41	1.9881	1.18743	3.75500	1.86	3.4596	1.36382	4.31277
1.42	2.0164	1.19164	3.76829	1.87	3.4969	1.36748	4.32435
1.43	2.0449	1.19583	3.78153	1.88	3.5344	1.37113	4.33590
1.44	2.0736	1.20000	3.79473	1.89	3.5721	1.37477	4.34741

Table XI Squares and Square Roots.
(*continued*)

N	N²	√N	√10N	N	N²	√N	√10N
1.90	3.6100	1.37840	4.35890	2.35	5.5225	1.53297	4.84768
1.91	3.6481	1.38203	4.37035	2.36	5.5696	1.53623	4.85798
1.92	3.6864	1.38564	4.38178	2.37	5.6169	1.53948	4.86826
1.93	3.7249	1.38924	4.39318	2.38	5.6644	1.54272	4.87852
1.94	3.7636	1.39284	4.40454	2.39	5.7121	1.54596	4.88876
1.95	3.8025	1.39642	4.41588	2.40	5.7600	1.54919	4.89898
1.96	3.8416	1.40000	4.42719	2.41	5.8081	1.55242	4.90918
1.97	3.8809	1.40357	4.43847	2.42	5.8564	1.55563	4.91935
1.98	3.9204	1.40712	4.44972	2.43	5.9049	1.55885	4.92950
1.99	3.9601	1.41067	4.46094	2.44	5.9536	1.56205	4.93964
2.00	4.0000	1.41421	4.47214	2.45	6.0025	1.56525	4.94975
2.01	4.0401	1.41774	4.48330	2.46	6.0516	1.56844	4.95984
2.02	4.0804	1.42127	4.49444	2.47	6.1009	1.57162	4.96991
2.03	4.1209	1.42478	4.50555	2.48	6.1504	1.57480	4.97996
2.04	4.1616	1.42829	4.51664	2.49	6.2001	1.57797	4.98999
2.05	4.2025	1.43178	4.52769	2.50	6.2500	1.58114	5.00000
2.06	4.2436	1.43527	4.53872	2.51	6.3001	1.58430	5.00999
2.07	4.2849	1.43875	4.54973	2.52	6.3504	1.58745	5.01996
2.08	4.3264	1.44222	4.56070	2.53	6.4009	1.59060	5.02991
2.09	4.3681	1.44568	4.57165	2.54	6.4516	1.59374	5.03984
2.10	4.4100	1.44914	4.58258	2.55	6.5025	1.59687	5.04975
2.11	4.4521	1.45258	4.59347	2.56	6.5536	1.60000	5.05964
2.12	4.4944	1.45602	4.60435	2.57	6.6049	1.60312	5.06952
2.13	4.5369	1.45945	4.61519	2.58	5.6564	1.60624	5.07937
2.14	4.5796	1.46287	4.62601	2.59	6.7081	1.60935	5.08920
2.15	4.6225	1.46629	4.63681	2.60	6.7600	1.61245	5.09902
2.16	4.6656	1.46969	4.64758	2.61	6.8121	1.61555	5.10882
2.17	4.7089	1.47309	4.65833	2.62	6.8644	1.61864	5.11859
2.18	4.7524	1.47648	4.66905	2.63	6.9169	1.62173	5.12835
2.19	4.7961	1.47986	4.67974	2.64	6.9696	1.62481	5.13809
2.20	4.8400	1.48324	4.69042	2.65	7.0225	1.62788	5.14782
2.21	4.8841	1.48661	4.70106	2.66	7.0756	1.63095	5.15752
2.22	4.9284	1.48997	4.71169	2.67	7.1289	1.63401	5.16720
2.23	4.9729	1.49332	4.72229	2.68	7.1824	1.63707	5.17687
2.24	5.0176	1.49666	4.73286	2.69	7.2361	1.64012	5.18652
2.25	5.0625	1.50000	4.74342	2.70	7.2900	1.64317	5.19615
2.26	5.1076	1.50333	4.75395	2.71	7.3441	1.64621	5.20577
2.27	5.1529	1.50665	4.76445	2.72	7.3984	1.64924	5.21536
2.28	5.1984	1.50997	4.77493	2.73	7.4529	1.65227	5.22494
2.29	5.2441	1.51327	4.78539	2.74	7.5076	1.65529	5.23450
2.30	5.2900	1.51658	4.79583	2.75	7.5625	1.65831	5.24404
2.31	5.3361	1.51987	4.80625	2.76	7.6176	1.66132	5.25357
2.32	5.3824	1.52315	4.81664	2.77	7.6729	1.66433	5.26308
2.33	5.4289	1.52643	4.82701	2.78	7.7284	1.66733	5.27257
2.34	5.4756	1.52971	4.83735	2.79	7.7841	1.67033	5.28205

TABLE XI. Squares and Square Roots
             (*continued*)

N	N^2	$\sqrt{N}$	$\sqrt{10N}$	N	N^2	$\sqrt{N}$	$\sqrt{10N}$
2.80	7.8400	1.67332	5.29150	3.25	10.5625	1.80278	5.70088
2.81	7.8961	1.67631	5.30094	3.26	10.6276	1.80555	5.70964
2.82	7.9524	1.67929	5.31037	3.27	10.6929	1.80831	5.71839
2.83	8.0089	1.68226	5.31977	3.28	10.7584	1.81108	5.72713
2.84	8.0656	1.68523	5.32917	3.29	10.8241	1.81384	5.73585
2.85	8.1225	1.68819	5.33854	3.30	10.8900	1.81659	5.74456
2.86	8.1796	1.69115	5.34790	3.31	10.9561	1.81934	5.75326
2.87	8.2369	1.69411	5.35724	3.32	11.0224	1.82209	5.76194
2.88	8.2944	1.69706	5.36656	3.33	11.0889	1.82483	5.77062
2.89	8.3521	1.70000	5.37587	3.34	11.1556	1.82757	5.77927
2.90	8.4100	1.70294	5.38516	3.35	11.2225	1.83030	5.78792
2.91	8.4681	1.70587	5.39444	3.36	11.2896	1.83303	5.79655
2.92	8.5264	1.70880	5.40370	3.37	11.3569	1.83576	5.80517
2.93	8.5849	1.71172	5.41295	3.38	11.4244	1.83848	5.81378
2.94	8.6436	1.71464	5.42218	3.39	11.4921	1.84120	5.82237
2.95	8.7025	1.71756	5.43139	3.40	11.5600	1.84391	5.83095
2.96	8.7616	1.72047	5.44059	3.41	11.6281	1.84662	5.83952
2.97	8.8209	1.72337	5.44977	3.42	11.6964	1.84932	5.84808
2.98	8.8804	1.72627	5.45894	3.43	11.7649	1.85203	5.85662
2.99	8.9401	1.72916	5.46809	3.44	11.8336	1.85472	5.86515
3.00	9.0000	1.73205	5.47723	3.45	11.9025	1.85742	5.87367
3.01	9.0601	1.73494	5.48635	3.46	11.9716	1.86011	5.88218
3.02	9.1204	1.73781	5.49545	3.47	12.0409	1.86279	5.89067
3.03	9.1809	1.74069	5.50454	3.48	12.1104	1.86548	5.89915
3.04	9.2416	1.74356	5.51362	3.49	12.1801	1.86815	5.90762
3.05	9.3025	1.74642	5.52268	3.50	12.2500	1.87083	5.91608
3.06	9.3636	1.74929	5.53173	3.51	12.3201	1.87350	5.92453
3.07	9.4249	1.75214	5.54076	3.52	12.3904	1.87617	5.93296
3.08	9.4864	1.75499	5.54977	3.53	12.4609	1.87883	5.94138
3.09	9.5481	1.75784	5.55878	3.54	12.5316	1.88149	5.94979
3.10	9.6100	1.76068	5.56776	3.55	12.6025	1.88414	5.95819
3.11	9.6721	1.76352	5.57674	3.56	12.6736	1.88680	5.96657
3.12	9.7344	1.76635	5.58570	3.57	12.7449	1.88944	5.97495
3.13	9.7969	1.76918	5.59464	3.58	12.8164	1.89209	5.98331
3.14	9.8596	1.77200	5.60357	3.59	12.8881	1.89473	5.99166
3.15	9.9225	1.77482	5.61249	3.60	12.9600	1.89737	6.00000
3.16	9.9856	1.77764	5.62139	3.61	13.0321	1.90000	6.00833
3.17	10.0489	1.78045	5.63028	3.62	13.1044	1.90263	6.01664
3.18	10.1124	1.78326	5.63915	3.63	13.1769	1.90526	6.02495
3.19	10.1761	1.78606	5.64801	3.64	13.2496	1.90788	6.03324
3.20	10.2400	1.78885	5.65685	3.65	13.3225	1.91050	6.04152
3.21	10.3041	1.79165	5.66569	3.66	13.3956	1.91311	6.04979
3.22	10.3684	1.79444	5.67450	3.67	13.4689	1.91572	6.05805
3.23	10.4329	1.79722	5.68331	3.68	13.5424	1.91833	6.06630
3.24	10.4976	1.80000	5.69210	3.69	13.6161	1.92094	6.07454

Table XI.   Squares and Square Roots
(*continued*)

N	N^2	$\sqrt{N}$	$\sqrt{10N}$	N	N^2	$\sqrt{N}$	$\sqrt{10N}$
3.70	13.6900	1.92354	6.08276	4.15	17.2225	2.03715	6.44205
3.71	13.7641	1.92614	6.09098	4.16	17.3056	2.03961	6.44981
3.72	13.8384	1.92873	6.09918	4.17	17.3889	2.04206	6.45755
3.73	13.9129	1.93132	6.10737	4.18	17.4724	2.04450	6.46529
3.74	13.9876	1.93391	6.11555	4.19	17.5561	2.04695	6.47302
3.75	14.0625	1.93649	6.12372	4.20	17.6400	2.04939	6.48074
3.76	14.1376	1.93907	6.13188	4.21	17.7241	2.05183	6.48845
3.77	14.2129	1.94165	6.14003	4.22	17.8084	2.05426	6.49615
3.78	14.2884	1.94422	6.14817	4.23	17.8929	2.05670	6.50384
3.79	14.3641	1.94679	6.15630	4.24	17.9776	2.05913	6.51153
3.80	14.4400	1.94936	6.16441	4.25	18.0625	2.06155	6.51920
3.81	14.5161	1.95192	6.17252	4.26	18.1476	2.06398	6.52687
3.82	14.5924	1.95448	6.18061	4.27	18.2329	2.06640	6.53452
3.83	14.6689	1.95704	6.18870	4.28	18.3184	2.06882	6.54217
3.84	14.7456	1.95959	6.19677	4.29	18.4041	2.07123	6.54981
3.85	14.8225	1.96214	6.20484	4.30	18.4900	2.07364	6.55744
3.86	14.8996	1.96469	6.21289	4.31	18.5761	2.07605	6.56506
3.87	14.9769	1.96723	6.22093	4.32	18.6624	2.07846	6.57267
3.88	15.0544	1.96977	6.22896	4.33	18.7489	2.08087	6.58027
3.89	15.1321	1.97231	6.23699	4.34	18.8356	2.08327	6.58787
3.90	15.2100	1.97484	6.24500	4.35	18.9225	2.08567	6.59545
3.91	15.2881	1.97737	6.25300	4.36	19.0096	2.08806	6.60303
3.92	15.3664	1.97990	6.26099	4.37	19.0969	2.09045	6.61060
3.93	15.4449	1.98242	6.26897	4.38	19.1844	2.09284	6.61816
3.94	15.5236	1.98494	6.27694	4.39	19.2721	2.09523	6.62571
3.95	15.6025	1.98746	6.28490	4.40	19.3600	2.09762	6.63325
3.96	15.6816	1.98997	6.29285	4.41	19.4481	2.10000	6.64078
3.97	15.7609	1.99249	6.30079	4.42	19.5364	2.10238	6.64831
3.98	15.8404	1.99499	6.30872	4.43	19.6249	2.10476	6.65582
3.99	15.9201	1.99750	6.31664	4.44	19.7136	2.10713	6.66333
4.00	16.0000	2.00000	6.32456	4.45	19.8025	2.10950	6.67083
4.01	16.0801	2.00250	6.33246	4.46	19.8916	2.11187	6.67832
4.02	16.1604	2.00499	6.34035	4.47	19.9809	2.11424	6.68581
4.03	16.2409	2.00749	6.34823	4.48	20.0704	2.11660	6.69328
4.04	16.3216	2.00998	6.35610	4.49	20.1601	2.11896	6.70075
4.05	16.4025	2.01246	6.36396	4.50	20.2500	2.12132	6.70820
4.06	16.4836	2.01494	6.37181	4.51	20.3401	2.12368	6.71565
4.07	16.5649	2.01742	6.37966	4.52	20.4304	2.12603	6.72309
4.08	16.6464	2.01990	6.38749	4.53	20.5209	2.12838	6.73053
4.09	16.7281	2.02237	6.39531	4.54	20.6116	2.13073	6.73795
4.10	16.8100	2.02485	6.40312	4.55	20.7025	2.13307	6.74537
4.11	16.8921	2.02731	6.41093	4.56	20.7936	2.13542	6.75278
4.12	16.9744	2.02978	6.41872	4.57	20.8849	2.13776	6.76018
4.13	17.0569	2.03224	6.42651	4.58	20.9764	2.14009	6.76757
4.14	17.1396	2.03470	6.43428	4.59	21.0681	2.14243	6.77495

Table XI.   Squares and Square Roots
(continued)

N	N²	√N	√10N	N	N²	√N	√10N
4.60	21.1600	2.14476	6.78233	5.05	25.5025	2.24722	7.10634
4.61	21.2521	2.14709	6.78970	5.06	25.6036	2.24944	7.11337
4.62	21.3444	2.14942	6.79706	5.07	25.7049	2.25167	7.12039
4.63	21.4369	2.15174	6.80441	5.08	25.8064	2.25389	7.12741
4.64	21.5296	2.15407	6.81175	5.09	25.9081	2.25610	7.13442
4.65	21.6225	2.15639	6.81909	5.10	26.0100	2.25832	7.14143
4.66	21.7156	2.15870	6.82642	5.11	26.1121	2.26053	7.14843
4.67	21.8089	2.16102	6.83374	5.12	26.2144	2.26274	7.15542
4.68	21.9024	2.16333	6.84105	5.13	26.3169	2.26495	7.16240
4.69	21.9961	2.16564	6.84836	5.14	26.4196	2.26716	7.16938
4.70	22.0900	2.16795	6.85565	5.15	26.5225	2.26936	7.17635
4.71	22.1841	2.17025	6 86294	5.16	26.6256	2.27156	7.18331
4.72	22.2784	2.17256	6.87023	5.17	26.7289	2.27376	7.19027
4.73	22.3729	2.17486	6.87750	5.18	26.8324	2.27596	7.19722
4.74	22.4676	2.17715	6.88477	5.19	26.9361	2.27816	7.20417
4.75	22.5625	2.17945	6.89202	5.20	27.0400	2.28035	7.21110
4.76	22.6576	2.18174	6.89928	5.21	27.1441	2.28254	7.21803
4.77	22.7529	2.18403	6.90652	5.22	27.2484	2.28473	7.22496
4.78	22.8484	2.18632	6.91375	5.23	27.3529	2.28692	7.23187
4.79	22.9441	2.18861	6.92098	5.24	27.4576	2.28910	7.23878
4.80	23.0400	2.19089	6.92820	5.25	27.5625	2.29129	7.24569
4.81	23.1361	2.19317	6.93542	5.26	27.6676	2.29347	7.25259
4.82	23.2324	2.19545	6.94262	5.27	27.7729	2.29565	7.25948
4.83	23.3289	2.19773	6.94982	5.28	27.8784	2.29783	7.26636
4.84	23.4256	2.20000	6.95701	5.29	27.9841	2.30000	7.27324
4.85	23.5225	2.20227	6.96419	5.30	28.0900	2.30217	7.28011
4.86	23.6196	2.20454	6.97137	5.31	28.1961	2.30434	7.28697
4.87	23.7169	2.20681	6.97854	5.32	28.3024	2.30651	7.29383
4.88	23.8144	2.20907	6.98570	5.33	28.4089	2.30868	7.30068
4.89	23.9121	2.21133	6.99285	5.34	28.5156	2.31084	7.30753
4.90	24.0100	2.21359	7.00000	5.35	28.6225	2.31301	7.31437
4.91	24.1081	2.21585	7.00714	5.36	28.7296	2.31517	7.32120
4.92	24.2064	2.21811	7.01427	5.37	28.8369	2.31733	7.32803
4.93	24.3049	2.22036	7.02140	5.38	28.9444	2.31948	7.33485
4.94	24.4036	2.22261	7.02851	5.39	29.0521	2.32164	7.34166
4.95	24.5025	2.22486	7.03562	5.40	29.1600	2.32379	7.34847
4.96	24.6016	2.22711	7.04273	5.41	29.2681	2.32594	7.35527
4.97	24.7009	2.22935	7.04982	5.42	29.3764	2.32809	7.36206
4.98	24.8004	2.23159	7.05691	5.43	29.4849	2.33024	7.36885
4.99	24.9001	2.23383	7.06399	5.44	29.5936	2.33238	7.37564
5.00	25.0000	2.23607	7.07107	5.45	29.7025	2.33452	7.38241
5.01	25.1001	2.23830	7.07814	5.46	29.8116	2.33666	7.38918
5.02	25.2004	2.24054	7.08520	5.47	29.9209	2.33880	7.39594
5.03	25.3009	2.24277	7 09225	5.48	30.0304	2.34094	7.40270
5.04	25.4016	2.24499	7.09930	5.49	30.1401	2.34307	7.40945

Table XI.   Squares and Square Roots
(*continued*)

N	N²	√N	√10N	N	N²	√N	√10N
5.50	30.2500	2.34521	7.41620	5.95	35.4025	2.43926	7.71362
5.51	30.3601	2.34734	7.42294	5.96	35.5216	2.44131	7.72010
5.52	30.4704	2.34947	7.42967	5.97	35.6409	2.44336	7.72658
5.53	30.5809	2.35160	7.43640	5.98	35.7604	2.44540	7.73305
5.54	30.6916	2.35372	7.44312	5.99	35.8801	2.44745	7.73951
5.55	30.8025	2.35584	7.44983	6.00	36.0000	2.44949	7.74597
5.56	30.9136	2.35797	7.45654	6.01	36.1201	2.45153	7.75242
5.57	31.0249	2.36008	7.46324	6.02	36.2404	2.45357	7.75887
5.58	31.1364	2.36220	7.46994	6.03	36.3609	2.45561	7.76531
5.59	31.2481	2.36432	7.47663	6.04	36.4816	2.45764	7.77174
5.60	31.3600	2.36643	7.48331	6.05	36.6025	2.45967	7.77817
5.61	31.4721	2.36854	7.48999	6.06	36.7236	2.46171	7.78460
5.62	31.5844	2.37065	7.49667	6.07	36.8449	2.46374	7.79102
5.63	31.6969	2.37276	7.50333	6.08	36.9664	2.46577	7.79744
5.64	31.8096	2.37487	7.50999	6.09	37.0881	2.46779	7.80385
5.65	31.9225	2.37697	7.51665	6.10	37.2100	2.46982	7.81025
5.66	32.0356	2.37908	7.52330	6.11	37.3321	2.47184	7.81665
5.67	32.1489	2.38118	7.52994	6.12	37.4544	2.47386	7.82304
5.68	32.2624	2.38328	7.53658	6.13	37.5769	2.47588	7.82943
5.69	32.3761	2.38537	7.54321	6.14	37.6996	2.47790	7.83582
5.70	32.4900	2.38747	7.54983	6.15	37.8225	2.47992	7.84219
5.71	32.6041	2.38956	7.55645	6.16	37.9456	2.48193	7.84857
5.72	32.7184	2.39165	7.56307	6.17	38.0689	2.48395	7.85493
5.73	32.8329	2.39374	7.56968	6.18	38.1924	2.48596	7.86130
5.74	32.9476	2.39583	7.57628	6.19	38.3161	2.48797	7.86766
5.75	33.0625	2.39792	7.58288	6.20	38.4400	2.48998	7.87401
5.76	33.1776	2.40000	7.58947	6.21	38.5641	2.49199	7.88036
5.77	33.2929	2.40208	7.59605	6.22	38.6884	2.49399	7.88670
5.78	33.4084	2.40416	7.60263	6.23	38.8129	2.49600	7.89303
5.79	33.5241	2.40624	7.60920	6.24	38.9376	2.49800	7.89937
5.80	33.6400	2.40832	7.61577	6.25	39.0625	2.50000	7.90569
5.81	33.7561	2.41039	7.62234	6.26	39.1876	2.50200	7.91202
5.82	33.8724	2.41247	7.62889	6.27	39.3129	2.50400	7.91833
5.83	33.9889	2.41454	7.63544	6.28	39.4384	2.50599	7.92465
5.84	34.1056	2.41661	7.64199	6.29	39.5641	2.50799	7.93095
5.85	34.2225	2.41868	7.64853	6.30	39.6900	2.50998	7.93725
5.86	34.3396	2.42074	7.65506	6.31	39.8161	2.51197	7.94355
5.87	34.4569	2.42281	7.66159	6.32	39.9424	2.51396	7.94984
5.88	34.5744	2.42487	7.66812	6.33	40.0689	2.51595	7.95613
5.89	34.6921	2.42693	7.67463	6.34	40.1956	2.51794	7.96241
5.90	34.8100	2.42899	7.68115	6.35	40.3225	2.51992	7.96869
5.91	34.9281	2.43105	7.68765	6.36	40.4496	2.52190	7.97496
5.92	35.0464	2.43311	7.69415	6.37	40.5769	2.52389	7.98123
5.93	35.1649	2.43516	7.70065	6.38	40.7044	2.52587	7.98749
5.94	35.2836	2.43721	7.70714	6.39	40.8321	2.52784	7.99375

Table XI. Squares and Square Roots
(*continued*)

N	N²	√N	√10N	N	N²	√N	√10N
6.40	40.9600	2.52982	8.00000	6.85	46.9225	2.61725	8.27647
6.41	41.0881	2.53180	8.00625	6.86	47.0596	2.61916	8.28251
6.42	41.2164	2.53377	8.01249	6.87	47.1969	2.62107	8.28855
6.43	41.3449	2.53574	8.01873	6.88	47.3344	2.62298	8.29458
6.44	41.4736	2.53772	8.02496	6.89	47.4721	2.62488	8.30060
6.45	41.6025	2.53969	8.03119	6.90	47.6100	2.62679	8.30662
6.46	41.7316	2.54165	8.03741	6.91	47.7481	2.62869	8.31264
6.47	41.8609	2.54362	8.04363	6.92	47.8864	2.63059	8.31865
6.48	41.9904	2.54558	8.04984	6.93	48.0249	2.63249	8.32466
6.49	42.1201	2.54755	8.05605	6.94	48.1636	2.63439	8.33067
6.50	42.2500	2.54951	8.06226	6.95	48.3025	2.63629	8.33667
6.51	42.3801	2.55147	8.06846	6.96	48.4416	2.63818	8.34266
6.52	42.5104	2.55343	8.07465	6.97	48.5809	2.64008	8.34865
6.53	42.6409	2.55539	8.08084	6.98	48.7204	2.64197	8 35464
6.54	42.7716	2.55734	8.08703	6.99	48.8601	2.64386	8.36062
6.55	42.9025	2.55930	8.09321	7.00	49.0000	2.64575	8.36660
6.56	43.0336	2.56125	8.09938	7.01	49.1401	2.64764	8.37257
6.57	43.1649	2.56320	8.10555	7.02	49.2804	2.64953	8.37854
6.58	43.2964	2.56515	8.11172	7.03	49.4209	2.65141	8.38451
6.59	43.4281	2.56710	8.11788	7.04	49.5616	2.65330	8.39047
6.60	43.5600	2.56905	8.12404	7.05	49.7025	2.65518	8.39643
6.61	43.6921	2.57099	8.13019	7.06	49.8436	2.65707	8.40238
6.62	43.8244	2.57294	8.13634	7.07	49.9849	2.65895	8.40833
6.63	43.9569	2.57488	8.14248	7.08	50.1264	2.66083	8.41427
6.64	44.0896	2.57682	8.14862	7.09	50.2681	2.66271	8.42021
6.65	44.2225	2.57876	8.15475	7.10	50.4100	2.66458	8.42615
6.66	44.3556	2.58070	8.16088	7.11	50.5521	2.66646	8.43208
6.67	44.4889	2.58263	8.16701	7.12	50.6944	2.66833	8.43801
6.68	44.6224	2.58457	8.17313	7.13	50.8369	2.67021	8.44393
6.69	44.7561	2.58650	8.17924	7.14	50.9796	2.67208	8.44985
6.70	44.8900	2.58844	8.18535	7.15	51.1225	2.67395	8.45577
6.71	45.0241	2.59037	8.19146	7.16	51.2656	2.67582	8.46168
6.72	45.1584	2.59230	8.19756	7.17	51.4089	2.67769	8.46759
6.73	45.2929	2.59422	8.20366	7.18	51.5524	2.67955	8.47349
6.74	45.4276	2.59615	8.20975	7.19	51.6961	2.68142	8.47939
6.75	45.5625	2.59808	8.21584	7.20	51.8400	2.68328	8.48528
6.76	45.6976	2.60000	8.22192	7.21	51.9841	2.68514	8.49117
6.77	45.8329	2.60192	8.22800	7.22	52.1284	2.68701	8.49706
6.78	45.9684	2.60384	8.23408	7.23	52.2729	2.68887	8.50294
6.79	46.1041	2.60576	8.24015	7.24	52.4176	2.69072	8.50882
6.80	46.2400	2.60768	8.24621	7.25	52.5625	2.69258	8.51469
6.81	46.3761	2.60960	8.25227	7.26	52.7076	2.69444	8.52056
6.82	46.5124	2.61151	8.25833	7.27	52.8529	2.69629	8.52643
6.83	46.6489	2.61343	8.26438	7.28	52.9984	2.69815	8.53229
6.84	46.7856	2.61534	8.27043	7.29	53.1441	2.70000	8.53815

TABLE XI. Squares and Square Roots
(*continued*)

N	N²	√N	√10N	N	N²	√N	√10N
7.30	53.2900	2.70185	8.54400	7.75	60.0625	2.78388	8.80341
7.31	53.4361	2.70370	8.54985	7.76	60.2176	2.78568	8.80909
7.32	53.5824	2.70555	8.55570	7.77	60.3729	2.78747	8.81476
7.33	53.7289	2.70740	8.56154	7.78	60.5284	2.78927	8.82043
7.34	53.8756	2.70924	8.56738	7.79	60.6841	2.79106	8.82610
7.35	54.0225	2.71109	8.57321	7.80	60.8400	2.79285	8.83176
7.36	54.1696	2.71293	8.57904	7.81	60.9961	2.79464	8.83742
7.37	54.3169	2.71477	8.58487	7.82	61.1524	2.79643	8.84308
7.38	54.4644	2.71662	8.59069	7.83	61.3089	2.79821	8.84873
7.39	54.6121	2.71846	8.59651	7.84	61.4656	2.80000	8.85438
7.40	54.7600	2.72029	8.60233	7.85	61.6225	2.80179	8.86002
7.41	54.9081	2.72213	8.60814	7.86	61.7796	2.80357	8.86566
7.42	55.0564	2.72397	8.61394	7.87	61.9369	2.80535	8.87130
7.43	55.2049	2.72580	8.61974	7.88	62.0944	2.80713	8.87694
7.44	55.3536	2.72764	8.62554	7.89	62.2521	2.80891	8.88257
7.45	55.5025	2.72947	8.63134	7.90	62.4100	2.81069	8.88819
7.46	55.6516	2.73130	8.63713	7.91	62.5681	2.81247	8.89382
7.47	55.8009	2.73313	8.64292	7.92	62.7264	2.81425	8.89944
7.48	55.9504	2.73496	8.64870	7.93	62.8849	2.81603	8.90505
7.49	56.1001	2.73679	8.65448	7.94	63.0436	2.81780	8.91067
7.50	56.2500	2.73861	8.66025	7.95	63.2025	2.81957	8.91628
7.51	56.4001	2.74044	8.66603	7.96	63.3616	2.82135	8.92188
7.52	56.5504	2.74226	8.67179	7.97	63.5209	2.82312	8.92749
7.53	56.7009	2.74408	8.67756	7.98	63.6804	2.82489	8.93308
7.54	56.8516	2.74591	8.68332	7.99	63.8401	2.82666	8.93868
7.55	57.0025	2.74773	8.68907	8.00	64.0000	2.82843	8.94427
7.56	57.1536	2.74955	8.69483	8.01	64.1601	2.83019	8.94986
7.57	57.3049	2.75136	8.70057	8.02	64.3204	2.83196	8.95545
7.58	57.4564	2.75318	8.70632	8.03	64.4809	2.83373	8.96103
7.59	57.6081	2.75500	8.71206	8.04	64.6416	2.83549	8.96660
7.60	57.7600	2.75681	8.71780	8.05	64.8025	2.83725	8.97218
7.61	57.9121	2.75862	8.72353	8.06	64.9636	2.83901	8.97775
7.62	58.0644	2.76043	8.72926	8.07	65.1249	2.84077	8.98332
7.63	58.2169	2.76225	8.73499	8.08	65.2864	2.84253	8.98888
7.64	58.3696	2.76405	8.74071	8.09	65.4481	2.84429	8.99444
7.65	58.5225	2.76586	8.74643	8.10	65.6100	2.84605	9.00000
7.66	58.6756	2.76767	8.75214	8.11	65.7721	2.84781	9.00555
7.67	58.8289	2.76948	8.75785	8.12	65.9344	2.84956	9.01110
7.68	58.9824	2.77128	8.76356	8.13	66.0969	2.85132	9.01665
7.69	59.1361	2.77308	8.76926	8.14	66.2596	2.85307	9.02219
7.70	59.2900	2.77489	8.77496	8.15	66.4225	2.85482	9.02774
7.71	59.4441	2.77669	8.78066	8.16	66.5856	2.85657	9.03327
7.72	59.5984	2.77849	8.78635	8.17	66.7489	2.85832	9.03881
7.73	59.7529	2.78029	8.79204	8.18	66.9124	2.86007	9.04434
7.74	59.9076	2.78209	8.79773	8.19	67.0761	2.86182	9.04986

TABLE XI.   Squares and Square Roots
             (*continued*)

N	N^2	$\sqrt{N}$	$\sqrt{10N}$	N	N^2	$\sqrt{N}$	$\sqrt{10N}$
8.20	67.2400	2.86356	9.05539	8.65	74.8225	2.94109	9.30054
8.21	67.4041	2.86531	9.06091	8.66	74.9956	2.94279	9.30591
8.22	67.5684	2.86705	9.06642	8.67	75.1689	2.94449	9.31128
8.23	67.7329	2.86880	9.07193	8.68	75.3424	2.94618	9.31665
8.24	67.8976	2.87054	9.07744	8.69	75.5161	2.94788	9.32202
8.25	68.0625	2.87228	9.08295	8.70	75.6900	2.94958	9.32738
8.26	68.2276	2.87402	9.08845	8.71	75.8641	2.95127	9.33274
8.27	68.3929	2.87576	9.09395	8.72	76.0384	2.95296	9.33809
8.28	68.5584	2.87750	9.09945	8.73	76.2129	2.95466	9.34345
8.29	68.7241	2.87924	9.10494	8.74	76.3876	2.95635	9.34880
8.30	68.8900	2.88097	9.11043	8.75	76.5625	2.95804	9.35414
8.31	69.0561	2.88271	9.11592	8.76	76.7376	2.95973	9.35949
8.32	69.2224	2.88444	9.12140	8.77	76.9129	2.96142	9.36483
8.33	69.3889	2.88617	9.12688	8.78	77.0884	2.96311	9.37017
8.34	69.5556	2.88791	9.13236	8.79	77.2641	2.96479	9.37550
8.35	69.7225	2.88964	9.13783	8.80	77.4400	2.96648	9.38083
8.36	69.8896	2.89137	9.14330	8.81	77.6161	2.96816	9.38616
8.37	70.0569	2.89310	9.14877	8.82	77.7924	2.96985	9.39149
8.38	70.2244	2.89482	9.15423	8.83	77.9689	2.97153	9.39681
8.39	70.3921	2.89655	9.15969	8.84	78.1456	2.97321	9.40213
8.40	70.5600	2.89828	9.16515	8.85	78.3225	2.97489	9.40744
8.41	70.7281	2.90000	9.17061	8.86	78.4996	2.97658	9.41276
8.42	70.8964	2.90172	9.17606	8.87	78.6769	2.97825	9.41807
8.43	71.0649	2.90345	9.18150	8.88	78.8544	2.97993	9.42338
8.44	71.2336	2.90517	9.18695	8.89	79.0321	2.98161	9.42868
8.45	71.4025	2.90689	9.19239	8.90	79.2100	2.98329	9.43398
8.46	71.5716	2.90861	9.19783	8.91	79.3881	2.98496	9.43928
8.47	71.7409	2.91033	9.20326	8.92	79.5664	2.98664	9.44458
8.48	71.9104	2.91204	9.20869	8.93	79.7449	2.98831	9.44987
8.49	72.0801	2.91376	9.21412	8.94	79.9236	2.98998	9.45516
8.50	72.2500	2.91548	9.21954	8.95	80.1025	2.99166	9.46044
8.51	72.4201	2.91719	9.22497	8.96	80.2816	2.99333	9.46573
8.52	72.5904	2.91890	9.23038	8.97	80.4609	2.99500	9.47101
8.53	72.7609	2.92062	9.23580	8.98	80.6404	2.99666	9.47629
8.54	72.9316	2.92233	9.24121	8.99	80.8201	2.99833	9.48156
8.55	73.1025	2.92404	9.24662	9.00	81.0000	3.00000	9.48683
8.56	73.2736	2.92575	9.25203	9.01	81.1801	3.00167	9.49210
8.57	73.4449	2.92746	9.25743	9.02	81.3604	3.00333	9.49737
8.58	73.6164	2.92916	9.26283	9.03	81.5409	3.00500	9.50263
8.59	73.7881	2.93087	9.26823	9.04	81.7216	3.00666	9.50789
8.60	73.9600	2.93258	9.27362	9.05	81.9025	3.00832	9.51315
8.61	74.1321	2.93428	9.27901	9.06	82.0836	3.00998	9.51840
8.62	74.3044	2.93598	9.28440	9.07	82.2649	3.01164	9.52365
8.63	74.4769	2.93769	9.28978	9.08	82.4464	3.01330	9.52890
8.64	74.6496	2.93939	9.29516	9.09	82.6281	3.01496	9.53415

Table XI.   Squares and Square Roots
          (*concluded*)

N	N²	√N	√10N	N	N²	√N	√10N
9.10	82.8100	3.01662	9.53939	9.55	91.2025	3.09031	9.77241
9.11	82.9921	3.01828	9.54463	9.56	91.3936	3.09192	9.77753
9.12	83.1744	3.01993	9.54987	9.57	91.5849	3.09354	9.78264
9.13	83.3569	3.02159	9.55510	9.58	91.7764	3.09516	9.78775
9.14	83.5396	3.02324	9.56033	9.59	91.9681	3.09677	9.79285
9.15	83.7225	3.02490	9.56556	9.60	92.1600	3.09839	9.79796
9.16	83.9056	3.02655	9.57079	9.61	92.3521	3.10000	9.80306
9.17	84.0889	3.02820	9.57601	9.62	92.5444	3.10161	9.80816
9.18	84.2724	3.02985	9.58123	9.63	92.7369	3.10322	9.81326
9.19	84.4561	3.03150	9.58645	9.64	92.9296	3.10483	9.81835
9.20	84.6400	3.03315	9.59166	9.65	93.1225	3.10644	9.82344
9.21	84.8241	3.03480	9.59687	9.66	93.3156	3.10805	9.82853
9.22	85.0084	3.03645	9.60208	9.67	93.5089	3.10966	9.83362
9.23	85.1929	3.03809	9.60729	9.68	93.7024	3.11127	9.83870
9.24	85.3776	3.03974	9.61249	9.69	93.8961	3.11288	9.84378
9.25	85.5625	3.04138	9.61769	9.70	94.0900	3.11448	9.84886
9.26	85.7476	3.04302	9.62289	9.71	94.2841	3.11609	9.85393
9.27	85.9329	3.04467	9.62808	9.72	94.4784	3.11769	9.85901
9.28	86.1184	3.04631	9.63328	9.73	94.6729	3.11929	9.86408
9.29	86.3041	3.04795	9.63846	9.74	94.8676	3.12090	9.86914
9.30	86.4900	3.04959	9 64365	9.75	95.0625	3.12250	9.87421
9.31	86.6761	3.05123	9.64883	9.76	95.2576	3.12410	9.87927
9.32	86.8624	3.05287	9.65401	9.77	95.4529	3.12570	9.88433
9.33	87.0489	3.05450	9.65919	9.78	95.6484	3.12730	9.88939
9.34	87.2356	3.05614	9.66437	9.79	95.8441	3.12890	9.89444
9.35	87.4225	3.05778	9.66954	9.80	96.0400	3.13050	9.89949
9.36	87.6096	3.05941	9.67471	9.81	96.2361	3.13209	9.90454
9.37	87.7969	3.06105	9.67988	9.82	96.4324	3.13369	9.90959
9.38	87.9844	3.06268	9.68504	9.83	96.6289	3.13528	9.91464
9.39	88.1721	3.06431	9.69020	9.84	96.8256	3.13688	9.91968
9.40	88.3600	3.06594	9.69536	9.85	97.0225	3.13847	9.92472
9.41	88.5481	3.06757	9.70052	9.86	97.2196	3.14006	9.92975
9.42	88.7364	3.06920	9.70567	9.87	97.4169	3.14166	9.93479
9.43	88.9249	3.07083	9.71082	9.88	97.6144	3.14325	9.93982
9.44	89.1136	3.07246	9.71597	9.89	97.8121	3.14484	9.94485
9.45	89.3025	3.07409	9.72111	9.90	98.0100	3.14643	9.94987
9.46	89.4916	3.07571	9.72625	9.91	98.2081	3.14802	9.95490
9.47	89.6809	3.07734	9.73139	9.92	98.4064	3.14960	9.95992
9.48	89.8704	3.07896	9.73653	9.93	98.6049	3.15119	9.96494
9.49	90.0601	3.08058	9.74166	9.94	98.8036	3.15278	9.96995
9.50	90.2500	3.08221	9.74679	9.95	99.0025	3.15436	9.97497
9.51	90.4401	3.08383	9.75192	9.96	99.2016	3.15595	9.97998
9.52	90.6304	3.08545	9.75705	9.97	99.4009	3.15753	9.98499
9.53	90.8209	3.08707	9.76217	9.98	99.6004	3.15911	9.98999
9.54	91.0116	3.08869	9.76729	9.99	99.8001	3.16070	9.99500

Review Exercises

The set of exercises in this Appendix cover the methods developed in Chapters 3 to 13 and are designed to give the reader experience in deciding which of the concepts and techniques from those chapters to apply in each exercise. All of the exercises in this Appendix are different from those in the chapters of the book.

1. Given the following frequency distribution:

Class boundaries	Frequencies
5– 9	6
9–13	9
13–17	20
17–21	10
21–25	8
25–29	3

Find, accurate to two decimal places,
(a) the mean, using the definition;
(b) the mean, using an assumed mean;
(c) the median;
(d) the modal class;
(e) the variance, using an assumed mean.

2. Given the following frequency distribution:

Class boundaries	Frequencies
5–15	3
15–25	8
25–35	11
35–45	15
45–55	16
55–65	10
65–75	7

Answer the questions given for Exercise 1.

3. In a collection of twenty numbers, seven are 1's, three are 2's, nine are 3's, and the other is a 4.
   (a) Find the mean of these numbers.
   (b) Find their median.
   (c) Find their standard deviation.
   (d) If all possible samples of size 10 are taken from this population and the mean of each of these samples is calculated, find the exact value of the standard deviation of this set of means.

4. A group of 100 athletes is transported across the country on two airplanes. The first plane carries 40 of these athletes, the second the remaining 60. It is known that the average weight of the 100 athletes is 186.3 lbs and that the average weight of the athletes on the second plane is 10 lbs less than the average of those on the first plane. What is the average weight of the athletes on each of the two planes?

5. An examination was given to a class of 68 students. The mean of the examination was 70.6. The mean grade for all girls in the class was 75.0, the one for all boys was 65.0. How many girls and how many boys were in the class?

6. The probabilities that three men hit a target are $\frac{1}{6}$, $\frac{1}{4}$, and $\frac{1}{3}$, respectively. If all of them shoot at the target, what is the probability that exactly one of them hits the target?

7. A student is required to answer 8 out of 10 questions on an examination.
   (a) How many possible choices does he have?
   (b) How many, if he must answer the first three questions?
   (c) How many, if he must answer at least 4 of the first 5 questions (selecting the remaining questions from the last 5 questions)?

8. If 600 missions of bombers, each consisting of 7 planes, are sent out on a bombing mission, and if the probability is $\frac{1}{3}$ that a bomber will be shot down, how many of these missions would you expect to return with at least four planes?
   (a) Use the binomial distribution.
   (b) Use the Poisson distribution.
   (c) Use the normal distribution.

9. One bag contains 4 white and 6 black balls, and a second bag contains 3 white and 5 black balls.
   (a) If one ball is drawn from each bag, find the probability that both will be of the same color.
   (b) If a ball is drawn from a bag selected at random, find the probability that it will be white.

10. A teacher has told the students in a class that on an examination he will given an equal number of A's, B's, C's, D's, and F's. Four students compare their grades after the examinations have been returned. What is the probability that
    (a) all four have received different grades;
    (b) at least three of the four students have received the same grade?

11. An urn contains 60 balls, each of which is colored red or white and numbered 1 or 2. The exact composition of the urn is given below

	Red	White
Numbered 1	10	25
Numbered 2	20	5

   (a) If two balls are drawn at random from this urn without replacement, what is the probability that exactly one is white?
   (b) If, one at a time, two balls are drawn at random from this urn and neither is returned to the urn, what is the probability that a white ball is drawn on the second draw, if a ball numbered 1 is drawn on the first draw?
   (c) One by one, balls are drawn from this urn and then returned. The color of each is noted. What is the probability that when the hundredth ball is drawn, a white ball will have been drawn at least 55 times?

12. Sixty percent of the citizens of a particular town are over age 60. Thirty percent of the citizens are unemployed or retired. Twenty percent of the citizens are over 60 and unemployed or retired.
   (a) If a citizen of that town is selected at random, what is the probability that the individual is over 60 and working?
   (b) If the citizen selected is under 60, what is the probability that this person is unemployed or retired?

13. The works of a famous composer are available in 11 different record albums. If a person owns four of the albums, what is the probability that, if you buy him three more of the composer's albums, you select all three of them so that he does not already have them?

14. The top drawer of a dresser contains six black socks and a pair of red and grey argyles. If two socks are picked at random and they match, what is the probability that the argyle socks have been picked?

15. If it has been determined that 30% of the voters favor a certain candidate, what is the probability that, if we ask 2000 voters, at least 550 will favor this candidate?

16. A random selection of 20,000 steel balls was made from an assembly line and the diameter of each was measured. The mean diameter was 4.30 cm, and the standard deviation was 0.01 cm. What is the probability that the next ball chosen from the assembly line will have a diameter less than 4.28 cm, assuming that the that the diameters satisfy a normal distribution.

17. The lifetime of a particular brand of electric light bulbs is normally distributed with a mean of 1100 hours and a standard deviation of 25 hours. Of 2000 light bulbs of this brand, how many can be expected to last
   (a) more than 1120 hours;
   (b) at most 1150 hours;
   (c) between 1120 and 1140 hours;
   (d) 1100 hours, when recorded to the nearest hour;
   (e) 1100 hours exactly?

18. A particular disease affects 3% of all human beings. In a group of 100 people, what is the probability that
    (a) exactly three people have this disease;
    (b) two, three, or four have the disease.

19. Eighteen female mice underwent tubal ligation. Nine of them (the control group) were allowed to run through a specially constructed maze; they did this in an average of 28 seconds with a variance of 10. The other nine (the experimental group) were inseminated with live sperm and allowed to run the same maze; they did this in an average time of 25 seconds with a variance of 8. Does this indicate that the mean times for treated and untreated mice can be considered the same (5% level)?

20. If among manufactured articles of a certain brand the true proportion of defectives is known to be 0.01, find the probability that in a sample of 100 articles there are
    (a) no defectives;
    (b) exactly two defectives;
    (c) fewer than 5 defectives;
    (d) at least 5 defectives.

21. A commuter reads a report of a traffic survey that claims the mean time to cross a bridge during a rushhour is 8.2 minutes with a standard deviation of 2.1 minutes. He notes that his own time for ten trips (each made at the same time each day) has an average of 9.6 minutes. Assuming that the time for crossing the bridge is normally distributed, does this indicate that his time is significantly different from that claimed in the report (5% level)?

22. Loaded trucks arrive at a cannery at an average rate of 4 per hour. Because the trucks come from different fields, they arrive independently and at random times. What is the probability that no truck arrives in a one-half-hour period?

23. Find the 98% confidence limits for the mean of a normally distributed population with a standard deviation of 3.19, if a sample taken from that population has the following variates: 13.6, 10.9, 8.2, 8.5, 9.0, 12.8.

24. Two brands of orange juice are rated for quality and the scores on a 10-point scale are shown below. On the basis of these data is there a significant difference between the mean scores of the two brands (5% level)?

Brand A	Brand B
7	10
6	9
8	7
5	6
4	8

25. Wire being manufactured by process A shows a mean breaking strength of 10.00 lbs, with a standard deviation of 2.50 lbs. Wire being manufactured by process B shows a mean breaking strength of 9.20 lbs, with a standard deviation of 2.30 lbs. Someone purchases 45 pieces of wire manufactured by process A and 50 pieces by process B. What is the probability that the average breaking strength of the 50 pieces of process B is less than that of the 45 pieces of process A?

26. A clerk entering data makes an average of one error in two pages. What is the probability that there are less than two errors in 6 pages?

27. Preceding an election in which there are two candidates, A and B, it is believed from preliminary indications that 52% of the population favors candidate A. In order to determine the winner before the election, a poll must be able to predict the winner within 2%. How large a sample of people needs to be surveyed in order to predict the winner with 99% confidence?

28. The upper 98% confidence limit for the mean of a normal population as determined from a sample with variates 13.6, 10.8, 8.2, 8.5, 9.0, and 12.8 was found to be 15.11 when the true value of the population standard deviation was used. What was the true value of the population standard deviation?

29. Under standard manufacturing conditions, 1.5% of all light bulbs manufactured by a particular company are known to be defective. If 400 light bulbs are examined, what is the probability that at most 10 will be defective?

30. The standard deviation of a binomial distribution is 6, and the probability that any variate is equal to or exceeds 65 is 0.226. Use the normal curve approximation to find p and n.

31. The grade-point averages of 800 college freshmen follow a normal distribution with a mean of 2.12 and a standard deviation of 1.10. How many of these freshmen would you expect to have a grade-point average between 1.50 and 2.00?

32. The following data give the length of time, Y, in hours, for a headache to disappear after administration of different dosages, X, in grains of aspirin.

Dosage:	1	2	3	4	5	6
Time:	1.5	1.3	1.8	1.0	1.0	0.9

Find the regression line for predicting from the dosage the time when the headache will disappear. (Calculate the slope to three decimal places.)

33. Of two boxes, one box contains 3 white balls and 5 black balls; the other contains 9 white and 3 black balls. A die is thrown and if it shows a six a ball is drawn from the first box; otherwise a ball is drawn from the second box. Find the probability that the ball drawn is white.

34. A researcher wishes to poll a sample in order to know public opinion on a certain issue. He will ask his sample to rate their confidence in a President's Administration on a scale from 0 to 100. Assuming a normal distribution for these confidence ratings with a standard deviation of 20, how many people should he poll in order to estimate the mean confidence of the population within 5 units with a probability of 90%?

35. An agency that regulates hunting makes a survey to compare hunter success by various categories. One such comparison is the following:

	Number successful	Number unsuccessful
Urban hunters	20	40
Rural hunters	30	30

Is there evidence of a difference in success for urban and rural hunters (5% level)?

36. Two types of paint are to be tested. Type I is somewhat cheaper than Type II. The test consists of giving scores to the paints after they have been exposed to certain weather conditions for a period of 6 months. Five samples of Type I are scored and another five samples of Type II are scored with the following results:

Type I:	85	87	92	80	84
Type II:	89	89	90	84	88

We should like to adopt Type I, the cheaper one, unless we have reason to believe that Type II is better. Is there reason to believe that Type II is better (5% level)?

37. One of the 98% confidence limits for the mean of a population as determined from a sample of 25 variates is 22.55. If, for the sample, $\Sigma(X - \bar{X})^2 = 600$, between what limits can one be 98% sure that the mean of the population will fall?

38. A widely used measure of a person's reading ability is the Iowa Silent Reading Test. A group of slow readers has been given class time to study with a remedial reading teacher over a period of weeks. At the end of that time the teacher estimates that the average improvement in scores on the Iowa Silent Reading Test will be 12 points. A sample of five students is selected and given the Iowa Test. Their scores on the Iowa Test, compared with the scores they made before beginning the experiment, are as follows:

Student	Before remedial reading	After remedial reading
1	98	105
2	93	108
3	95	103
4	95	104
5	99	110

Do these data confirm the teacher's claim that the average gain in the Iowa Test is 12 points (5% level)?

39. Prove that for a given set of N numbers $X_1, X_2, \ldots, X_N$, whose mean is m:

$$\sum_{i=1}^{N}(X_i - m)(X_i + 2m) - \sum_{i=1}^{N}(X_i^2 + 1) + \sum_{i=1}^{N}(X_i - 1)^2 = \sum_{i=1}^{N} X_i^2 - mN(m + 2).$$

40. For what values of r would the correlation coefficient of a sample of size 32 be considered significantly different from zero on the basis of the 1% level of significance?

41. In a sample of 160 people in a community, 40 were men and 120 women. Of these, 60 favored school busing, and 100 were against it. Of the men, 19 favored school busing. Do these data indicate that in this community there is a relationship between sex and opinion on school busing (5% level)?

42. Test at the 1% level of significance the hypothesis that $\rho = 0.70$, if in a sample of 67 pairs the correlation coefficient was found to be equal to 0.50.

43. If four people meet, what is the probability that they all have birthdays in different months? Assume that all months are equally likely for a birthday.

44. Of students who took the entrance examination at a certain law school, random samples were drawn from those whose college majors differed. Performance on the examination is summarized in the following table:

	Above median	Below median
Social sciences	70	80
Natural sciences	40	10
Humanities	40	60

Do these data indicate that performance on the examination is dependent on college major (5% level)?

45. A person tosses two fair coins. What is the probability that he has tossed *exactly* one head if it is known that he has tossed *at least* one head.

46. A certain baseball player is at bat 400 times in season I and gets 100 hits. In season II, in 300 times at bat, he also gets 100 hits. Does this indicate that his batting ability is the same in the two seasons (5% level)?

47. From two samples of the same size taken from two normally distributed populations, we wish to calculate the 90% confidence limits for the difference of the means of these populations. If the variances of the two populations are 16 and 20, respectively, and if the width of the confidence interval must be equal to 1.98, what must be the sample size?

48. The mean life of electric light bulbs produced by a company has in the past been 1120 hours with a standard deviation of 125 hours. A sample of newly-produced bulbs is taken with the following results: 975, 895, 1155, and 935 hours. Test the hypothesis that the mean life of the bulbs has not changed. Use the 5% level of significance.

49. In a large community only 45% of the registered voters are in favor of a bond issue. If only 600 voters go to the polls and vote, what is the probability that the issue will pass? Assume that only a simple majority is needed for the bond issue to pass.

50. "Number-ten" cans of peaches are supposed to hold 72 ounces. The canner adjusts the machine to fill the can with, on the average, 72.6 oz. The distribution of fill weights is approximately normal with a standard deviation of 0.4 oz.
    (a) How many of 100,000 cans will have less than 72 oz?
    (b) The manufacturer feels that this is too many. Assuming the standard deviation remains unchanged, what mean should the manufacturer use if he wants no more than 300 out of the 100,000 cans to contain less than 72 oz?

51. In a large orchard 10% of the apples are wormy. If five apples are selected at random, what is the probability that
    (a) exactly three will be wormy (use an exact method, not an approximation);
    (b) the first and last apples selected will be free of worms, whereas the other three are all wormy?

52. If a group consists of five men and five women, in how many ways can a committee of four be selected if
    (a) there are no restrictions on the number of men and women on the committee?
    (b) there must be at least one of each sex?

53. A poll on a campus asks students to name their favorite of four soft drinks. The results are as follows:

Brand:	A	B	C	D
Number preferring:	190	198	187	225

    Do these data show that there are no differences in the preferences for these brands (5% level)?

54. Milk is judged according to cream percent. In a certain area, the average is 4.30 with a standard deviation of 1.25. At the county fair, of all samples entered, 20 percent were rated I (highest cream percent), 20 percent were rated II, 35 percent were rated III, 15 percent were rated IV, and 10 percent were rated V. Assuming that the cream content follows a normal distribution, find the lowest and highest cream content in the IV range.

55. Two chemical treatments were applied to random samples of seeds. After the treatments, germination tests were conducted. It was found that among the 150 seeds that received chemical treatment A, 80 percent germinated, while among the 125 seeds that received chemical treatment B, 88 percent germinated. Do these data indicate that the chemicals differ in their effects on germination, using the 5% level of significance?

56. A sample of size 52 is taken from a population in which the correlation coefficient is 0.72. Find the probability that the correlation coefficient of the sample is between 0.62 and 0.82.

57. In a recent election in Davis, California, 75% of those voting voted against making Davis a Charter City. A group of nine persons who were serving currently on the Davis City Council or who had served on the prior council was asked how each voted on this issue. Assuming these nine people represent a random group of Davis residents, what is the probability that at least 6 of them voted in favor of making Davis a Charter City? (Give *exact* probability, no approximation.)

58. Find the correlation coefficient (to three decimal places) for the following sample of values of X and Y:

$$\begin{array}{ccccccc}
X: & 6 & 4 & 2 & 1 & 0 & -1 \\
Y: & 10 & 4 & -1 & -4 & -5 & -10
\end{array}$$

59. The 95% confidence limits for the mean of a population, as determined from a sample of 9 variates from the population, are 15.47 and 38.53. What is the value of $\Sigma(X - \bar{X})^2$ for the sample?

60. A survey of 450 randomly selected persons requested information about the amount of their education and the amount of their income. The results are as follows:

Education	High	Medium	Low
	Income		
Less than high school	30	50	70
High school graduate	80	100	120

Are income level and education independent (5% level)?

61. Under the hypothesis that 20% of a certain kind of seed germinates, what is the probability that in an experiment at most 100 seeds out of 600 germinate
(a) using the normal distribution;
(b) using adjusted chi-square?

62. A psychologist gives a test to four blue-collar workers and five while-collar workers to see if there is any significant difference at the 5% level of significance in their scores as groups. The blue-collar workers made scores of 23, 18, 22, 21 and the white-collar workers made scores of 17, 22, 19, 18, 20. Is there a significant difference in the average scores of the two groups?

63. A candy bar manufacturer has a machine that is supposed to make 2.34-ounce bars. It is known, however, that the weights vary slightly but are normally distributed. To test quality, a random sample of four bars is drawn from the day's output, and the four bars are found to have weights of 2.24 oz, 2.34 oz, 2.28 oz, and 2.30 oz. Does this indicate that the machine performs as it is supposed to (5% level)? Use a two-tailed test.

64. A large-scale cancer research study is to be undertaken on persons none of whom currently shows any sign of cancer. If 5% of all people become cancer patients, how large a sample should be taken so that the researchers can be 99% sure that it includes at least 50 persons who will get cancer?

65. Prove that $C_{n-1,r} + C_{n-1,r-1} = C_{n,r}$.

66. In a test of the yields of two varieties of wheat, 5 blocks of land are each divided into two plots of equal size. In each block one plot is planted with variety A and the other with variety B. The yields in bushels are shown in the following table. On the basis of these data, is variety A superior in yield to variety B at the 5% level of significance?

Block	Variety A	Variety B
1	50	44
2	45	42
3	60	55
4	55	57
5	60	52

67. A manufacturer of woolen cloth claims that the average number of flaws in his product is one per 2 square yards. A sample square yard of his product, selected at random, shows 3 flaws. What is the probability of obtaining 3 or more flaws in any one square yard if the manufacturer's claim is valid?

68. To test the effectiveness of a new tuberculosis drug in pill form, it was decided to use it in 8 tuberculosis institutions on the West Coast of the United States; in each of these institutions two individuals with tuberculosis were selected, one of which was given the drug, the other an ineffective pill. The number of days until recovery for both types of patients are listed below. Does this indicate that the cure was effective (5% level)?

Recovery Time for Patients

Institution	Treated with new drug	Treated with ineffective pill
1	36	40
2	38	35
3	48	48
4	28	29
5	25	30
6	35	41
7	35	42
8	38	42

(a) Solve without using the Wilcoxon test.
(b) Solve by using the Wilcoxon test.

69. Prove that $\hat{s}_e = \sqrt{\hat{s}_y^2 - b\hat{s}_{xy}}$.

70. Two populations have means $m_x = 15$ and $m_y = 14$, respectively. Both have the same standard deviation, namely 4. What is the probability that the mean of a sample of 64 variates from the X-population is greater than the mean of a sample of 36 variates from the Y-population?

71. In a class election there are two parties. John is nominated by one party, and the probability that he will be elected is $\frac{3}{4}$, provided William is not nominated by the opposition. The probability that William will not be nominated is $\frac{1}{3}$ and the probability that he will be elected if nominated is $\frac{2}{3}$. What is the probability that John will be elected?

72. Do the following data give us reason to believe that the fouling of boat bottoms can be reduced by the use of antifouling paint (1% level)?

	Paint	
	Antifouling	Standard
No fouling	37	15
Some fouling	53	50
Much fouling	10	35

73. If it is assumed that of the automobile drivers in a certain state 75% are accident free, what is the probability that in a sample of 200 drivers at least 62 are involved in one or more accidents?
    (a) Solve by using the normal distribution.
    (b) Solve by using adjusted χ^2.

74. Verify Formula (8-4) for $N = 5$ and $n = 2$.

75. Verify Formula (8-5) for $N = 5$ and $n = 2$.

76. In a manufacturing process it is known from long experience that 2% of the articles manufactured are defective. On a particular day 5 defective articles appeared in 100 manufactured. At the 3% level, is this a significantly large number of defects?

77. In a study of identical twins, one twin is given a drug and then an intelligence test while under the influence of the drug; the other twin is given an intelligence test but is not given the drug. The data are as follows:

Pair	Twin with drug	Twin without drug
1	110	114
2	122	135
3	149	143
4	116	125
5	148	158
6	130	130

Do these data indicate that the drug has an effect on the intelligence test score (5% level)?

78. A state department of motor vehicles claimed that 60% of the cars on the road are in violation of the state inspection laws. To test this claim a research organization took a random sample of 900 cars and found that 500 of them failed to pass inspection. Does this cast doubt on the claim of the department of motor vehicles (5% level)?

79. In a normal population 61% of the means of samples of 25 variates deviate by more than 1.02 in absolute value from the population mean, and 96% of the variates in the population exceed 2.5. Find the mean of the population.

80. In testing the yields of two varieties of wheat one variety was planted on 7 plots and the other on 9 plots. The difference between the means of the plot yields of the two varieties was 8.58 pounds and was just significant at the 5% level. Find the estimate of the variance (s^2) of the populations from which the samples came.

81. A researcher is 90% confident that a sample of 40 pairs of values he is working with came from a bivariate population with a correlation coefficient of 0.70, but he believes that there is a 10% chance that it came from another population with a correlation coefficient of 0.30. What is the probability that the correlation coefficient of the sample is between 0.30 and 0.70?

82. In a regression equation, if the regression coefficient b is greater than 1, show that Σy^2 must be greater than Σx^2.

83. In a 2×2 contingency table giving the preference between a scientific and a non-scientific career among 15 men and women entering college, one third preferred a scientific career. If the value of adjusted χ^2 is 1.35, what are the observed values?

84. Two classes in statistics take the same final examination. The following information is available for the two classes:

	Number of students	Sum of squares of deviations from mean
Class A	25	430
Class B	16	350

What is the least difference between the means of the class scores that would just be significant at the 1% level?

85. Suppose that the dollar value of the accounts receivable of a certain corporation is normally distributed with a mean of $10,000 and a standard deviation of $2000.
   (a) If one account is randomly selected, what is the probability that its value will be between $9000 and $12,000?
   (b) If a random sample of 400 accounts is selected, what is the probability that its mean will be between $10,100 and $10,200?

86. On a certain transcontinental highway, on the average, 12 traffic deaths occur every 3 months. What is the probability that in any one month
    (a) there are more than 4 traffic deaths;
    (b) there are exactly 4 deaths?

87. At the 5% level of significance, what is the complete range of values for which the coefficient of correlation of a sample of 19 pairs of values is significantly different from a population value of 0.77?

88. If at the 1% level of significance one confidence limit for the coefficient of correlation of a population is 0.77, as determined from a sample of 147 pairs of values, what can you say about the other confidence limit?

89. From long experience it is known that workers perform a particular job in an average time of 8.3 hours with a standard deviation of 2.0 hours. Four workers do the job in the following times (in hours): 9, 6, 5, 6. On the basis of the average time required for these workers to do the job, would you say that this is a superior group of workers (5% level)?

90. A doctor asserts that he has a method that is 80% accurate in determining, several months before birth, the sex of an unborn child. He is challenged to prove his claim, by being asked to predict the sex of 40 unborn children. Use the 5% level of significance and a one-tailed test to determine the least number of children for which he must predict the sex correctly so that his claim can be believed?

91. A committee of 4 is selected at random from 6 lawyers, 7 engineers, and 2 doctors. What is the probability that all on the committee are of the same profession?

92. An examination is taken by 80 girls and 60 boys. The class mean is 71.86, and the mean of the girls in this class is 5.00 points higher than that of the boys. Find the mean of the girls and that of the boys.

93. A football player completed 60% of his passes this season. Assuming that he maintains the same record next year, what is the probability that he will complete at least 5 of his first 8 passes? (Use an *exact* method, not an approximation).

94. Three people, A, B, and C, in that order, toss two dice until one obtains a total of 4 on the two dice; that is, A, B, and C do not stop until someone obtains a 4. What is the probability that C will be the first to obtain a 4?

95. A jar contains a large number of colored beads. It is assumed that the colors red, yellow, green, and blue are in the proportions 8 : 7 : 3 : 2. If a sample of 400 contains 180 red, 120 yellow, 40 green, and 60 blue beads, does this confirm the assumption (5% level)?

96. Under the hypothesis that the sex ratio in the population is 1 : 1 for chickens, find the probability that in a sample of 40 chickens at least 25 males are obtained.
    (a) Use the normal distribution.
    (b) Use adjusted χ^2.

97. From a sample of 18 variates, the 95% confidence limits for the mean of the population whose standard deviation is unknown were determined to be 28.2 and 32.6. Find the 99% confidence limits for the mean of the same population.

98. A group of students working on a summer project with a social agency took a random sample of 120 families in a well-defined "poverty area" of a large city in order to determine the mean annual family income in this area. If the sample gave a mean family income of $2810 and a standard deviation of $780, what are the 99% confidence limits for the mean income of all families in this poverty area?

99. The following are percentages of fat found in samples of two types of meat. Do the meats have significantly different fat contents (5% level)?

| Meat A: | 30 | 26 | 26 | 19 | 25 | 37 | 27 | 32 |
| Meat B: | 40 | 34 | 42 | 29 | 35 | 36 | 33 | 37 |

(a) Solve without using the Wilcoxon test.
(b) Solve by using the Wilcoxon test.

100. A coach claims that his eighth-grade boys can do an average of 10 pull-ups on the overhead bar. A sample of ten boys is selected at random and each does, respectively, 9, 9, 11, 10, 8, 10, 9, 9, 8, 10 pull-ups. Do these data confirm the coach's claim (5% level)?

101. For what range of values can be the correlation coefficient of a sample of 15 pairs of X, Y-values be considered significantly different from zero at the 2% level of significance?

102. An urn contains two white, three red, and five black balls. A second urn contains one white, six red, and three black balls. One ball is selected at random from each urn. Find the probability (as a fraction in lowest terms) that
(a) both are the same color;
(b) at least one is red.

103. A utility table is assembled from three pieces: a tray, a leg, and a stand. These pieces are manufactured separately, and any one table contains pieces drawn at random from each of the three stocks. One in twenty of the trays, one in a hundred of the legs, and one in fifty of the stands are flawed. The finished tables are examined and rejected if and only if at least two of the three pieces are flawed. Find the probability (as a fraction in lowest terms) that a finished utility table, selected at random, will be rejected.

104. A church has been led by experience to expect that about $\frac{1}{3}$ of the 300 members will attend a church supper. How many meals should be planned if the church does not want to incur a risk of more than 1% of providing for too few?

105. Ten pairs of brothers made the following scores on a test. Do these results indicate that in general the older of two brothers will make a higher score (5% level)?

Brother	1	2	3	4	5	6	7	8	9	10
Older	86	79	67	92	65	75	61	77	91	69
Younger	80	81	60	87	71	72	56	75	93	67

Pair

(a) Solve without using the Wilcoxon test.
(b) Solve using the Wilcoxon test.

106. In a factory 10% of all die castings have been found to be defective. A new operator made 64 die castings of which 10 were found to be defective. He was fired on the grounds that the sample of his work examined contained a significantly higher than average number of defective castings. What can you say about the level of significance that must have been used?

Answers to Odd-Numbered Exercises

CHAPTER 3

1(a). $X_2 + X_3 + X_4 + X_5 + X_6$.   **1(b).** $X_7^2 + X_8^2 + X_9^2$.

1(c). $(X_1 + X_2 + X_3 + X_4 + X_5 + X_6 + X_7)^2$.

1(d). $(X_1 - 2)^2 + (X_2 - 2)^2 + (X_3 - 2)^2$.

1(e). $[(X_1 + 3)X_1 + (X_2 + 3)X_2 + (X_3 + 3)X_3 + (X_4 + 3)X_4]^2$.

1(f). $cX_1 + cX_2 + cX_3 + cX_4$.   **1(g).** $(Y_2 + 7) + (Y_3 + 7) + (Y_4 + 7) + (Y_5 + 7)$.

1(h). $W_1^3 + W_2^3 + W_3^3 + W_4^3$.   **1(i).** $f_1 X_1^2 + f_2 X_2^2 + f_3 X_3^2 + f_4 X_4^2$.

3(a). 140.   **3(b).** 104.   **5(a).** 334.   **5(b).** 7.   **7(a).** 49.   **7(b).** 29.

9(a). 4.   **9(b).** 210.   **11(a).** 68.   **11(b).** $\sqrt{21}$.

CHAPTER 4

1(a). 8.50.   **1(b).** 8.95.   **1(c).** 1.45.   **1(d).** 1.92.   **3(a).** 24.60.

3(b). 24.60.   **3(c).** 25.00.

3(d). Class with boundaries 24 and 30.   **3(e).** 6.54.   **3(f).** 7.91.   **3(g).** 7.91.

5(a). 2.87.   **5(b).** 1.16.   **7.** Use class boundaries 5–8, 8–11, etc.

7(a). 16.96.   **7(b).** 16.96.   **7(c).** 16.31.

7(d). Class with boundaries 11 and 14.   **7(e).** 4.67.   **7(f).** 5.60.   **7(g).** 5.60.

9(a). 63.04.   **9(b).** 67.58.   **9(c).** 65.40.   **11.** 51.55 gals.   **13.** 120 and 80.

15. 60.28.   **17.** 170, 160, 155 lbs.   **19.** 74.04.   **21.** 56% and 44%.

23. \$2205.00.   **25.** 55 m.p.h.   **31.** 1.

CHAPTER 5

1(a). $\frac{5}{36}$.   **1(b).** $\frac{13}{18}$.   **1(c).** $\frac{5}{12}$.   **3(a).** $\frac{2}{13}$.   **3(b).** $\frac{3}{13}$.   **3(c).** $\frac{4}{13}$.

5. $\frac{5}{12}$.   **7.** $\frac{2}{3}$.   **9.** $\frac{5}{8}$.   **11(a).** $\frac{3}{52}$.   **11(b).** $\frac{29}{52}$.   **11(c).** $\frac{23}{52}$.   **11(d).** $\frac{5}{13}$.

13. $\frac{5}{288}$.   **15(a).** $\frac{11}{1530}$.   **15(b).** $\frac{1}{33,150}$.   **17(a).** $\frac{1}{2000}$.   **17(b).** $\frac{59}{1000}$.

17(c). $\frac{1881}{2000}$.   **19.** \$1.38.   **21.** $\frac{699}{715}$.   **23.** $\frac{7}{12}$.   **25.** $\frac{91}{648}$.   **27(a).** $\frac{1}{216}$.

27(b). $\frac{1}{36}$.   **27(c).** $\frac{5}{72}$.   **27(d).** $\frac{5}{12}$.   **27(e).** $\frac{5}{9}$.   **29.** \$1.20.   **31.** $\frac{3}{10}$.

33. $\frac{41}{59}$.   **35.** $\frac{90}{589}$.   **37.** $\frac{45}{173}$.   **39.** $\frac{1}{15}$.   **41.** 220.   **43.** 350.

45(a). 1280.   **45(b).** 2047.   **47.** 560.   **49.** 5760.   **51(a).** 495.

51(b). 450.   **51(c).** 255.   **53.** 8400.   **55(a).** 1152.   **55(b).** 648.

57. 115, 165, 670, 400.   **59(a).** $\frac{280}{2187}$.   **59(b).** $\frac{232}{243}$.   **61(a).** $\frac{216}{625}$.
61(b). $\frac{1053}{3125}$.   **63.** $\frac{6,385,729}{6,400,000}$.   **65.** $\frac{11}{64}$.   **67.** $\frac{5}{16}$.   **69.** $\frac{75}{154}$.   **71.** $\frac{160}{729}$.
73. $\frac{16}{27}$.   **75.** $\frac{3}{20}$.   **77.** $\frac{87}{260}$.   **79.** $\frac{1}{225}$.   **81.** $\frac{61}{216}$.   **83.** $\frac{5}{16}$.
85. $\frac{25}{324}$.   **87.** $\frac{391}{16,807}$.   **89.** $\frac{53}{3125}$.   **91.** $\frac{20,480}{1,594,323}$.   **93.** $\frac{1225}{65,536}$.
95. $\frac{3}{32}$.   **97(a).** $\frac{1}{11}$.   **97(b).** $\frac{6}{11}$.   **99(a).** $\frac{5}{16}$.   **99(b).** $\frac{15}{32}$.   **99(c).** $\frac{3}{16}$.
101. $\frac{25}{1296}$.   **103.** $\frac{1}{5}$.   **105.** $\frac{841}{1225}$.   **107.** $\frac{18}{35}$.   **109.** $\frac{2}{105}$.   **111.** $\frac{2}{7}, \frac{1}{7}$.
113. $\frac{12}{23}, \frac{11}{23}$.   **115.** $\frac{3}{5}$.   **117.** $\frac{4}{9}$.   **119.** $\frac{244}{495}$.

CHAPTER 6

1(a). $\frac{15,625}{46,656}$, $\frac{18,750}{46,656}$, $\frac{9375}{46,656}$, $\frac{2500}{46,656}$, $\frac{375}{46,656}$, $\frac{30}{46,656}$, $\frac{1}{46,656}$.
1(b). 15,625, 18,750, 9375, 2500, 375, 30, 1.   **3(a).** 94, 469, 938, 938, 469, 94.
3(c). 1500.   **3(f).** 2.50, 1.12.   **3(g).** 2.50, 1.12.   **5(a).** 0, 3, 12, 28, 32, 15.
5(b). 3.50, 1.02.

CHAPTER 7

1(a). 0.0796.   **1(b).** 0.0287.   **3(a).** 17.   **3(b).** 17.   **5.** 0.0287.
7. 0.0885.   **9.** 0.2266.   **11(a).** 0.018.   **11(b).** 0.053.   **13(a).** 0.030.
13(b). 0.003.   **15.** 0.0250.   **17.** 46.78%.   **19.** 9.   **21.** 22.   **23.** 790.
25(a). 1151.   **25(b).** 148.   **25(c).** 0.   **25(d).** 66.12 in., 71.88 in.
27. 3.82 oz, 5.63 oz.   **29.** 4.00.   **31.** 68.55 in., 3.00 in.   **33.** 1800.
35. 4 ft. 11.36 in.   **37(a).** 22.73.   **37(b).** 15.33.   **39.** 662.   **41.** 237.
43. 0.2409.   **45.** 218.

CHAPTER 8

9. 0.1131.   **11.** 0.0062.   **13(a).** 0.9773.   **13(b).** 0.0227.   **15.** 81.
17. 0.0146.   **19.** 0.0294.   **21.** 0.1131.   **23.** 18.69 to 21.31.   **25.** 5.42.
27. 0.5987.   **29.** 166.   **31.** 167.   **33.** 41.

CHAPTER 9

1. Yes, since $P = 0.0465$.   **3.** Yes, since $P = 0.0073$.   **5.** $X \leq 3$ and $X \geq 12$.
7. No, since $P = 0.0192$.   **9.** No, since $P = 0.0537$.
11. Yes, since $P = 0.2628$.   **13.** Yes, since $P = 0.0008$.
15. Yes, since $P = 0.0055$.   **17.** Yes, since $P = 0.0062$.   **19.** Yes, since $z = 3.26$.
21. 5%.   **23.** 40 ± 0.69.   **25.** 68.52 ± 0.64.   **27.** 36 : 25.   **29.** 514.
31. 6.37.   **33.** 16,247.   **35.** 3384.   **37.** 3394.

CHAPTER 10

1. 26.3 ± 1.02.   **3(a).** 16.90 ± 2.56.   **3(b).** 16.90 ± 2.58.
5(a). 12.15 ± 0.25 **5(b).** Yes.   **7(a).** Yes, since $z = 2.25$.
7(b). Yes, since $t = 2.38$.   **9.** No, since $t = -3.00$.   **11.** 15.00 oz, 1.52 oz.
13. No, since $t = -1.04$.   **15.** No, since $t = 1.89$.
17. No significant difference, since $t = -1.04$.   **19.** Yes, since $t = 7.42$.
21. Yes, since $t = 2.40$.   **23.** 7.10.   **25.** No, since $t = 1.73$.
27. No, since $t = -1.28$.   **29.** Yes, since $t = -2.01$.   **31.** 5.50 ± 4.54.
33. 8.36 or 12.04.

CHAPTER 11

1. No significant difference, since $P = 0.0649$ for a two-tailed test.
3. No, since $P = 0.3524$.   **5.** Yes, since $P = 0.0318$.
7. No significant difference, since $P = 0.0512$.
9. Same average scores, since $P = 0.2748$.   **11.** Yes, since $P = 0.0032$.
13. 7.   **15.** 12.   **17.** No, since $P = 0.0644$.   **19.** No, since $P = 0.0781$.
21. No, since $P = 0.2188$.

CHAPTER 12

1(a). $Y_e = 6.46 - 3.52X$.   **1(c).** -7.62.   **1(d).** 0.97.   **1(e).** -0.992.
5(a). \$37.05.   **5(b).** -0.95.   **7.** $Y_e = 8.15X$.   **9.** Yes, since $t = 2.89$.
11. ± 0.40.   **13.** $r > 0.25$ or $r < -0.25$.   **15(a).** $Y_e = -0.862 + 0.729X$.
15(b). 0.996.   **15(c).** No.   **15(d).** 0.7.   **17.** 0.50, 0.73.   **19.** 100.
21. Yes, since $z = 0.92$.   **23.** 61.   **25.** 0.1344.   **27.** 66.   **29.** 25.
31. 46.

CHAPTER 13

1. Can be considered honest die, since $\chi^2 = 4.51$.   **3.** Yes, since $\chi^2 = 1.50$.
5(a). 8.00.   **5(b).** Yes.   **5(c).** 0.047.   **5(d).** 0.0512.   **5(e).** 0.052.
7(a). 0.1841.   **7(b).** 0.1926.   **9(a).** 0.1210.   **9(b).** 0.123.
11. Yes, since $\chi^2 = 1.44$.   **13.** No, since adj. $\chi^2 = 9.80$.
15. Yes, since adj. $\chi^2 = 13.32$.   **17.** No, since adj. $\chi^2 = 1.30$.
19. No, since adj. $\chi^2 = 3.54$.   **21.** No, since adj. $\chi^2 = 1.25$.   **23.** 0.044.
25. 20, 10, 10, 20, or 10, 20, 20, 10.   **29.** 7.24.   **31.** No, since $\chi^2 = 11.00$.
33. Yes, since $\chi^2 = 3.25$.   **35.** Yes, since $\chi^2 = 16.01$.   **37.** Yes, since $\chi^2 = 5.59$.

CHAPTER 14

1(a). 100, 104.6, 110.3, and so on.   **1(b).** 95.6, 100, 105.4, and so on.
3. \$13,256.   **5(a).** 100.7, 103.6.   **5(b).** 102.3, 106.4.   **5(c).** 19.5.

CHAPTER 15

1(a). $Y_e = 2.60 + 0.247x$.   **1(c).** 1.36, 1.61, 1.86, and so on.
3. Specific seasonals for 1973: 112, 104, 103, and so on. Typical seasonals: 109 for Jan., 104 for Feb., 102 for Mar., and so on.   **5(a).** 59, 94, 174, 67, and so on.
5(b). 65, 83, 170, 83.   **5(c).** 1337, 1558, 1374, 1047, and so on.
5(d). $Y'_e = 1711.1 + 78.4x$, $Y'_e = 2240.9 - 19.9x$.
5(e). 1368, 1388, 1407, 1427, and so on, where the trend values are calculated for the middle of the quarters.   **5(f).** -31, 170, -33, -380, and so on.

CHAPTER 16

1(a). 0.098.   **1(b).** 0.179.   **1(c).** 0.455.   **3.** Yes, since $F = 1.95$.
5. Not homogeneous, since $F = 19.03$.   **7.** 5.58 and 18.31.   **9.** $n_2 \geq 11$.

CHAPTER 17

1(a). $\frac{1}{1,000,000}$.   **1(b).** $\frac{177,147}{500,000}$.   **1(c).** $\frac{468,559}{1,000,000}$.   **1(d).** $\frac{199,989}{200,000}$.
3(a). 190.00.   **3(b).** 105.00.   **3(c).** 85.00.   **5(a).** 142.00.   **5(b).** 87.30.
5(c). 54.70.   **7(a).** Yes, since $F = 8.24$.   **7(b).** AB, BC, and BD.
7(c). AB, BC, and BD.   **9(a).** Yes, since $F = 11.96$.   **9(b).** AB and BC.
11. $F = 1.85$.
13. Highly significant difference between feeds, since $F = 135.92$; pairs with highly significantly different means are AC, AD, BC, BD, and CD.   **15(a).** 50.
15(b). 4.4.
17. Significant difference for original data, since $F = 60.85$; significant difference for transformed data, since $F = 70.93$.
19(a). Significant difference for original data, since $F = 18.31$; significant difference for transformed data, since $F = 16.78$.
19(b). AB, AC, AD, AE, BE, CE, and DE for both sets of data.
21(a). Yes, since $F = 23.28$.   **21(b).** Yes, since $F = 12.37$.   **21(c).** 1 2 3 4 $\overline{5}$.

21(d). $D\ C\ B\ A$.   **23(a).** Yes, since $F = 2.56$.   **23(b).** 3 5 2 6 $\overline{4\ 1}$.

23(c). 3 5 2 6 $\overline{4\ 1}$.   **23(d).** Yes, since $F = 13.59$.

23(e). 1964 $\overline{1963}$ 1968 1969 1962 1961 1965 $\overline{1966\ 1960\ 1967}$ 1959.

23(f). Same as part e.

REVIEW EXERCISES (APPENDIX)

1(a). 16.00.   **1(b).** 16.00.   **1(c).** 15.60.
1(d). Class with boundaries 13 and 17.   **1(e).** 28.14.   **3(a).** 2.20.
3(b). 2.50.   **3(c).** 0.98.   **3(d).** 0.22.   **5.** 38, 30.   **7(a).** 45.   **7(b).** 21.
7(c). 35.   **9(a).** $\frac{21}{40}$.   **9(b).** $\frac{31}{80}$.   **11(a).** $\frac{30}{59}$.   **11(b).** $\frac{205}{413}$.

11(c). 0.1841.   **13.** $\frac{7}{33}$.   **15.** 0.9931.   **17(a).** 424.   **17(b).** 1955.
17(c). 314.   **17(d).** 32.   **17(e).** 0.   **19.** No, since $t = 2.13$.
21. Yes, since $P = 0.0340$.   **23.** 10.5 ± 3.03.   **25.** 0.9485.   **27.** 4161
29. 0.9678.   **31.** 135.   **33.** $\frac{11}{16}$.   **35.** No, since adj. $\chi^2 = 2.78$.
37. 17.57, 27.53.   **41.** No, since adj. $\chi^2 = 1.74$.   **43.** $\frac{55}{96}$.   **45.** $\frac{2}{3}$.
47. 99.   **49.** 0.0062.   **51(a).** $\frac{81}{10,000}$.   **51(b).** $\frac{81}{100,000}$.
53. Yes, since $\chi^2 = 4.49$.   **55.** No, since adj. $\chi^2 = 2.68$.   **57.** $\frac{655}{65,536}$.
59. 1800.   **61(a).** 0.0233.   **61(b).** 0.0238.   **63.** Yes, since $t = -2.38$.
67. 0.014.   **71.** $\frac{17}{36}$.   **73(a).** 0.0300.   **73(b).** 0.0321.
77. No, since $t = 1.72$.   **79.** 20.00.   **81.** 0.4997.   **83.** 7, 3, 8, 12, or 3, 7, 12, 8.
85(a). 0.5328.   **85(b).** 0.1360.   **87.** $r \le 0.48$ and $r \ge 0.91$.
89. Yes, since $P = 0.0359$.   **91.** $\frac{10}{273}$.   **93.** $\frac{46,413}{78,125}$.
95. No, since $\chi^2 = 22.03$.   **97.** 30.4 ± 3.01.   **99(a).** Yes, since $t = -3.38$.
99(b). Yes, since $P = 0.0087$.   **101.** $r < -0.592$ and $r > 0.592$.   **103.** $\frac{21}{12,500}$.
105(a). No, since $t = 1.52$.   **105(b).** No, since $P = 0.0884$.

Index